THE
BIOGENESIS OF
MITOCHONDRIA

Academic Press Rapid Manuscript Reproduction

Proceedings of the
International Conference on the Biogenesis of Mitochondria
Bari, Italy, June 25–28, 1973

THE BIOGENESIS OF MITOCHONDRIA

Transcriptional, Translational and Genetic Aspects

edited by

A. M. Kroon

Laboratory of Physiological Chemistry
State University, Groningen, The Netherlands

C. Saccone

Institute of Biological Chemistry
University of Bari, Bari, Italy

Academic Press, Inc.   *1974*
A Subsidiary of Harcourt Brace Jovanovich, Publishers

ACADEMIC PRESS, INC.
111 Fifth Avenue, New York, New York 10003

United Kingdom Edition published by
ACADEMIC PRESS, INC. (LONDON) LTD.
24/28 Oval Road, London NW1

Library of Congress Cataloging in Publication Data

International Conference on the Biogenesis of Mitochondria,
        Rosa Marina, Italy, 1973.
        The biogenesis of mitochondria.

        1.   Mitochondria–Congresses.   2.   Protein biosyn-
thesis–Congresses.   I.   Kroon, A. M., ed.   II.   Sac-
cone, C., ed.   III.   Title.   [DNLM: 1.   Mitochondria–
Metabolism–Congresses.   QH603.M5 I601b 1973]
QH603.M5I55 1973   574.8'734   73-5314
ISBN 0–12–426750–5

PRINTED IN THE UNITED STATES OF AMERICA

Contents

CONTENTS

CONTENTS

CONTENTS

Part III.   Synthesis of Mitochondrial Proteins

CONTENTS

List of Contributors

A. Adoutte, Laboratoire de Génétique, Université Paris-Sud, Orsay, France.

E. Agsteribbe, Laboratory of Physiological Chemistry, State University, Groningen, The Netherlands.

A. Akai, The Public Health Research Institute of the City of New York, New York, N.Y., U.S.A.

A. A. Algeri, Institute of Genetics, University of Parma, Parma, Italy.

I. Ali, Max-Planck-Institute für experimentelle Medizin, Abteilung Chemie, Göttingen, Germany.

V. F. Allison, Department of Biochemistry, the University of Texas Southwestern Medical School and Dept. of Biology, Southern Methodist University, Dallas, Texas, U.S.A.

J. André, Laboratoire de Biologie Cellulaire 4, Université de Paris XI, Orsay, France.

A. J. Arendzen, Laboratory of Physiological Chemistry, State University, Groningen, The Netherlands.

G. Attardi, Division of Biology, California Institute of Technology, Pasadena, California, U.S.A.

C. J. Avers, Department of Biological Sciences, Douglass Campus, Rutgers University, New Brunswick, New Jersey, U.S.A.

W. Bandlow, Institute for Genetics, University of Munich, Munich, Germany.

D. S. Beattie, Department of Biochemistry, Mount Sinai School of Medicine of the City University of New York, New York, N.Y., U.S.A.

J. S. Beckmann, Department of Biochemistry, The Weizmann Institute of Science, Rehovot, Israel.

C. Blossey, Max-Planck-Institute für experimentelle Medizin, Abteilung Chemie, Göttingen, Germany.

W. F. Bodmer, Genetics Laboratory, Dept. of Biochemistry, University of Oxford, Oxford, Great Britain.

P. Borst, Section for Medical Enzymology, Laboratory of Biochemistry, University of Amsterdam, Amsterdam, The Netherlands.

H. B. Bosmann, Department of Pharmacology and Toxicology, University of Rochester School of Medicine and Dentistry, Rochester, New York, U.S.A.

R. A. Butow, Department of Biochemistry, the University of Texas Southwestern Medical School and Department of Biology, Southern Methodist University, Dallas, Texas, U.S.A.

J. Casey, Department of Medicine, University of Chicago, Chicago, Illinois, U.S.A.

A. O. S. Chiu, Department of Biology, University of Pennsylvania, Philadelphia, Pa., U.S.A.

N. Chiu, Department of Biology, University of Pennsylvania, Philadelphia, Pa., U.S.A.

O. Ciferri, Institutes of Plant Physiology and Genetics, University of Pavia, Pavia, Italy.

H. G. Coon, Department of Embryology, Carnegie Institution of Washington, Baltimore, Maryland, U.S.A.

C. S. Cooper, Biology Department, Massachusetts Institute of Technology, Cambridge, Massachusetts, U.S.A.

P. Costantino, Division of Biology, California Institute of Technology, Pasadena, California, U.S.A.

I. W. Craig, Genetics Laboratory, Department of Biochemistry, University of Oxford, Oxford, Great Britain.

R. S. Criddle, Department of Biochemistry and Biophysics, University of California, Davis, California, U.S.A.

J-J. Curgy, Laboratoire de Biologie Cellulaire 4, Université de Paris XI, Orsay, France.

R. Datema, Laboratory of Experimental Phytomorphology, State University, Groningen, The Netherlands.

I. B. Dawid, Department of Embryology, Carnegie Institution of Washington, Baltimore, Maryland, U.S.A.

K. Dawidowicz, Department of Chemistry, Indiana University, Bloomington, Indiana, U.S.A.

N. D. Denslow, Department of Biochemistry, J. Hillis Miller Health Center, University of Florida, Gainsville, Florida, U.S.A.

E. Ebner, Section of Biochemistry, Molecular and Cell Biology, Cornell University, Ithaca, New York, U.S.A.

T. R. Eccleshall, Department of Biochemistry and Biophysics, University of California, Davis, California, U.S.A.

J. van Etten, Department of Plant Physiology, University of Nebraska, Lincoln, Nebraska, U.S.A.

I. Evans, Department of Botany and Microbiology, University College, London, England.

G. Faye, Centre de Génétique Moléculaire du C.N.R.S., Gif-sur-Yvette 91190, France.

F. Feldman, Department of Chemistry, Indiana University, Bloomington, Indiana, U.S.A.

H. Fukuhara, Centre de Génétique Moléculaire du C.N.R.S., Gif-sur-Yvette 91190, France.

M. N. Gadaleta, Institute of Biological Chemistry, University of Bari, Bari, Italy.

R. Gallerani, Institute of Biological Chemistry, University of Bari, Bari, Italy.

G. S. Getz, Department of Medicine, University of Chicago, Chicago, Illinois, U.S.A.

E. B. Gingold, Biochemistry Department, Monash University, Clayton, Victoria, Australia.

P. Gordon, Department of Medicine, University of Chicago, Chicago, Illinois, U.S.A.

M. Greco, Institute of Biological Chemistry, University of Bari, Bari, Italy.

D. E. Griffiths, Department of Molecular Sciences, University of Warwick, Coventry, England.

L. A. Grivell, Section for Medical Enzymology, Laboratory of Biochemistry, University of Amsterdam, Amsterdam, The Netherlands.

G. S. P. Groot, Section for Medical Enzymology, Laboratory of Biochemistry, University of Amsterdam, Amsterdam, The Netherlands.

H. Grossfeld, Department of Biochemistry, The Weizmann Institute of Science, Rehovot, Israel.

P. Hamill, Department of Chemistry, Indiana University, Bloomington, Indiana, U.S.A.

V. van Heyningen, Genetics Laboratory, Department of Biochemistry, University of Oxford, Oxford, Great Britain.

I. Horak, Department of Embryology, Carnegie Institution of Washington, Baltimore, Maryland, U.S.A.

R. L. Houghton, Department of Molecular Sciences, University of Warwick, Coventry, England.

N. Howell, Biochemistry Department, Monash University, Clayton, Victoria, Australia.

H-J. Hsu, Department of Medicine, University of Chicago, Chicago, Illinois, U.S.A.

R. E. Kellems, Department of Biochemistry, the University of Texas Southwestern Medical School and Department of Biology, Southern Methodist
University, Dallas, Texas, U.S.A.

W. Kleinow, Institut für Physiologische Chemie und Physikalische Biochemie der
Universität München, München, German Federal Republic.

A. M. Kroon, Laboratory of Physiological Chemistry, State University, Groningen,
The Netherlands.

H. Küntzel, Max-Planck-Institut für experimentelle Medizin, Abteilung Chemie,
Göttingen, Germany.

Y. Kuriyama, Laboratory of Cell Biology, The Rockefeller University, New York,
N.Y. U.S.A.

W. E. Lancashire, Department of Molecular Sciences, University of Warwick,
Coventry, England.

G. Lazar, Biological Research Centre, Hungarian Academy of Sciences, Szeged,
Hungary.

J. Lazowska, Centre de Génétique Moléculaire du C.N.R.S., Gif-sur-Yvette
91190, France.

G. Ledoigt, Laboratoire de Biologie Cellulaire 4, Universite de Paris XI, Orsay,
France.

L-F. H. Lin, Department of Biochemistry, Mount Sinai School of Medicine of
the City University of New York, New York, U.S.A.

A. W. Linnane, Biochemistry Department, Monash University, Clayton, Victoria,
Australia.

D. Linstead, Department of Botany and Microbiology, University College,
London, England.

U. Z. Littauer, Department of Biochemistry, The Weizmann Institute of Science,
Rehovot, Israel.

J. Locker, Department of Medicine, University of Chicago, Chicago, Illinois,
U.S.A.

D. J. L. Luck, Laboratory of Cell Biology, The Rockefeller University, New York, N.Y. U.S.A.

H. B. Lukins, Biochemistry Department, Monash University, Clayton, Victoria, Australia.

W. Machleidt, Institut für Physiologische Chemie und Physikalische Biochemie der Universität München, München, German Federal Republic.

H. R. Mahler, Department of Chemistry, Indiana University, Bloomington, Indiana, U.S.A.

G. R. Martin, Department of Biochemistry, J. Hillis Miller Health Center, University of Florida, Gainsville, Florida, U.S.A.

T. L. Mason, Section of Biochemistry, Molecular and Cell Biology, Cornell University, Ithaca, New York, U.S.A.

G. Michaelis, Centre de Génétique Moléculaire du C.N.R.S., Gif-sur-Yvette 91190, France.

F. Michel, Centre de Génétique Moléculaire du C.N.R.S., Gif-sur-Yvette 91190, France.

R. Michel, Institut für Physiologische Chemie und Physikalische Biochemie der Universität München, München, German Federal Republic.

F. Miller, Institut für Zellbiologie der Universität München, München, German Federal Republic.

M. W. Myers, Department of Pharmacology and Toxicology, University of Rochester School of Medicine and Dentistry, Rochester, New York, U.S.A.

Ph. Nagley, Biochemistry Department, Monash University, Clayton, Victoria, Australia.

M. M. K. Nass, Department of Therapeutic Research, University of Pennsylvania School of Medicine, Philadelphia, Pa., U.S.A.

W. Neupert, Institut für Physiologische Chemie und Physikalische Biochemie der Universität München, München, German Federal Republic.

T. W. O'Brien, Department of Biochemistry, J. Hillis Miller Health Center, University of Florida, Gainsville, Florida, U.S.A.

D. Ojala, Division of Biology, California Institute of Technology, Pasadena, California, U.S.A.

B. Ono, Section of Biochemistry, Molecular and Cell Biology, Cornell University, Ithaca, New York, U.S.A.

J. Otto, Institut für Physiologische Chemie und Physikalische Biochemie der Universität München, München, German Federal Republic.

L. Packer, Department of Physiology-Anatomy, University of California, Berkeley, California, U.S.A.

G. Pepe, Institute of Biological Chemistry, University of Bari, Bari, Italy.

E. Petrochilo, Centre de Génétique Moléculaire du C.N.R.S., Gif-sur-Yvette 91190, France.

S. H. Phan, Department of Chemistry, Indiana University, Bloomington, Indiana, U.S.A.

R. O. Poyton, Section of Biochemistry, Molecular and Cell Biology, Cornell University, Ithaca, New York, U.S.A.

P. P. Puglisi, Institute of Genetics, University of Parma, Parma, Italy.

M. Rabinowitz, Department of Medicine, University of Chicago, Chicago, Illinois, U.S.A.

A. Reitsema, Laboratory of Physiological Chemistry, State University, Groningen, The Netherlands.

E. Ross, Section of Biochemistry, Molecular and Cell Biology, Cornell University, Ithaca, New York, U.S.A.

M. S. Rubin, The Public Health Institute of the City of New York, New York, N.Y., U.S.A.

C. Saccone, Institute of Biological Chemistry, University of Bari, Bari, Italy.

G. Schatz, Section of Biochemistry, Molecular and Cell Biology, Cornell University, Ithaca, New York, U.S.A.

G. L. Scherphof, Laboratory of Physiological Chemistry, State University, Groningen, The Netherlands.

H. Schmitt, Department of Biochemistry, The Weizmann Institute of Science, Rehovot, Israel.

A. J. Schwab, Institut für Physiologische Chemie und Physikalische Biochemie der Universität München, München, German Federal Republic.

B. C. van Schijndel, Laboratory of Physiological Chemistry, State University, Groningen, The Netherlands.

A. H. Scragg, National Institute for Medical Research, Mill Hill, London NW7, Great Britain.

W. Sebald, Institut für Physiologische Chemie und Physikalische Biochemie der Universität München, München, German Federal Republic.

P. Slonimski, Centre de Génétique Moléculaire du C.N.R.S., Gif-sur-Yvette 91190, France.

B. J. Stevens, Laboratoire de Biologie Cellulaire 4, Université de Paris XI, Orsay, France.

R. N. Stuchell, Department of Biochemistry, Mount Sinai School of Medicine of the City University of New York, New York, N.Y., U.S.A.

Y. Suyama, Department of Biology, University of Pennsylvania, Philadelphia, Pa., U.S.A.

O. Tiboni, Institutes of Plant Physiology and Genetics, University of Pavia, Pavia, Italy.

A. Tzagoloff, The Public Health Research Institute of the City of New York, New York, N.Y., U.S.A.

H. de Vries, Laboratory of Physiological Chemistry, State University, Groningen, The Netherlands.

H. Weiss, Institut für Physiologische Chemie und Physikalische Biochemie der Universität München, München, German Federal Republic.

S. Werner, Institut für Physiologische Chemie und Physikalische Biochemie der Universität München, München, German Federal Republic.

D. Wilkie, Department of Botany and Microbiology, University College, London, England.

L. Worthington, Physiology Research Laboratory, VA Hospital, Martinez, California, U.S.A.

G-J. Wu, Department of Embryology, Carnegie Institution of Washington, Baltimore, Maryland, U.S.A.

B. Ziganke, Institut für Physiologische Chemie und Physikalische Biochemie der Universität München, München, German Federal Republic.

Preface

The International Conference on the Biogenesis of Mitochondria, held in Rosa Marina near Bari, Italy in June 1973, was the ninth in the series of annual meetings on mitochondria organized by members of the Laboratory of Biochemistry of the University of Bari in collaboration with their colleagues from other Italian Universities and from abroad. Within this series it was the second completely devoted to the different aspects of mitochondrial biogenesis. Since 1967, the interest in, and the number of laboratories working on, the biogenesis of mitochondria has grown tremendously. For this reason, it was decided to put some restriction on the topics of the conference. Thus, the physical characteristics of mitochondrial DNA and the mechanism of mitochondrial DNA replication were not discussed. As a consequence, the mechanism of mitochondrial translation and transcription could be dealt with in more detail, as could the genetic approaches to the unraveling of the complicated interplay between the nuclear and mitochondrial genetic systems in the assembly of mitochondria within the cell.

All but one of the contributions to the symposium appear in this volume. An abstract of the missing contribution appears. In the last chapter of the book we have tried to summarize the main points of consensus reached during the conference and have indicated a few of the unsolved or controversial problems raised.

We want to seize this opportunity to thank all those who have contributed to the organization of the conference. The members of the advisory board, P. Borst, Th. Bücher, H. Küntzel, A. W. Linnane, and G. Schatz, gave valuable advice for drawing up the scientific program. We thank Professor Quagliariello, Rector of the University of Bari, who rendered the organization possible and contributed in facilitating its direction. Finally we thank our own scientific and technical staffs for their cooperation before, during, and after the conference.

PREFACE

The conference was sponsored by the following organizations: Ministero della Publica Instruzione; Consiglio Nazionale delle Ricerche; Università di Bari; Provincie di Bari e Brindisi; Boehringer, Mannheim; Lepetit, Milano.

INTRODUCTION

E. Quagliariello, *Professor of Biochemistry and Rector Magnificus of the University of Bari, Bari, Italy.*

Already far back in 1965 during the first in the series of meetings on aspects of the biochemistry of mitochondria organized by the University of Bari, some of the papers presented concerned the existence of mitochondrial nucleic acids and of a mitochondrial machinery for protein synthesis. I remember that the subject was then very controversial and that the discussion on bacterial contamination and on contamination with other cell constituents of the mitochondrial preparations was very lively. It was soon clear that mitochondrial molecular biology requested a special meeting.

So, in 1967, we held the first specialized meeting on mitochondrial biogenesis entitled: Biochemical Aspects of Mitochondrial Biogenesis. The problem of bacterial and other contaminations had been definetely clarified and the semi-autonomy of the mitochondrion in the cell completely accepted. The meeting dealt with all the possible aspects of the problem at that time: histochemical evidence of the existence of DNA and RNA in mitochondria, properties of mitochondrial nucleic acids and protein synthesis and genetic aspects of mitochondrial biogenesis. The field of mitochondrial biogenesis has since then expended enormously. It is now one of the most active in cell biology and this fact has forced the organizers this year to restrict the area of interest of the meeting of which this volume holds the proceedings. The topics of this International Conference on Biogenesis of Mitochondria are only two: mitochondrial transcription and mitochondrial translation. But as always happens in Science, it is difficult or impossible to put restrictions to topics, unless one wishes to have a cold assembly

1

of papers which differ only in experimental details and contain few speculations. Thus it was thought better to include the role of mitochondrial DNA and the genetic aspects of mitochondrial biogenesis in relation to the main themes.

The physical properties of isolated mitochondrial DNA which was one of the main subjects in 1967 will not be discussed here. This because, as I have already said, the field is too large at the moment: on the other hand many of the properties of mtDNA are now known in details and have been extensively discussed in previous symposia and in a number of reviews.

The physical properties of animal mitochondrial DNA make it a highly suitable model system for the study of DNA replication and the knowledge on the replication of mitochondrial DNA is accumulating so rapidly that it seemed appropriate to leave this topic out as well. On the contrary the role of mitochondrial DNA in the transcription mechanism will be particularly stressed. After the interesting experiments on mitochondrial RNA synthesis in HeLa cells, the problem of the mechanism of mitochondrial transcription became one of the most exciting in the field of mitochondrial biogenesis. It has been found that, at least in HeLa cells the transcription of mitochondrial DNA is symmetric. It seems that both strands of DNA are completely transcribed, but that afterwards the product of the L strand is almost completely (98 %) degraded. This means that in the organelle, the problem of strand selection could be achieved by a posttranscriptional control mechanism. In recent years the enzyme responsible for transcription, the DNA-dependent RNA polymerase present in the organelle, has been purified from various organisms (*Neurospora*, yeast, rat liver). Although some functional differences have been described in mitochondrial polymerases, there is general agreement regarding their size. The molecular weight of the enzyme consisting of one single polypeptide chain, seems to be about 60,000 daltons. Mitochondrial transcription can, therefore now be studied also *in vitro*, using the isolated template and the homologous enzyme.

Another interesting point which will be discussed in relation to the synthesis of mitochondrial RNA will be that of the presence of polyadenilic acid sequences in the mitochondrial messenger RNA. One of the hypotheses put forward to explain the function of the poly-A tracts in the extramitochondrial messenger RNAs is that the poly-A sequences protect the messenger RNA during nuclear export to the cytoplasm. Since this function is meaningless for the RNA made inside the mitochondria the functional significance of poly-A sequences in RNA has probably to be revised completely.

All mitochondrial DNAs studied up to now contain genes for ribosomal RNA and transfer RNA. Genes for ribosomal RNA are represented in only one copy per mitochondrial genome and this property differentiates mitochondrial DNA from all other cellular DNAs which contain more copies of ribosomal genes per genome.

The genetic function of mitochondrial DNA is extensively studied also by carrying out genetic experiments. Yeast is the most favoured organism for these experiments and it is contibuously giving important informations on the biogenesis of mitochondria. The existence of cytoplasmic mutants and the study of the their behaviour have permitted the construction of a map of mitochondrial genes.

Another part of these proceedings will deal with the characteristics of the mitochondrial protein synthetic machinery. Here we should say that many of the properties of this machinery are well known at the present. Mitochondrial ribosomes from lower organisms have been well characterized: they appear to be in the 70-80 S size range, but completely different in physicochemical properties from their cytoplasmic counterparts. In several cases they retain *in vitro* a high activity in protein synthesis. Although the general occurrence, of 55S ribosomes in animal mitochondria has been widely reported, complete characterization of mitochondrial ribosomes from animal cells was still lacking.

Recently, in a number of laboratories, ribosomes

highly active in protein synthesis *in vitro* have been obtained and these will probably be of use in clearing up this point. Also studies regarding the enzymes and factors involved in mitochondrial protein synthesis are now greatly expanding. The factors required for mitochondrial protein synthesis differ also from their cytoplasmic counterparts and in several cases it has been demonstrated that they are interchangeable with factors from *E. coli*.

With regard to the sensitivity of the mitochondrial protein synthesizing system to antibiotics, it is generally accepted that the translational apparatus of the organelle is of "prokaryotic type". It has been reported that the ribosomal systems of mammalian and yeast mitochondria differ in their antibiotic sensitivity. This functional difference was ultimately correlated to the physical differences between mammalian ribosomes and ribosomes of lower organisms and was ascribed to a phylogenetic difference between the two systems, This is, however, still a matter of controversy which - we may hope - will be clarified in the near future.

A main part of these proceedings will be dealing with the synthesis of mitochondrial proteins. The majority of mitochondrial proteins are made in the extramitochondrial space on cytoplasmic ribosomes, a minority is made within the mitochondria themselves on mitochondrial ribosomes. The interrelation between these two systems is one of the most important points of mitochondrial biogenesis. The contribution of the mitochondrial protein synthesizing system for the synthesis of several enzymes of the inner mitochondrial membrane has been demonstrated. It has been found that several components of large mitochondrial enzyme systems, like cytochrome *c* oxidase. ATPase and cytochrome *b* are synthesized within the mitochondria.

For other proteins of inner mitochondrial membrane, like cytochrome c_1, it has been demonstrated that although they are made on cytoplasmic ribosomes, they still require the presence of an active mitochondrial protein synthesizing system in order to be rightly in-

tegrated in the respiratory chain. Also some glycoproteins seem to be made in the mitochondria. I hope
that the present proceedings of the International Conference on the Biogenesis of Mitochondria, may show
that at least a few of the problems put above are either clarified or brought closer to a final solution.
If so, the Conference and also these Proceedings have
served their main purpose.

Part I

**Mitochondrial Transcription:
The Mechanism and Products, the Role of Mitochondrial
DNA and Some Genetic Aspects**

MOLECULAR APPROACHES TO THE DISSECTION OF THE MITOCHONDRIAL GENOME IN HELA CELLS

Giuseppe Attardi, Paolo Constantino and Deanna Ojala

Division of Biology, California Institute of Technology, Pasadena, California 91109, U.S.A.

Against the traditional rules of the game, the deciphering of the genetic content of the mitochondrial genome in animal cells has been so far the lonely task of biochemistry, and probably will be so for some time to come. This is by no means, however, a motive for pessimism. On the contrary, there are good reasons to believe that the biochemical approach will go a long way towards the final goal of understanding the informational role of mitochondrial DNA (mtDNA) and the control of its expression. By the variety and sophistication of the techniques available today, it is possible to identify the primary gene products of mtDNA, to test their function in *in vitro* systems, and to use them for mapping the corresponding genes in the RNA with a resolution which approaches or surpasses that of the finest genetic techniques.

GENETIC MAP OF HELA CELL MITOCHONDRIAL DNA 1973.

Fig. 1 shows the up-to-date genetic map of HeLa cell mtDNA, as constructed by electron microscopy of RNA-DNA and DNA-DNA hybrids, by a collaborative effort of our laboratory and Norman Davidson's laboratory (ref. 1 and unpublished observations). The two mitochondrial ribosomal RNA (rRNA) genes in the heavy (H) strand were recognized from the positions of the duplex regions corresponding to the 12S and 16S rRNA-DNA hybrids. The 4S RNA sites were identified from the positions of binding of ferritin-coupled 4S hybridized with the complementary sequences in DNA. There are nine such sites (H1 to H9) on the H strand, which were mapped relative to the rRNA genes, and three (L1 to L3) on the light (L) strand. In order to provide a reference point for the identification of the position of the three 4S RNA sites on the L strand relative to the genes on the H strand, C. Tu in our laboratory purified the mitochondrial rRNA genes by annealing, under saturation conditions, rRNA with the H strand, digesting single-stranded DNA segments with the *Aspergillus* S1 nuclease and non-hybridized RNA with pancreatic RNAse, and isolating the RNA-DNA hybrids in a Cs$_2$SO$_4$ density gradient. The purified coding sequences for rRNA were used by M. Wu in Norman Davidson's laboratory as a marker for the identification of the anti-rRNA sites on the L strand. A preliminary assignment of the position of these sites is indicated in the map of the L strand in Fig. 1. The relative orientation of the anti-12S and anti-16S RNA sequences in this region with respect to the three 4S RNA sites still remains to be determined.

CODING CAPACITY OF MITOCHONDRIAL DNA FOR tRNA.

As appears from Fig. 1, three of the 4S RNA coding sequences on the H strand, H1, H2, and H3, are very close to the two rRNA genes. By analogy with bacterial systems, this suggests the possibility that one of these specifies a small RNA which plays the role of a 5S RNA equivalent in the mitochondrial ribosomes structure. Such a possibility is at present under investigation in our laboratory.

There is evidence from RNA-DNA hybridization data obtained with rat liver material that animal cell mtDNA contains information for at least four tRNA species (2). On the other hand, the RNA-DNA hybridization results from which the map of Fig. 1 was constructed have set an upper limit of 12 genes for the coding capacity for tRNA of HeLa mtDNA. In *Xenopus* mtDNA, the reported hybridization plateau for 4S RNA corresponds to 15 cistrons (3). These observations raise the question of whether animal cell mitochondria utilize for protein synthesis an incomplete set of endogenous tRNA

10

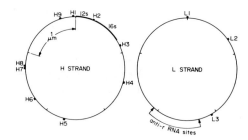

Fig. 1. Circular map of the positions of the complementary sequences for 4S RNAs on the H and L strands of HeLa mtDNA and of the 12S and 16S rRNA genes on the H strand. (Modified from ref. 1). See text for details.

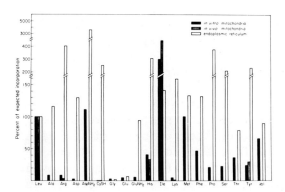

Fig. 2. Percent incorporation of various amino acids by HeLa cell mitochondria or endoplasmic reticulum, relative to that expected for the synthesis of an "average" HeLa cell protein. It is assumed that the relative rates of incorporation for the various amino acids into this "average" protein are proportional to their mole percent in whole HeLa cell protein (5). Taken from ref. 4.

species, or whether, on the contrary, the missing tRNA species are imported from the cytoplasm. With the aim of obtaining information on this question, recently, experiments have been carried out in our laboratory to test the capacity of HeLa mitochondria, either isolated or in the intact cell, to incorporate different labelled amino acids into proteins (4). As appears in Fig. 2, eight amino acids (alanine, arginine, aspartic acid, cysteine, glutamic acid, glutamine, glycine and lysine), which include most of the charged polar ones, showed a very low level, if any at all, of chloramphenicol-sensitive incorporation relative to that expected for an "average" HeLa cell protein. By contrast, the most hydrophobic amino acids (leucine, isoleucine, valine, phenylalanine and methionine) were the most actively incorporated by HeLa mitochondria. Furthermore, the various labelled amino acids appeared to be utilized by the HeLa mitochondria system in proportion strikingly different from that observed for protein synthesis by endoplasmic reticulum polysomes. The available evidence suggests that pool effects cannot account for this general pattern of utilization of amino acids; furthermore, this pattern is in good agreement with the known hydrophobic properties of proteins synthesized in mitochondria.

Whether the apparent constraint which exists in the pattern of amino acid utilization by the HeLa cell mitochondrial protein synthesizing system reaches the extreme limit of a complete lack of utilization of a certain set of amino acids, including most of the charged polar ones, and whether it can be correlated with the existence in animal cell mitochondria of an incomplete set of tRNA genes (1,3), are important questions which remain to be answered. *Neurospora* mitochondria apparently contain a complete or almost complete set of tRNA species (6), and in yeast mtDNA the presence of at least 20 4S RNA genes has been recently reported (7,8). It will be interesting to see whether in these organisms the pattern of amino acid utilization for mitochondrial protein synthesis is different from that described here for HeLa cells, and,

if so, whether this difference can be correlated with
and evolutionary change in the spectrum of the proteins synthesized in mitochondria. In another context,
the striking difference observed in the pattern of
amino acid utilization by the mitochondrial and the
cytoplasmic protein synthesizing systems in HeLa cells
will represent a valuable tool for the identification
of the proteins synthesized in animal mitochondria in
the absence of inhibitors.

MITOCHONDRIAL DNA CODED MESSENGER RNA.

The genes identified so far in HeLa mtDNA account
for about 25 % of the potential information contained
in a 5 μ long DNA molecule with no major internal repetitions (9). This estimate ignores of course the
remote, though not to be completely excluded possibility, which is suggested by the extensive symmetric
transcription of HeLa mtDNA (10), of a meaningful information being coded on both strands of mtDNA at the
same site. We have previously shown that the H strand
of HeLa mtDNA is completely transcribed (11), and
that the L strand is also transcribed over a major
portion, if not the entirely of its length (10). One
might ask whether the RNA sequences which are transcribed from mtDNA, other than those which correspond
to the rRNA and 4S RNA genes, represent spacers or
ancillary sequences necessary for the processing or
regulation of synthesis of the above mentioned discrete RNA species, or whether on the contrary other
informational RNA molecules, in particular messenger
RNA (mRNA) molecules, are coded for by mtDNA.

In the rest of this paper we aim to present some
evidence which strongly suggests that the last possibility I have just raised, namely the existence of mt-
DNA coded mRNA species is indeed true.

The poly(A) story. A strong indication of the
existence of mRNA in animal cell mitochondria and at
the same time a handle for its isolation have been
provided by the demonstration of the occurrence of po-
ly(A) sequences of a distinctive size in HeLa mitochondrial RNA (12,13) and of a poly(A) synthesizing

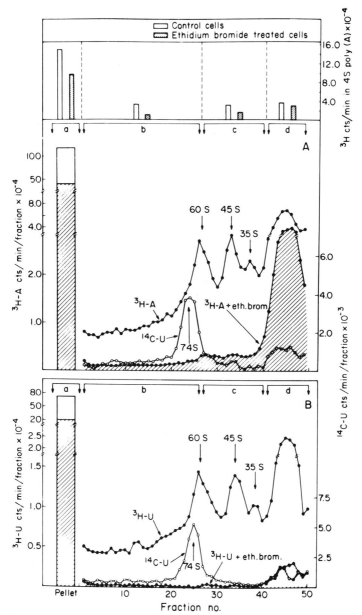

Fig. 3. Sedimentation patterns of the components of the mitochondrial Triton X-
100 lysates from cells labeled for 2 h with 8[³H]adenosine (A) or 5[³H]uridine
(B), in the presence of 0.1 μg actinomycin D/ml and in the absence or presence
of 1 μg ethidium bromide/ml, and distribution of labeled 4S$_E$ poly(A) among different components of the lysate. The 8[³H]adenosine or 5[³H]uridine labeled cells
had been mixed with cells labeled for 24 h with 2[¹⁴C]uridine to provide sedimentation markers. The lysates were run through 15 to 30 % sucrose gradients in 0.05
M Tris buffer, pH 6.7, 0.1 M KCl, 0.01 M MgCl₂, in the Spinco SW27 rotor at
19,000 rev./min for 11 h.

enzyme in rat liver mitochondria (14,15).

We have investigated the distribution of labelled poly(A) sequences among the different centrifugal components of a mitochondrial lysate from HeLa cells labelled for 2 h with 8-[^{3}H]adenosine in the presence of 0.1 µg/ml actinomycin D. RNA was extracted from each of the various cuts of the gradients shown in Fig. 3a, passed through a poly(T)-cellulose column to select for poly(A)-containing RNA, and the fraction retained and eluted from poly(T)-cellulose was digested with DNAse and with pancreatic and T$_1$ RNAse, precipitated with ethanol and analyzed by polyacrylamide gel electrophoresis. Fig. 4a shows the electrophoretic profile of the digest of the RNA from the mitochondrial polysome region (74-180S). One can see a relatively narrow peak of labelled molecules which migrate slightly faster than mitochondrial 4S RNA, and which, ignoring conformational effects, can be tentatively assigned a length corresponding to 60-80S residues; this peak has an advancing edge of faster moving segments extending up to the large peak near the buffer front, which represents ethanol precipitable non-specific digestion products of RNA. That the 4S$_E$[†] 8-[^{3}H]adenosine labelled segments resistant to pancreatic and T$_1$ RNAse are indeed poly(A) stretches is shown by the fact that they are sensitive to alkali (and, therefore, are RNA chains), and that they do not contain any appreciable amount of uridine (and, therefore, are not double-stranded RNA segments): in fact, as shown in Fig. 4a, there is no S$_E$ peak in the digest of 5-[^{3}H]uridine labelled RNA isolated from the mitochondrial polysome region of cells labelled for 2 h with 5-[^{3}H]uridine (Fig. 3b). Poly(A) isolated under the same conditions from 8-[^{3}H]adenosine labelled mRNA of free cytoplasmic polysomes exhibited in polyacrylamide gel a peak at about 7S$_E$ (Fig. 4b).

If the 2 h labelling of HeLa cells with 8-[^{3}H]adenosine is done in the presence of ethidium bromide at

[†] The symbol S$_E$ is used to indicate the S-value estimated from the relative electrophoretic mobility of the RNA (16).

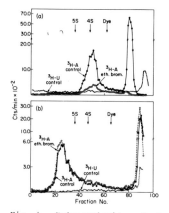

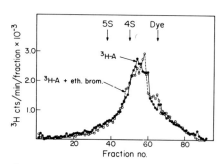

Fig. 4. Polyacrylamide gel electrophoresis pattern of poly(A) present in mitochondrial RNA (a) or cytoplasmic mRNA (b). RNA was extracted from the polysome region [cut (b)] of the sucrose gradient sedimentation patterns shown in Fig. 3a and 3b and from free polysomes isolated in the same experiment, and passed through poly(T)-cellulose columns; the RNA components retained and eluted from the columns were subjected to RNase-DNase digestion, and then analyzed by electrophoresis through 12 % polyacrylamide gels.

Fig. 5. Detection of "free" $4S_E$ poly(A) in mitochondrial RNA. RNA extracted from the slowest sedimenting components of the mitochondrial lysate [cut (d)] in the sedimentation runs of Fig. 3a, and selected for poly(A) content by poly(T)-cellulose chromatography, was analyzed by polyacrylamide gel electrophoresis without RNAse-DNase digestion.

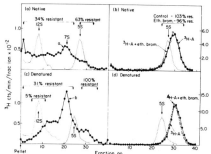

Fig. 6. Sedimentation analysis under native and denaturing conditions and tests of RNAse resistance of RNA from the mitochondrial polysome region [cut (b)] and from the slowest sedimenting components [cut (d)] of the gradients of Fig. 3a, after selection for poly(A) content by poly(T)-cellulose chromatography. The RNA from cut (b) had been passed twice through poly(T)-cellulose. (a) and (b) centrifugation through 5 to 20 % sucrose gradients in 0.01 M sodium acetate buffer, pH 5.0, 0.1 M NaCl, in the Spinco SW41 rotor at 37,000 rev./min for 22h. (c) and (d) the samples were denatured in formaldehyde (25) and run through sucrose gradients in the presence of formaldehyde (26) in the SW41 rotor at 41,000 rev./min for 27 h. The [³H] labeled mitochondrial 12S RNA and cytoplasmic 5S RNA were treated in the same way as the experimental samples and run in parallel gradients. The RNA from the indicated fractions was tested for RNAse resistant acid precipitable radioactivity (2 μg/ml heated pancreatic RNAse and 100 units/ml heated T_1 RNAse, 30 min at 37° C) after extensive dialysis versus 0.01 M Tris buffer, pH 7.4, 10^{-3} M $MgCl_2$, 0.25 M NaCl.

16

1 µg/ml, a concentration which inhibits more than 96 %
of the synthesis of the RNA sedimenting in the mitochondrial polysome region (Fig. 3b), the labelling of
the $4S_E$ poly(A) stretches isolated from this RNA is
inhibited by about 70 % (Fig. 4a). The same concentration of the drug has no effect on the synthesis of
poly(A) associated with cytoplasmic mRNA (Fig. 4b).
This observation suggests that the RNA molecules covalently attached to $4S_E$ poly(A) stretches are coded
for by mtDNA. We will come back to this question below.

Poly(A) stretches of a size corresponding to $4S_E$
are not restricted to the RNA isolated from the mitochondrial polysome region of the gradient. On the contrary, as shown in the upper part of Fig. 3, about
60 % of the $4S_E$ poly(A) labelled during a 2 h 8-[^{3}H]-
adenosine pulse is associated with the RNA extracted
from the fast sedimenting components of the lysate
(>200S), while the remainder is approximately equally
distributed in the mitochondrial polysome region, in
the ribosome-subunit region and in the region of the
slower sedimenting material. The fast-sedimenting
structures contain only a relatively small part of
the mitochondria protein synthesizing structures, as
judged from the distribution of the radioactivity after a 4 min 4,5-[^{3}H]isoleucine pulse in the presence
of emetine (100 µg/ml) (an inhibitor of cytoplasmic
protein synthesis: cf. ref's. 17-19). It seems likely
that the poly(A) present in these structures is associated with precursor mRNA molecules in the course of
processing. It should be noticed that these fast-sedimenting structures contain the transcription complexes of mtDNA (20) and the bulk of pulse-labelled
mitochondrial RNA (21).

The partial ethidium bromide resistance of the synthesis of mitochondrial $4S_E$ poly(A), which is as high
as 85 % for the poly(A) present in the slow sedimenting components of the lysate (Fig. 3), strongly suggests that poly(A) continues to be synthesized in the
presence of ethidium bromide and added, presumably at
the 3'-end, to mitochondrial RNA molecules synthesized

17

prior to the addition of the drug and, therefore, unlabelled. In agreement with this idea are recent data
(22) on the binding of polyribonucleotides to H and
L strands of rat mtDNA, which indicate the absence of
long continuous stretches of T in this DNA. The situation is similar to that described for nuclear-cyto-
plasmic poly(A) (22,24).

A surprising finding in the course of the present
investigations was that the poly(T)-cellulose selected RNA from the slowly sedimenting components of the
mitochondrial lysate was completely or almost completely RNAse resistant (see, for example, below, Fig.
6b). This suggested the possibility that all or almost
all this RNA might be represented by "free"* poly(A)
sequences. This possibility turned out indeed to be
true. As shown in Fig. 5, the electrophoretic profile
of the undigested poly(T)-cellulose selected RNA from
the slowest sedimenting components (<25S) of the mitochondrial lysate of both the control and the ethidium bromide treated cells showed a labelled peak at
the same position (just ahead of mitochondrial 4S
RNA) as the poly(A) isolated from the same components
after RNAse-DNAse digestion, and with an almost identical profile, apart from a trailing edge of segments
moving slower than 5S RNA. Quantitatively, the radioactivity in the $4S_E$ material from the undigested RNA
accounted for all the poly(A) isolated after RNAse digestion.

Two interpretations can be proposed for the presence of "free" poly(A). The first one is that poly(A) in
HeLa cell mitochondria is not added stepwise to the
RNA molecules, but is synthesized separately and then
added as a preformed chain. The second possibility is
that "free" poly(A) represents the residue of the degradation of RNA carrying covalently bound poly(A) sequences. If the trailing edge of "free" poly(A) men-

* The expression "'free' poly(A)" is used to indicate
poly(A) sequences present in the undigested RNA sample with an electrophoretic mobility indistinguishable from that of poly(A) isolated after RNAse digestion.

tioned above represents poly(A) linked to short RNA molecules of various sizes, this would speak in favor of the second alternative. On the other hand, the presence of labelled poly(A) of identical size distribution and in approximately the same amount in the slow sedimenting components of the lysate from both the control and the ethidium bromide treated cells would rather favor the first alternative. This is also supported by the recent discovery in rat liver mitochondria (14,15) of a poly(A) synthesizing enzyme insensitive to DNAse, actinomycin D and ethidium bromide, and not requiring the bulk of mitochondrial RNA as a primer.

While all or the great majority of the poly(A) present in the slow sedimenting components of the lysate appeared to be "free", most of the poly(A) associated with the components sedimenting in the mitochondrial polysome region or with the faster sedimenting structures was shown to be covalently bound to RNA molecules. As appears from Fig. 6c, after formaldehyde treatment and centrifugation through a sucrose gradient in the presence of formaldehyde, the great majority of the 8-[^{3}H]adenosine labelled RNA from the polysome region, selected for poly(A) content by two passages through poly(T)-cellulose, sedimented in the region corresponding to >5S, and showed about the same proportion of RNAse resistance (31 %) as the components heavier than 5S RNA run in the native state (Fig. 6a). After denaturation, some molecules completely resistant to RNAse appeared to have accumulated in the region corresponding to "free" poly(A): these presumably represent "free" poly(A) which, in the native RNA preparation, cosedimented with heavier molecules, due to aggregation.

The sedimentation profile of the poly(A)-containing RNA from the mitochondrial polysome region showed, after formaldehyde denaturation, a pronounced, sharp peak (peak "b" in Fig. 6c) moving about 20 % faster than the denatured 5S RNA marker and 40 % slower than the denatured 12S RNA. In the sedimentation pattern of the native RNA, on the contrary, only a small peak

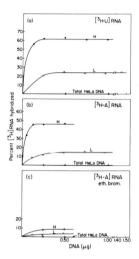

Fig. 7. Homology to separated mtDNA strands of 5[³H]uridine or 8[³H]adenosine labeled RNA extracted from the mitochondrial polysome region of the gradients of Fig. 3a and 3b, and selected for poly(A) by two successive poly(T)-cellulose chromatography runs.

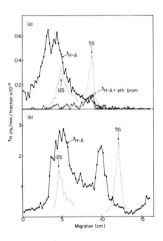

Fig. 8. Polyacrylamide gel electrophoresis in the native state (a), or after formaldehyde treatment and in the presence of formaldehyde (b), of 8[³H]adenosine labeled RNA from the mitochondrial polysome region of the gradient, selected for poly(A) content by passage through a poly(T)-cellulose column. Electrophoresis was carried out through 2.7 % polyacrylamide gels at 5 mA per 16 cm gel for 3.5 h (a) or 3 h (b). In (a), 0.04 M Tris buffer, pH 7.4 (25°C), 0.02 M sodium acetate, 0.002 M EDTA, containing 0.5 % sodium dodecyl sulfate; in (b), 0.02 M phosphate buffer, pH 7.8, containing 0.5 % sodium dodecyl sulfate and 3 % formaldehyde, was used as a buffer. The 12S and 5S RNA were treated as the experimental samples and run in parallel gels.

sedimenting at about 7S (peak "a" in Fig. 6a) could be seen to emerge over a background of heterogenous RNA, most of the material sedimenting faster, up to about 16S. This suggested the posibility that peak "b" represented the denatured form of the material of peak "a", and that a part of the components of peak "a" had sedimented in the heavier portion of the gradient in Fig. 6a, presumably due to aggregation. We will come back to this question below.

The partial ethidium bromide sensitivity of the labelling of poly(A) isolated from the mitochondrial polysome and ribosome-subunit regions of the gradient and from the fast-sedimenting components of the lysate, as opposed to the lack of effect of this drug on the synthesis of poly(A) associated with cytoplasmic mRNA (Fig. 4), suggested that the RNA molecules covalently linked to $4S_E$ poly(A) stretches are coded for by mtDNA. Unambiguous evidence in favor of this interpretation was obtained by analyzing, in RNA exhaustion hybridization tests, the homology with the separated H and L mtDNA strands of 5-[^{3}H]uridine or 8-[^{3}H]adenosine labelled poly(A)-containing mitochondrial RNA molecules. As shown in Fig. 7a, when 5-[^{3}H]-uridine labelled RNA from the mitochondrial polysome region of the sucrose gradient pattern from untreated cells, which had been selected for poly(A) content by two successice passages through poly(T)-cellulose, was used, about 2.5 times as many counts/min hybridized with the H strands as compared to those hybridized with the L strands. Assuming that different sequences hybridized with the H and L mtDNA strands, 86 % of the original labelled RNA formed complexes with mtDNA. On the contrary, when poly(T)-cellulose selected 8-[^{3}H]-adenosine labelled RNA molecules from the mitochondrial polysome cut of the gradient were used for hybridization, about 60 % hybridized one with or the other mtDNA strand, the relative proportion of hybridization with the two strands being, however, roughly maintained (Fig. 7b). The lower extent of hybridization obtained with 8-[^{3}H]adenosine labelled RNA very likely is due to the fact that the labelled poly(A)

stretches of such molecules did not hybridize with mt-
DNA, with the covalently linked poly(A) being in most
part cut off by the RNAse digestion of the hybrids.
Much less hybridization with mtDNA occurred when 8-
[^{3}H]adenosine labelled RNA from ethidium bromide
treated cells was used (Fig. 7c). The low level observed may be due to a small fraction of unlabelled
mtDNA-coded RNA molecules with still attached labelled
poly(A) stretches, which were not cut off by RNAse
treatment of the hybrids. No significant hybridization
occurred between any of the RNA samples and total denatured HeLa DNA.

*Identification of discrete poly(A)-containing RNA
components.* As discussed earlier, the poly(A)contai-
ning mitochondrial RNA from the polysome region of
the gradient showed in sucrose gradient a fairly uniform distribution of components covering the range
from 6 to 16S, with only a hint of a discrete component sedimenting at 7S. The evidence suggested that
the difficulty in resolving discrete components of po-
ly(A)-containing RNA may have been the tendency of
poly(A)-containing molecules to aggregate and cosedi-
ment with heavier molecules. The experiments which
will be described below indicate that this is indeed
the case, and that it is possible to identify discrete poly(A)-containing RNA components either in sucrose
or in gel by using denaturing conditions.

Fig. 8a shows the electrophoretic profile obtained
by running through a polyacrylamide gel, in the native
state, a sample of 8-[^{3}H]adenosine labelled RNA from
the mitochondrial polysome region of the gradient,
which had been selected for poly(A) content by passage through poly(T)-cellulose. The pattern shows a
broad band, partially resolved into two peaks, moving
slower than 12S RNA, with smaller amounts of faster
moving material extending up to the 5S region, without clearly defined components. The corresponding RNA
sample from ethidium bromide treated cells shows a very low level of radioactivity spread fairly uniformly
throughout the proximal half of the gel. After denaturation in formaldehyde and electrophoresis in the

presence of formaldehyde (Fig. 8b), the control RNA
exhibits a broad band of radioactivity in the region
10 to 13S_E, but the most striking fact is the appear-
ence of a pronounced peak at a position corresponding
to about 6.5S_E relative to the 5S RNA marker. The considerable sharpness of the peak, as compared to that
of the 5S RNA and 12S RNA, suggests that it consists
of sequences fairly uniform in length.

Fig. 9a shows the sedimentation profile in sucrose
gradient in low ionic strength buffer of a sample of
2 h 5-[^{3}H]uridine labelled RNA extracted from the polysome region of the sedimentation pattern of a mitochondrial lysate, and passed twice through poly(T)-
cellulose. The sample had been heat denatured in low
ionic strength buffer before the rerun through poly-
(T)-cellulose (in order to reduce to the minimum the
adventious retention by poly(T)-cellulose of non-po-
ly(A)-containing molecules due to base-pairing with
poly(A)-containing molecules), and again prior to the
sedimentation run. One recognized two sharp, partially resolved peaks, one sedimenting somewhat slower
than 5S RNA and the other sedimenting between the 5S
and 12S RNA markers,which emerge over a background of
heterogenous RNA extending on the heavy side of the
faster moving peak up to the 12S region. As appears
in Fig. 9b, the material of the peak sedimenting slower than 5S RNA in low ionic strength buffer, when rerun in acetate-NaCl buffer, migrated as a fairly sharp
peak moving faster than 5S RNA, with an estimated sedimentation coefficient of about 7S relative to the
5S and 12S RNA markers. The material corresponding to
the peak sedimenting between 5S and 12S RNA in low
ionic strength buffer showed in acetate-NaCl buffer a
main component, with an estimated sedimentation coefficient of about 9S; however, this was only partially
resolved from other faster and slower moving components (Fig. 9d). The material of the slower sedimenting peak in Fig. 9a, when run through a polyacrylamide gel in the presence of formaldehyde, migrated as
a sharp 6.5S_E peak.

Fig. 9c shows the sedimentation profile in sucrose

23

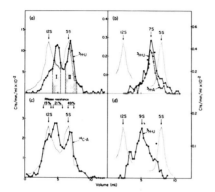

Fig. 9. Sedimentation analysis in low ionic strength buffer [(a) and (c)], or in acetate-NaCl buffer [(b) and (d)], of discrete components of poly(A)-containing RNA. RNA from the polysome region of the sedimentation pattern of a mitochondrial lysate from cells labeled for 2 h with 5[^{3}H]uridine (a) or 8[^{14}C]adenosine (c) was passed twice through poly(T)-cellulose, denatured by heating at 70° C for 6 min in 10^{-3} M Tris buffer, pH 7.4, 2.5 x 10^{-4} M EDTA, and fast cooling, and then run through a 5 to 20 % sucrose gradient in the same buffer in the SW41 rotor at 41,000 rev./min for 30 h. The RNA from the indicated cuts in (a) was collected by ethanol precipitation and centrifugation, and a portion of each sample was run in a sucrose gradient in acetate-NaCl buffer in the SW41 rotor at 41,000 rev./min for 18 h [(b), cut II, and (d), cut I]. In (b), the sedimentation profile of another 7S RNA sample, separated by first running heat-denatured poly(A)-containing 8[^{3}H]adenosine labeled RNA from the polysome region on a gel, and then rerunning the material from a 6-9S$_E$ shoulder in the gel pattern in a sucrose gradient in low ionic strength buffer, is also shown (o--o). The 12S and 5S RNA markers were treated as the experimental samples and run in parallel gradients. The RNA from the indicated cuts in (c) was tested for RNase resistance as explained in the legend of Fig. 6.

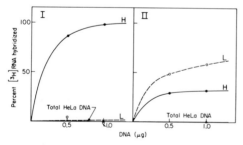

Fig. 10. Homology to separated mtDNA strands of 5[^{3}H]uridine labeled RNA from cuts I and II of Fig. 9a.

24

gradient in low ionic strength buffer of 2 h 8-[^{14}C]-adenosine labelled RNA from the mitochondrial polysome region, which had been passed twice through poly-(T)-cellulose. The pattern is very similar to that of Fig. 9a. The RNA from the indicated fractions was pooled, heat denatured, and subjected to DNAse and pancreatic and T$_1$ RNAse digestion, under the conditions normally used for isolation of poly(A) (13). As appears in Fig. 9c, the labelled material corresponding to the region of the slower and the faster sedimenting peak was, respectively, about 21 % and 49 % DNAse-RNAse resistant; under the same conditions, 12S RNA was made about 94 % acid soluble and [^{3}H]poly(A) was about 100 % RNAse resistant.

In order to investigate the homology to mtDNA of the two discrete RNA components resolved by sedimentation in low ionic strength buffer, aliquots of the material from the region of each of the two peaks in Fig. 9a were hybridized with increasing amounts of separated H or L mtDNA strands. As appears from Fig. 10, the material from the region of the faster moving peak hybridized exclusively with the H mtDNA strand, while the RNA corresponding to the slower moving peak hybridized in its majority (about two-thirds) with the L strand. These results strongly suggest that the 9S component is coded for by the H strand, while the 7S component is coded for by the L strand; the minor hybridization with the H strand of the material from the slower moving peak is presumably due to contaminating RNA sedimenting in the region of the 7S component in the sucrose gradient in low ionic strength buffer.

CONCLUDING REMARKS.

A critical factor for the identification of the two discrete components of poly(A)-containing RNA described above has proven to be the use of denaturing conditions for both the sedimentation and the electrophoretic analysis. This appears to be due to the tendency of the poly(A)-containing mitochondrial RNA molecules to aggregate. This tendency may reflect the existence of complementary sequences in these RNA molecules resulting from the symmetric transcription of mtDNA (10). However, the poly(A) stretch itself may also be involved in this aggregation, possibly due to its lack of secondary structure at neutral pH (27), which makes it available for base-pairing with complementary sequences.

The molecular size of the 7S and 9S components, as estimated from their sedimentation coefficients in sucrose gradient in 0.1 M salt relative to the 5S and 12S RNA markers, using a formula (28) which gives for 16S and 12S mitochondrial RNA values of mol. wt. only slightly lower than those estimated by electron microscopy (29), corresponds to about 8.5×10^4 and 1.5×10^5 daltons, respectively. From the sedimentation velocity of the slower component in sucrose gradient in the presence of formaldehyde (corresponding to about 0.6 times the sedimentation velocity of 12S RNA) one can estimate the molecular weight by using the formula $S \sim M^{0.40}$ (25): this gives 9.3×10^4 daltons. On the basis of the poly(A) content of the slower sedimenting component, estimated from the RNAse resistance data, and assuming a length of about 70 nucleotides for the poly(A) stretch and an A content of 32 % (30) for the mitochondrial RNA molecules to which it is attached, one can arrive at an estimate of about 9.5×10^4 for the size of this component, in good agreement wtth the sedimentation estimates.

As to the significance of the two discrete poly(A)-containing RNA components described here, the most plausible interpretation is that they represent mtDNA coded mRNA, possibly individual messenger species. Previous work from this laboratory (31) has shown that the *in vivo* and *in vitro* products of mitochondrial protein synthesis consist of a group of not well resolved components in the mol. wt. range from 12000 to 25000 daltons, and another group, more abundant, in the range from 40000 to 55000 daltons. However, it is possible that the larger components are polymeric forms of smaller units, as in yeast (32). It is interesting to mention that in the latter organisms the major product of mitochondrial protein synthesis appears to be a protein of molecular weight of 8000 to 9000 daltons (32; see also Tzagoloff *et al.* this volume): this is approximately the size of the protein(s) that would be expected to be coded for by the 7S mitochondrial RNA described here. Conclusive evidence concerning the messenger nature of the discrete poly(A)-containing RNA components identified here will have to come from the demonstration of their *in vitro* template activity for the synthesis of defined products of mitochondrial protein synthesis.

The observation made here that the 7S RNA hybridizes with the L mtDNA strand adds one, and possibly more, RNA species to the three 4S RNA species already known to be coded for by this strand (1,33), and a - plifies the informational significance of the extensive transcription of this strand (10).

ACKNOWLEDGEMENTS.

This work was supported by a grant from the U.S. Public Health Service (GM-11726). The valuable help of Mrs. B. Keeley and A. Drew and Miss G. Engel is gratefully acknowledged. The poly(T)-cellulose was a generous gift of Dr. Joseph Kates, and actinomycin D a gift of Merck, Sharp and Dohme.

REFERENCES.

1. M. Wu, N. Davidson, G. Attardi and Y. Aloni, J. Mol.Biol. 71, 81 (1972).
2. M.M.K. Nass and C.A. Buck, J.Mol.Biol. 54, 198 (1971).
3. I.B. Dawid, J.Mol.Biol. 63, 201 (1972).
4. P. Constantino and G. Attardi, Proc.Nat.Acad.Sci. U.S.A. 70, 1490 (1973).
5. L. Levintow and J.E. Darnell, J.Biol.Chem. 235, 70 (1960).
6. W.E. Barnett and D.H. Brown, Proc.Nat.Acad.Sci. U.S.A. 57, 452 (1967).
7. L. Reijnders and P. Borst, Biochem.Biophys.Res. Commun. 47, 126 (1972).
8. J. Casey, M. Cohen, M. Rabinowitz, H. Fukuhara and G.S. Getz, J.Mol.Biol. 63, 431 (1972).

9. P. Borst, Ann.Rev.Biochem. 41, 333 (1972).
10. Y. Aloni and G. Attardi, Proc.Nat.Acad.Sci.U.S.A. 68, 1757 (1971).
11. Y. Aloni and G. Attardi, J.Mol.Biol. 55, 251 (1971).
12. S. Perlman, H.T. Abelson and S. Penman, Proc.Nat. Acad.Sci.U.S.A. 70, 350 (1973).
13. D. Ojala and G. Attardi, J.Mol.Biol., submitted for publication.
14. S.T. Jacob, D.G. Schindler and H.P. Morris, Science 178, 639 (1972).
15. S.T. Jacob abd D.G. Schindler, Biochem.Biophys. Res.Commun. 48, 126 (1972).
16. P. Borst and L.A. Grivell, FEBS Letters 13, 73 (1971).
17. A.P. Grollman, Proc.Nat.Acad.Sci.U.S.A. 56, 1867 (1966).
18. S. Perlman and S. Penman, Biochem.Biophys.Res. Commun. 40, 941 (1970).
19. D. Ojala and G. Attardi, J.Mol.Biol. 65, 273 (1972).
20. Y. Aloni and G. Attardi, J.Mol.Biol. 70, 363 (1972).
21. Y. Aloni and G. Attardi, J.Mol.Biol. 70, 375 (1972).
22. P. Borst and G.J.C.M. Ruttenberg, Biochim.Biophys. Acta 259, 313 (1972).
23. J.E. Darnell, L. Philipson, R. Wall and M. Adesnik, Science 174, 507 (1971).
24. W. Jelinek, M. Adesnik, M. Salditt, D. Sheiness, R. Wall, G. Molloy, L. Philipson and J.E. Darnell, J.Mol.Biol. 75, 515 (1973).
25. H. Boedtker, J.Mol.Biol. 35, 61 (1968).
26. L. Hatlen, F. Amaldi and G. Attardi, Biochemistry 8, 4989 (1969).
27. A. Rich, D.R. Davies, F.H.C. Crick and J.D. Watson, J.Mol.Biol. 3, 71 (1961).
28. A.S. Spirin, Biokhimiya 26, 454 (1961).
29. D. Robberson, Y. Aloni, G. Attardi and N. Davidson, J. Mol. Biol. 60, 473 (1971).
30. B. Attardi and G. Attardi, J.Mol.Biol. 55, 231

(1971).

31. M. Lederman and G. Attardi, J.Mol.Biol. in press.

32. A. Tzagoloff and A. Akai, J.Biol.Chem. 247, 6517 (1972).

33. Y. Aloni and G. Attardi, J.Mol.Biol. 55, 271 (1971).

THE DNA-DEPENDENT RNA POLYMERASES FROM YEAST MITOCHON-DRIA

T.R. Eccleshall and R.S. Criddle

Department of Biochemistry and Biophysics, University of California, Davis, California (U.S.A.)

Studies on the DNA-dependent RNA-polymerases from mitochondria of various sources have led to some major differences in conclusions regarding the nature of these enzymes. The most thoroughly characterized of these RNA polymerase preparations to date is that from *Neurospora* mitochondria which has been studied by Küntzel and Schäfer (1). Some properties of this enzyme preparation are summarized in Table I. It has been found to have a low molecular weight, single po-lypeptide structure and to be inhibited by rifamycin but not by α-amanitin. Wu and Dawid have reported the existence of a mitochondrial RNA polymerase from *Xenopus'* eggs with a similar small molecular weight but which is not inhibited by rifamycin (2). Low molecular weight RNA-polymerase have also been reported in wheat leaf (3) (rifamycin insensitive) and rat liver mito-chondria (4) (rifamycin sensitive) as summarized in Table I.

In contrast to these studies, it has been demon-strated that the RNA-polymerases associated with yeast mitochondria are much larger in size and resemble more closely the nuclear polymerases. Thus, we have reported multiple RNA polymerases associated with yeast mito-chondrial preparations having molecular weights near 500,000 and showing no sensitivity to rifamycin (5). Similar reports have come from the laboratories of Wintersberger (6) and of Benson (7). Scragg has re-ported a yeast mitochondrial RNA-polymerase with mo-lecular weight greater than 200,000, based on Sepha-dex G-200 chromatography, that does show a rifamycin

TABLE I

MITOCHONDRIAL RNA POLYMERASE PREPARATIONS

Source	Molecular Weights Intact	Subunits	Rifamycin Sensitivity	α-amanitin Sensitivity	Reference
Neurospora		64,000	+	−	Küntzel & Schäfer (1)
			−	−	Wintersberger (6)
	~500,000		−	−	Woodward (9)
Rat liver		~67,000	+	−	Reid & Parsons (4)
			−	−	Wintersberger (6)
			±		Mukerjee & Goldfeder (11)
Xenopus eggs		~57,000	−	−	Wu and Dawid (2)
Yeast I	~500,000	2 large	−	−	Eccleshall, Tsai &
II	~500,000	3 large	−	+	Criddle
III	~500,000	2 large	−	−	(unpublished data)
	>200,000		+	−	Scragg (8)
			−	−	Wintersberger (6)
I(Ia)			−	−	
II			−	+	Benson (7)
	~500,000	2 large	−		Hallick, Hager & Rutter (21)
Ehrlich Ascites			−	−	Jackisch (10)

sensitivity (8). Wintersberger has not been able to find a rifamycin sensitive enzyme in *Neurospora* (6), while Woodward has isolated an enzyme from *Neurospora* mitochondria that is both high molecular weight and rifamycin insensitive (9). The mitochondrial RNA polymerase from Ehrlich ascites cells appears rifamycin insensitive (10) while Mukerjee and Goldfeder (11) report both sensitive and insensitive preparations from rat liver. The overall picture of mitochondrial polymerase is then one of either a great diversity of enzyme types associated with different organisms or else some differences in the isolation procedures employed in the various laboratories give rise to preparations with these differences.

PURIFICATION AND PROPERTIES OF POLYMERASES IN YEAST MITOCHONDRIAL PREPARATIONS

We have continued our studies on the yeast mitochondrial RNA-polymerases in an attempt to further characterize these enzymes. Also, since these enzymes are similar to the corresponding nuclear enzymes, we have carefully compared some of the properties of enzymes from the two sources. Finally, we have investigated isolation conditions to optimize recovery and to determine whether any low molecular weight, rifamycin sensitive form of RNA—polymerase can either be found to exist naturally in yeast or may be formed as a result of the isolation and purification procedures. Our general procedures employed in purification of the mitochondrial polymerases have been published previously (5). In general these include preparation of mitochondria by differential sedimentation or sucrose gradient techniques, rupture of the mitochondria and solubilization of the tightly bound membrane enzyme with 0.5 M KCl. The enzymes are further purified by ammonium sulfate fractionation, DEAE-Sephadex and DNA-cellulose chromatography.

Three fractions with RNA-polymerase activity are obtained following DEAE-Sephadex chromatography. This elution pattern has been used as the basis for naming the enzymes as I, II and III as shown in Fig. 1A. The relative amounts of enzymes I, II and III appear relatively independent of growth phase of the yeast cells over the limited conditions studied but are markedly altered by addition of inhibitors of protease activity to the extraction buffers. Fig. 1B shows that the proportion of enzyme chromatographing as peak I is greatly reduced when homogenization of cells is followed immediately by the addition of 0.5 mM DFP* and 0.5 mM PMSF*. Peak III on the other hand is increased.

A further large increase in the amount of peak III is noted when the antioxidant 2,6-di-t-butyl-4-hydroxymethyl phenol (BHP; 50 μg/ml) is added to the isolation buffers (Fig. 1C). This antioxidant has recently been suggested by Minssen and Munkres (12) to prevent lipid autooxidation and subsequent protein crosslinking among mitochondrial proteins. It was noted that addition of BHP to the mitochondrial preparations gave approximately a ten-fold increase in recovery of total enzyme activity. Thus, use of this reagent has

---

*DFP = diisopropyl fluorophosphate, PMSF = phenylmethyl sulfonyl fluoride.

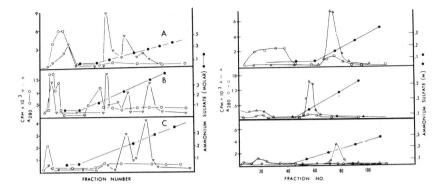

Fig. 1. Separation of RNA-polymerases from yeast mitochondrial preparations by
DEAE-Sephadex chromatography. A: enzymes prepared as described by Tsai *et al.* (5);
B: enzymes isolated in the presence of 0.5 mM PMSF and 0.5 mM DFP; C: enzyme isolated in the presence of PMSF and DFP and 50 μg/ml BHP. The yeast cells were grown
to late log phase (A_{660} = 20-22). Cells were harvested and broken in a glass bead
homogenizer. The homogenate was incubated for 15 min with 10 μg/ml bovine pancreatic DNAse and mitochondria isolated using the procedure of Criddle and Schatz (23)
and then frozen at -20° overnight. The mitochondria were lysed by again homogeni-
zing in the Braun mill for 60 sec. PMSF, when used, was added to a concentration
of 0.5-1 mM after the protein concentration was adjusted so that the A_{280} of a 1 to
100 dilution was between 1 and 2. DNAse (Worthington) was added (10 μg/ml) and the
homogenate stirred an additional 30 min. After centrifugation at 78,000 x g for 2
h, finely ground solid $(NH_4)_2SO_4$ was added to 25 % saturation. The supernatant was
then made 50 % saturated with respect to $(NH_4)_2SO_4$. The precipitate was dissolved
in TGMED containing 0.05 M $(NH_4)_2SO_4$ and dialyzed against the same buffer for 4 h.
This sample was chromatographed on a DEAE-Sephadex column.

After loading, the column was washed with 1-2 bed volumes of TGMED containing
0.05 M $(NH_4)_2SO_4$ and the RNA polymerase activities were eluted with a gradient of
0.05 to 0.50 M $(NH_4)_2SO_4$ in TGMED (6-7 bed volumes), (TGMED: 0.05 M Tris-HCl (pH
7.9), 5 mM $MgCl_2$, 0.1 mM EDTA, 25 % (v/v) glycerol, 0.5 mM DTT, 10 mM thioglycerol).
Assays for RNA-polymerase activity were carried out in a volume of 250 μl containing 25 μmoles Tris-HCl (pH 7.5), 40 mμmoles each of ATP, GTP, and CTP, 8 mμmoles
of UTP (unlabeled), 0.154 mμmoles of [³H]UTP (13-15 C/mmole), 20 μg native calf
thymus DNA, 1 μmole $MgCl_2$, 0.25 μmoles $MnCl_2$ and 1 μmole mercaptoethanol. The reaction mixture was incubated at 29° for 10 min. Reaction was stopped by the addition of 5 % (w/v) ice-cold TCA solution containing 0.1 M sodium pyrophosphate. The
reaction products were collected on cellulose nitrate filters and washed three times with 10 ml of the cold TCA solution. The filters were dried and counted in tolu-
ene scintillation fluid.

Fig. 2. DNA-cellulose chromatography of RNA-polymerase enzyme isolated from mitochondrial preparations. DNA-cellulose was prepared from single -stranded calf thymus DNA and cellulose by irradiating with UV light according to the method of Litman (24). A column (10-11 cm x 0.9 cm) of DNA-cellulose was equilibrated with TGMED
containing 0.05 M $(NH_4)_2SO_4$ (or 0.03 M $(NH_4)_2SO_4$ when peak II was to be chromatographed). The enzyme fraction was loaded at $\frac{1}{2}$ bed volume/h with frequent 15 min
periods when the flow was stopped for the enzyme to equilibrate with the DNA. After
washing with 1-2 bed volumes of the original buffer, 5-6 bed volume linear gradient
of 0.05 M $(NH_4)_2SO_4$ to 0.5 M $(NH_4)_2SO_4$ in TGMED was applied. Fractions were assayed
as indicated in the legend of Fig. 1.

34

allowed recovery of the yeast mitochondrial enzymes on a scale that now makes physical studies on the enzyme and its subunits possible.

After initial separation on DEAE-Sephadex, each of the three RNA polymerase fractions was rechromatograph-ed on DEAE-Sephadex and further purified on DNA-cellulose columns for structural analysis. The purified components were then analyzed by SDS disc gel electrophoresis. The final elution profiles of each of the enzyme fractions from DNA-cellulose are shown in Fig. 2. The corresponding SDS-disc gel patterns of subunits from each of these fractions are shown in Fig. 3. Using the method of Weber and Osborn (13), the molecular weights of each of the major subunit components of the RNA polymerases were determined. These are summarized in Table II. Both enzymes I and III have two major bands which are in approximately 1:1 stoichiometric ratios. Enzyme II was found to have three major subunit components, with the two larger components existing in molar ratios approximately one half that of the 150,000 molecular weight component. The existence of other low molecular weight components of the polymerases cannot be ruled out from these studies.

TABLE II

MOLECULAR WEIGHTS OF RNA-POLYMERASE SUBUNITS DETERMINED BY SDS-GEL ELECTROPHORESIS

	Molecular weights			Stoichiometry
Enzyme I	145,000	210,000		~1:1
Enzyme II	150,000	200,000	260,000	~1:0.4:0.5
Enzyme III	135,000	190,000		1:0.9

COMPARISON OF NUCLEAR AND MITOCHONDRIAL ENZYMES

RNA-polymerases have been isolated from our whole cell yeast preparations by modification of the technique of Wintersberger (14), for comparison with the enzymes isolated from our mitochondrial preparations. Separation of these components by DEAE-Sephadex chromatography indicates the existence of two enzymes in our preparation, while a third enzyme is noted by other workers (14,15). The first two of these enzymes, designated nuclear enzymes I and II, are present in

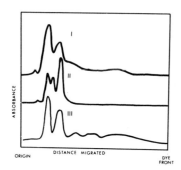

Fig. 3. SDS-gel electrophoresis of DNA-cellulose purified enzymes from yeast mitochondria. SDS-gel electrophoresis was performed using the method of Weber and Osborn
(13). The samples were prepared by boiling for 10 min in 1 % SDS and 2 % mercaptoethanol. The enzyme samples were stained with Coomassie blue stain and scanned with
a Gilford spectrophotometer at 549 nm. Sample migration is from left to right.

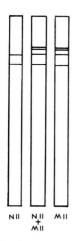

N II   N II   M II
        +
       M II

Fig. 4. SDS-gel electrophoresis for comparison of RNA-polymerases from nuclear enzyme preparation peak II and mitochondrial enzyme preparation peak II. Each enzyme
was electrophoresed separately and also combined for an electrophoretic analysis.
Protein was stained with Coomassie blue dye. Migration is from top to bottom.

large amounts relative to peak III. All three peaks
have elution properties from DEAE-Sephadex very similar to the corresponding enzymes isolated with the mitochondria and therefore further comparisons were made
(14,15). Co-electrophoresis of purified nuclear and
mitochondrial peak I enzymes on SDS gels indicated
that the subunit components from the two enzymes were
identical in size. These results suggest that peak I
enzyme found associated with the mitochondrial preparations may be a nuclear contaminant. Neither of these
enzyme preparations showed any sensitivity to inhibition by rifamycin or α-amanatin.

When nuclear enzyme II was electrophoresed on SDS
gels only two major bands were found. These had R_m
values identical to two of the bands from the mitochondrial enzyme preparation. As illustrated in Fig.
4, however, the slow moving ~ 250,000 molecular
weight component seems unique to the mitochondrial enzyme preparation. In the experiment of Fig. 4, the
high molecular weight band is split into major and minor components. This difference disappears when enzyme
is isolated in the presence of BHP. Since enzyme II
from both mitochondrial and nuclear preparations is
α-amanitin sensitive and similar in many properties,
doubt may be expressed concerning whether each is a
unique enzyme. These results definitely show that differences do exist between the isolated enzymes. Such
differences may or may not exist in the cell, as it
is possible that mitochondrial enzyme II is truly i-
dentical with nuclear enzyme II but is subsequently
modified during isolation so that roughly 50 % of the
enzyme has a higher molecular weight large component.
We have no means at present of knowing whether such a
proposed modification would be a physiological property of mitochondrial enzyme or whether it may simply
be an artifact of isolation which results in the formation of a higher molecular weight component. The
observed stoichiometric relationships suggest that the
additional gel component of mitochondrial enzyme II is
not just a non-enzyme contaminant of isolation.

An alternate explanation of the data would be to

propose that a distinct mitochondrial enzyme with high molecular weight subunit does exist. This may be contaminated in our preparations by as much as 50 % nuclear enzyme, thus yielding the three observed electrophoretic components. A final straight forward (but not necessarily correct) interpretation would be that mitochondrial peak II enzyme does indeed have three major subunit classes in a 1:1:2 stoichiometry, two of which are similar or identical to nuclear enzyme.

The results of Kedinger and Chambon (16) are interesting in this regard as they show two α-amanitin sensitive RNA-polymerases in preparations of calf thymus with electrophoretic properties very similar to what is noted here. Our results would lead us to speculate that these workers may in fact be seeing a mitochondrial as well as nuclear enzyme.

The data obtained upon isolation of RNA polymerases from mitochondrial preparations in the presence and absence of protease inhibitors and BHP indicates an inverse yield relationship between enzyme I and III. That is, when protease inhibitors and antioxidant are added, the yield of peak III increases greatly with respect to peak I and somewhat with respect to II. Two general schemes may be proposed to account for such observations. First, enzyme III could be converted into I in the absence of inhibitors. Such a change could result from proteolytic action on enzyme III or by lipid autooxidation covalently linking protein to some additional component to make a modified enzyme. A second scheme would propose that no interconversion takes place but that inhibitor and antioxidant function indirectly in either limiting the extraction of peak I enzyme or greatly favoring extraction of peak III enzyme during mitochondrial isolation.

If protease action on enzyme III gave rise to enzyme I, one would expect that the molecular weights of subunits of III would be greater than I. Just the opposite is found. Modification by lipid oxidation could possibly give the higher molecular weight forms, but since both major subunits would have to be altered to fit the data, this seems unlikely. When the yields

of RNA polymerase activities are analyzed at each step
of extraction it seems clear that addition of BHP during enzyme isolation allows a greatly enhanced extraction of peak III enzyme from mitochondria with virtually no effect on yield of enzyme I. Some increase in
enzyme II is also noted. The increased extraction appears to be a proportionate increase in both mass and
activity. BHP thus seems to have a direct effect upon
the binding of these enzymes to components of the mitochondrial membrane system. Thus conditions which
limit proteolytic digestion and conditions which favor
increased enzyme extraction from mitochondria both
yield increased total activity of enzymes II and III.
These observations support a conclusion that at least
peak III enzyme is specifically associated with mitochondria. Furthermore, comparison of the catalytic,
chromatographic and subunit properties of this enzyme
with those of the nuclear RNA polymerase enzymes from
yeast indicates that this enzyme is probably unique to
the mitochondria. It is less clearly shown but strongly suggested that a peak II enzyme may also be mitochondrial.

REACTION CHARACTERISTICS

Comparisons of RNA polymerases from different sources have frequently been facilitated by systematic
variations in reaction conditions. We have therefore
used this procedure to study each of the enzymes from
mitochondrial and nuclear preparations. Complete details of these results will not be presented here but
the results may be summarized as follows. The effect
of increasing ionic strength in the reaction mixture
by addition of ammonium sulfate shows all three mitochondrial enzymes to have optima in the range of 10-40
mM. The peak III mitochondrial enzyme also has an additional optimum at approximately 200 mM ammonium sulfate as shown in Fig. 5. Both peak I and peak II enzymes become markedly inhibited above 80 mM ammonium
sulfate. The two nuclear enzymes showed ionic strength
dependency similar to I and II, in agreement with the
results of Adman, Schultz and Hall (15).

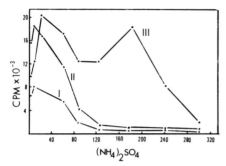

Fig. 5. Effect of ammonium sulfate on the activity of RNA-polymerases isolated from yeast mitochondrial preparations. The assays were carried out as described in the legend of Fig. 1 except for changes in ammonium sulfate concentrations. Ammonium sulfate concentrations are represented as milimolar.

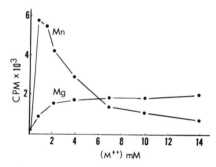

Fig. 6. Manganese and magnesium ion dependence of peak III enzyme from yeast mitochondria. Assays were carried out as described in the legend of Fig. 1 except for the changes in metal ion concentration.

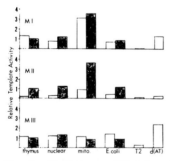

Fig. 7. Template specificities of RNA-polymerases from mitochondrial preparations. DNA preparations from calf thymus, yeast nuclei, yeast mitochondria, *E.coli*, T2 phage, and poly d(AT) were used as templates for the three enzyme preparations. Activities are reported relative to the activity of the enzyme using denatured calf thymus DNA as template. Solid bars indicate denatured templates, open bars indicate native DNA used as templates.

TABLE III

METAL ION DEPENDENCE OF YEAST RNA-POLYMERASES.
RNA-polymerase enzymes were assayed as described in the legend of Fig. 1, except that Mn^{++} and Mg^{++} concentrations were varied over the range 0 to 15 mM. The mitochondrial enzyme preparations used in these studies were purified by chromatography two times on DEAE-Sephadex. Data for nuclear preparations taken from the studies of Ponta, Ponta and Wintersberger (14).

	Optimum	
	Mn^{++}	Mg^{++}
Mitochondrial Preparations		
MI	2 mM	~10 mM (broad)
MII	4 mM	none
MIII	1 mM	> 3 mM
Nuclear Preparation		
NI	1.6 mM	~ 8 mM (broad)
NII	2 mM	none

Comparison of the metal ion dependence of the reactions of the various polymerases are shown in Table III. Again great similarities are noted between mitochondrial and nuclear enzyme preparations of peak I and II enzymes. The mitochondrial peak III enzyme uses Mn^{++} more effectively than Mg^{++} and has optimum at a lower level of added Mn^{++} as shown in Fig. 6.

When template specificities are compared for the various enzyme fractions using our assay conditions, again great similarities are noted between the nuclear and mitochondrial enzyme preparations. These data are shown in Fig. 7 for both native and denatured DNA templates from a number of sources. Peak II enzymes have a great preference for denatured DNA templates and work most efficiently with denatured mitochondrial DNA. Nuclear and mitochondrial peak I enzymes are equally effective with either native or denatured DNA templates but again use the mitochondrial DNA as the best template tested. The peak III enzyme, which appears to be specific to the mitochondrion, does not show any preference for mitochondrial DNA templates using our reaction conditions. Native and denatured DNAs are both used about equally well with the only preference noted being its greatly enhanced activity on poly d(AT) templates.

None of these purified RNA polymerases are sensitive to inhibition by rifamycin. All are inhibited to about the same extent by rifampin AF/013 and both peak

41

II preparations are inhibited by α-amanitin (Table IV).

TABLE IV

EFFECT OF INHIBITORS ON YEAST RNA-POLYMERASES.
Inhibitors were added to RNA-polymerase preparations in the assay system described in Fig. 1 using the method of Meilhac, Tysper and Chambon (22). Inhibitor concentrations were, α-amanitin 20 μg/ml, rifamycin 20 μg/ml, and rifampin AF/013 20 μg/ml. The data of the nuclear enzymes are taken from ref. 15.

Enzyme Source	% Inhibition		
	Rifamycin	α-Amanitin	Rifampin AF/103
Mitochondrial I	6	0	97
Mitochondrial II	3	100	100
Mitochondrial III	8	0	100
Nuclear I	0	0	92
Nuclear II	0	82	100
Nuclear (?) III	0	0	94

IS THERE A RIFAMYCIN SENSITIVE RNA-POLYMERASE IN YEAST?

Since rifamycin sensitive RNA polymerase preparations have been reported from yeast as well as from other mitochondrial sources, we have made a major effort at investigating other preparatory techniques to locate such activity in our yeast mitochondrial preparations. When we apply to yeast the procedures used successfully by Küntzel and Schäfer (1) in isolation of *Neurospora* mitochondrial enzyme, we cannot find a similar low molecular weight or inhibitor sensitive enzyme. When we follow precisely the isolation procedures of Scragg (8), who reported a high molecular weight rifamycin-sensitive enzyme from yeast, we were unable to show an enzyme which is inhibited by 40 μg/ml rifamycin.

It should be noted at this point that in no case where a low molecular weight mitochondrial RNA-polymerase has been reported was any mention made of the addition of protease inhibitors to the extraction medium. Also, the procedure of Scragg does not employ such inhibitors during extraction or further isolation. Preliminary experiments in our laboratory have indicated that limited hydrolysis of peak III enzyme with proteases causes breakdown of enzyme to distinct fragments observable on disc gel electrophoresis and that

this breakdown process appears to proceed more rapidly than loss of enzyme activity. We are not suggesting
at this time that proteases are responsible for formation of the low molecular weight polymerases, but
this could account for some of the major differences
observed in various laboratories. Proteases are certainly important in studies on the yeast system, since
the subunit molecular weights of peak II are distinct-
ly different and smaller in the absence of protease
inhibitors.

The differences between rifamycin sensitive and rifamycin resistant RNA-polymerase activity in yeast
may not be too large as Scher and Thomas (17) have
reported isolation of a rifamycin sensitive mutant of
yeast prepared from the normally resistant wild-type
strain. Genetic differences of this type could also
be used to explain some discrepancies between reports
of sensitive and insensitive polymerases.

RNA-POLYMERASE AS A MEMBRANE BOUND ENZYME

Initial reports on solubilization of RNA polymerase
activity from mitochondria described the use of detergents such as deoxycholate to release the enzyme from
the membrane system and led to suggestions that the
enzyme was tightly associated as a membrane complex.
We subsequently demonstrated solubilization of RNA
polymerase activity from mitochondria by treatment
with 0.5 M KCl. Repeated washings with lower levels
of KCl or other salts were not effective in solubilization of active enzyme. Both of these procedures support the conclusion that RNA polymerase may be considered part of a membrane complex. Further evidence
which would support such a conclusion is the demonstration in other systems of a lipid dependence of
some polymerase activities (18,19). Evidence for mitochondrial RNA-polymerase function in the lipid membrane complex has been discussed by Linnane and co-
workers (20). Our findings on the effect of an antioxidant shown to prevent lipid cross-linking of proteins in mitochondria, offer still another suggestion
of an intimate association of the enzyme with the mem-

43

brane matrix. It is quite clear that BHP allows a 10-fold increase in the amount of extractable RNA-polymerase using our conditions. While the mechanism of this release is unknown, a direct effect at the membrane level is strongly implied. The enzyme extracted from such surroundings may definitely be altered in its properties. Thus, it is quite possible that activity and reaction specificity of isolated enzyme may not be an accurate expression of *in vivo* activity.

CONCLUDING REMARKS

The major conclusions of these studies are that three separate RNA polymerase enzymes may be isolated from preparations of yeast mitochondria. Addition of protease inhibitors and the antioxidant BHP greatly affect the relative proportions and total yield of these enzymes. All three enzymes are insensitive to rifamycin. Peak I enzyme is found in much lower concentrations than II and III when isolation conditions are optimal. No differences between nuclear enzyme I and mitochondrial enzyme I have been noted in enzymatic or subunit structure. It is thus probable that this component in mitochondrial preparations represents contamination with nuclear enzyme.

Enzyme II also resembles the nuclear preparation of enzyme II in its enzymatic properties and inhibitor sensitivity. However, SDS gel electrophoresis of this preparation indicates that an additional major peptide component is found which differs from the nuclear enzyme preparation. It is difficult to state at this time whether all or part of this enzyme fraction is truly mitochondria-specific. The observations are consistent with two separate peak II enzymes, one nuclear and one mitochondrial or with one nuclear enzyme modified *in vivo* to function in the mitochondria. The large molecular weight polypeptides found in the mitochondrial preparations may alternatively be only an artifact of isolation which has resulted in addition of some 40,000 MW component to the large peptide of RNA polymerase yet which has no physiological meaning.

The enzyme of peak III appears to be specifically

an RNA polymerase of yeast mitochondria. It has catalytic and physical properties differing somewhat from
all nuclear counterparts and is found only in preparations containing mitochondria. It is greatly enriched
in all mitochondrial preparations but is released from
the mitochondria in high yields at high salt concentrations only in the presence of an antioxidant designed to prevent lipid autooxidation and cross-linking.

This mitochondrial enzyme resembles most commonly
reported eukaryote enzymes in size. It has two major
subunits of 135,000 and 190,000 molecular weight existing in 1:1 molar ratios and possibly other low molecular weight components as well. The enzyme utilizes denatured calf thymus DNA somewhat better than native
DNA for template and shows no great preference for mitochondrial as opposed to calf thymus DNA templates
under our assay conditions. The mitochondrial enzyme
does have a somewhat greater specificity for poly dAT
templates than other enzymes tested.

REFERENCES

1. H. Küntzel and K.P. Schäfer, Nature New Biol. 231,
   265 (1971).
2. G.-J. Wu and I.B. Dawid, Biochem. 11, 3589 (1972).
3. G.M. Polya, Arch.Biochem.Biophys. 155, 125 (1973).
4. B.D. Reid and P. Parsons, Proc.Natl.Acad.Sci.U.S.
   A. 68, 2830 (1971).
5. M.-J. Tsai, G. Michaelis and R.S. Criddle, Proc.
   Natl.Acad.Sci.U.S.A. 68, 473 (1971).
6. E. Wintersberger, Biochem.Biophys.Res.Commun. 48,
   1287 (1972).
7. R.W. Benson, Fed.Am.Soc.Exp.Biol., Abstr. 31, no.
   1450 (1972).
8. A.H. Scragg, Biochem.Biophys.Res.Commun. 45, 701
   (1971).
9. Woodward, personal communication.
10. R. Jackisch, A. Jung, W. Schlegel and D. Mayer,
    Hoppe-Seyler's Z.Physiol.Chem. 353, 1705 (1972).
11. H. Mukerjee and A. Goldfeder, Fed.Am.Soc.Exp.Biol.
    Abstr.31, no. 538 (1972).
12. M. Minssen and K.D. Munkres, Biochim.Biophys.Acta

$\underline{291}$, 398 (1973).

13. K. Weber and M. Osborn, J.Biol.Chem. $\underline{244}$, 4406 (1969).
14. H. Ponta, U. Ponta and E. Wintersberger, Europ.J. Biochem. $\underline{29}$, 10 (1972).
15. R. Adman, L.P. Schultz and B.D. Hall, Proc.Natl. Acad.Sci.U.S.A. $\underline{69}$, 1702 (1972).
16. C. Kedinger and P. Chambon, Europ.J.Biochem. $\underline{28}$, 283 (1972).
17. S. Scher and D. Thomas, J.Cell Biol. $\underline{55}$, 229a (Abstract 456) (1972).
18. I.A. Menon, Can.J.Biochem. $\underline{50}$, 807 (1972).
19. A. Lezius and B. Müller-Lornsen, Hoppe-Seyler's Z.Physiol.Chem. $\underline{353}$, 1872 (1972).
20. A.W. Linnane, J.M. Haslam and I.T. Forrester, in: Biochemistry and Biophysics of Mitochondrial Membranes, Eds., G.F. Azzone, E. Carafoli, A.L. Lehninger, E. Quagliariello and N. Siliprandi, Academic Press, New York, 523 (1972).
21. R.B. Hallick, G.L. Hager and W.J. Rutter, Fed.Am. Soc.Exp.Biol. Abstr. 32, no. 2299 (1973).
22. M. Meilhac, Z. Tysper and P. Chambon, Europ.J.Biochem. $\underline{28}$, 291 (1972).
23. R.S. Criddle and G. Schatz, Biochem. $\underline{8}$, 322 (1969).
24. R.M. Litman, J.Biol.Chem. $\underline{243}$, 6222 (1968).

A MITOCHONDRIAL DNA-DIRECTED RNA POLYMERASE FROM YEAST MITOCHONDRIA

Alan H. Scragg

National Institute for Medical Research, Mill Hill, London NW7 1AA, Great Britain.

Isolated mitochondria are capable of incorporating triphosphates into RNA (1-4) using an RNA polymerase(s) located within the mitochondrion (5). Recently a number of mitochondrial RNA polymerases have been isolated in significantly pure states. Those from *Neurospora crassa* (6), and rat liver mitochondria (7) have been characterized by their sensitivity to rifampicin, an inhibitor of bacterial RNA polymerase (8), and insensitivity to α-amanitin, an inhibitor of one of the nuclear polymerases (9). In contrast, the mitochondrial RNA polymerase from *Xenopus laevis* (10) is insensitive to rifampicin. With regard to the yeast mitochondrial RNA polymerase reports are confusing both rifampicin sensitivity (11), and insensitivity (12,13) having been reported. In this paper I report the isolation, and characterization of a rifampicin-sensitive RNA polymerase from yeast mitochondria which is clearly different from the nuclear polymerases.

METHODS.

Saccharomyces cerevisiae (strain 239 N.C.Y.C.) was grown and mitochondria prepared as described previously (11). The mitochondrial polymerase was solubilized from the isolated mitochondria by sonication at full power for 1 min buffer I (0.01 M Tris-HCl buffer, pH 7.8, containing 2 mM $MgCl_2$, 0.5 mM EDTA, 5 mM dithiothreitol and 15 % glycerol), or by incubation with digitonin (final concentration of 1 %) for 30 min at 2° C. The debris was removed by centrifugation at 15000 x g for 30 min and the supernatant referred to a crude

47

extract.

Assay of RNA polymerase activity. The reaction
mixture contained in a total volume of 50 μl, 50 mM
Tris-HCl buffer (pH 7.5), 20 mM magnesium acetate, 1
mM $MnCl_2$, 1 mM each of GTP and CTP, 2 mM ATP, 0.33
μCi [^{3}H]UTP (specific activity) 10.5 Ci/μmole, 10 μg
denatured calf thymus DNA and enzyme. Incubation was
at 37° C, after 20 min the 50 μl was removed and applied to a Whatmann No. 1 filter paper disc (2.4 cm
diameter) and treated as described by Bollum (14).

Polyacrylamide gel electrophoresis of RNA. The
gels were prepared and run as described by Dingman and
Peacock (15). The gels were sliced into 1 mm discs
and the discs digested with 0.2 ml 0.2 M NaOH at room
temperature overnight and counted in 10 ml dioxane
scintillant (16).

*SDS gel electrophoresis of in vitro and in vivo
products.* The cell-free system was prepared, run and
analyzed as described using 10 % gels (17). The *in vivo* products were labelled and prepared as described
by Thomas and Williamson (18).

Self-annealing of the transcription products was
carried as described by Schäfer *et al.* (19).

The yeast nuclear RNA polymerases were isolated as
described by Ponta, Ponta and Wintersberger (20).

MATERIALS.

Purified yeast mitochondrial DNA was kindly supplied by Dr. D.H. Williamson and T4 DNA by Dr. A.A.
Travers. [^{3}H]labelled rifampicin was kindly supplied
by Gruppo Lepetit, Milano and α-amanitin was a gift
from Professor T. Wieland (Heidelberg).

RESULTS.

Solubilization and isolation of the mitochondrial
RNA polymerase from a number of organisms has proved
difficult as many of the standard procedures for enzyme purification cause inactivation (6,21). Of the
methods tried for solubilization of the enzyme from
yeast mitochondria only sonication in low salt or
treatment with digitonin gave consistent results, the

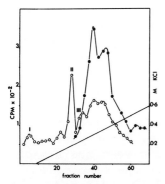 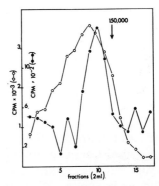

Fig. 1. DEAE-cellulose chromatography of the crude mitochondrial extract on a 2x12 cm column. 2 mg of protein was applied and elution was with a linear gradient (60 + 60 ml; 0.0-0.7 M KCl in buffer 1). 2.5 ml fractions were collected and assayed for polymerase activity as described in Methods (0—0). The pattern obtained for the nuclear enzyme (prepared as described in Methods) run under similar conditions is also shown (●—●).

Fig. 2. Density gradient centrifugation of enzyme activity II (after DEAE-cellulose, Fig. 1.). 2 µl of enzyme solution was layered onto a 32 ml linear 10-30 % glycerol gradient in buffer I containing 0.05 M KCl and centrifuged for 16 h at 25,000 g in an SW25/1 rotor. Parallel gradients were run with total nuclear polymerase and bovine gamma globulin (mol wt 150,000) as markers. 2 ml fractions were collected and assayed for enzyme activity (see Methods). (●—●): mitochondrial RNA polymerase; (0—0): nuclear RNA polymerase.

TABLE I

EFFECT OF INHIBITORS ON NUCLEAR AND MITOCHONDRIAL POLYMERASES.
The mitochondrial polymerases were those described in Fig. 1 and assayed as described in Methods. The nuclear polymerases were prepared and assayed as described by Ponta, Ponta and Wintersberger (18). The figures in brackets indicate the percentage inhibition obtained with the two antibiotics.

Additions	cpm/0.05 ml		
	peak I	peak II	peak III mito.
no DNA	176	58	60
denatured calf thymus DNA	260	131	112
rifampicin 15 µg/ml	188 (28)	70 (47)	65 (42)
amanitin 3.3 µg/ml	254 (2)	125 (5)	106 (5)
	pol. 1	pol. 2 nuclear	
denatured calf thymus DNA	577	2238	
rifampicin 20 µg/ml	565 (2)	2242 (0)	
amanitin 2 µg/ml	607 (0)	986 (56)	

digitonin treatment giving a better yield. Enzyme activity and rifampicin sensitivity were low and variable in crude extracts probably due to the presence of
nucleases. The enzyme activity is also unstable, being
lost after 2-3 days at -20° C, even when high concentrations of glycerol and dithiothreitol are used in
the storage medium.

Fig. 1 shows the chromatography of a crude mitochondrial extract on DEAE-cellulose. Three peaks are
obtained, the first eluting with the wash has a high
endogenous activity and is probably due to overloa-
ding. The main peak II elutes at 0.18 - 0.2 M KCl
prior to the nuclear polymerases (a nuclear enzyme
profile is shown in Fig. 1) eluting at 0.3 M and 0.37
M KCl respectively. Four nuclear RNA polymerase peaks
cen be obtained if the KCl is replaced by ammonium
sulphate. All three mitochondrial peaks show sensitivity to rifampicin and insensitivity to α-amanitin
in contrast to the nuclear RNA polymerases (Table I).

Fig. 2 shows the glycerol gradient centrifugation
of peak II from Fig. 1. Parallel gradients were run
with total nuclear RNA polymerase and bovine gamma
globulin as markers. Under these conditions of low
salt the enzyme appears to have a molecular weight of
about 200,000. However SDS gel electrophoresis of the
glycerol gradient peak yields a single main band which
when co-run with markers gave a molecular weight of
59-63,000 (Fig. 3). The effect of rifampicin and rifamycin SV on the further purified enzyme is shown in
Fig. 4. 10 µg/ml rifampicin was sufficient to give 95
% inhibition whereas rifamycin SV was less active giving 75 % inhibition at 10 µg/ml. Both inhibitors had
little effect on the nuclear polymerases 1 and 2 except at very high concentrations.

As stated before it is difficult to demonstrate rifampicin sensitivity in crude mitochondrial extracts
due probably to high nuclease activity. However, the
binding of [3H]rifampicin to the enzyme in crude extracts can be demonstrated by Sephadex G-25 chromatography (Fig. 5). A peak of [3H]rifampicin is found at
the void volume with both crude and partially purified

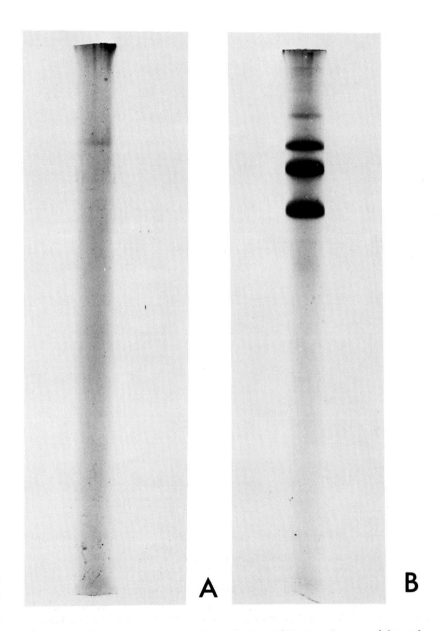

Fig. 3. SDS gel electrophoresis of the mitochondrial RNA polymerase (glycerol gradient enzyme) through 10 % gels as described in Methods. A: mitochondrial RNA polymerase (peak from glycerol gradient Fig. 2.); B: mitochondrial RNA polymerase with three markers, catalase (mol wt 53,000), ovalbumin (mol wt 40,000) and lactic dehydrogenase (mol wt 33,500).

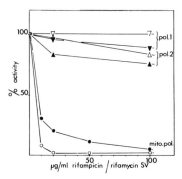

Fig. 4. The effect of rifampicin and rifampicin SV on the mitochondrial and nuclear polymerases. The mitochondrial enzyme was purified (90 %) by chromatography on a DNA-cellulose column as described by Alberts and Herrick (26), after DEAE-cellulose chromatography. Nuclear polymerases 1 and 2 were prepared and assayed as described (20). Rifampicin (open symbols); rifampicin SV (closed symbols).

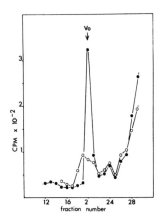

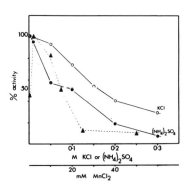

Fig. 5. Analysis of [³H]rifampicin complexes in crude and partially purified mitochondrial polymerase preparations. 0.5 µCi [³H]rifampicin was mixed with crude extract (12 µg protein) or DEAE-cellulose peak (800 µg protein) and applied to an 0.8 x 25 cm Sephadex G-25 column. The column was equilibrated and developed with buffer I. 0.25 µl fractions were collected and counted in 10 ml dioxane scintillant (15). Crude extract (O—O), DEAE-cellulose enzyme (●—●).

Fig. 6. The effect of salt concentration on the mitochondrial polymerase (DEAE-cellulose enzyme). KCl (O—O); ammonium sulphate (●—●); manganese chloride (▲-----▲).

enzyme. This peak is not found when the enzyme is omitted or substituted by nuclear RNA polymerase 1 (results not shown).

As found for the *Xenopus* mitochondrial RNA polymerase (10) increasing concentrations of KCl and ammonium sulphate cause inhibition of the yeast mitochondrial enzyme (Fig. 6). A concentration of 0.1 M KCl or ammonium sulphate gave inhibition of 25 % and 50 % respectively. The mitochondrial enzyme has an optimum for manganese chloride of 1-3 mM at 20 mM magnesium acetate.

The yeast mitochondrial RNA polymerase can use both yeast mitochondrial DNA and T4 DNA as templates, but functions well only with denatured calf thymus or yeast nuclear DNA. Fig. 7 shows polyacrylamide gel analysis of the products of the mitochondrial enzyme with either mitochondrial or nuclear DNA as the template. The mitochondrial DNA produces RNA 20-4S in size whereas the nuclear DNA's main product is of 4-5S in size.

Although in HeLa cells, rat liver and *Xenopus* (21) it is clear that virtually all the haevy strand of the mitochondrial DNA is transcribed and at least some of L strand, in yeast this remains unanswered. The RNA transcribed by the mitochondrial polymerase from mitochondrial DNA has been self-annealed and some 70 % remains ribonuclease sensitive (Fig. 8).

E.coli cell-free system capable of transcribing and translating small DNAs such as T7, ФX174 have been developed (22,23). To such a system the RNA transcribed from mitochondrial DNA by the mitochondrial polymerase has been added. The protein products have been analyzed by precipitation with antisera prepared against the mitochondrial membrane proteins, and SDS gel electrophoresis of the precipitates (17). This has been compared with those membrane proteins labelled *in vivo* in the presence of cycloheximide (18). The three main proteins labelled *in vivo* are found in the *in vitro* products (Fig. 9). In addition proteins of different and higher molecular weights are found in the *in vitro* products which represent minor species found only *in*

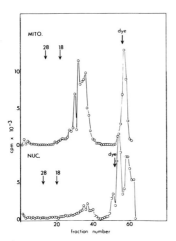

Fig. 7. Polyacrylamide gel analysis of the products of the mitochondrial polymerase and nuclear or mitochondrial DNA. The gels were run and analyzed as described in Methods. Larger reaction mixtures were used (250 μl total), and included 0.3 μCi of GTP, CTP and ATP. Ribosomal RNA was run in parallel as marker.

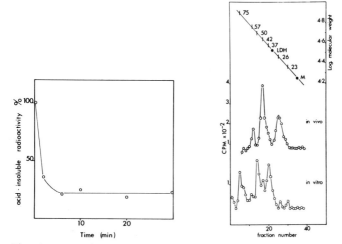

Fig. 8. Analysis of RNAse sensitivity of the transcription products of mitochondrial polymerase and DNA after self-annealing. Larger reaction mixtures were
used as described in Fig. 7. and analysis performed as described in Methods.
Without self-annealing some 10 % of the transcription products remained RNAse
resistant.

Fig. 9. SDS polyacrylamide gel analysis of *in vivo* and *in vitro* products. The
cell-free system was run and gels analyzed as described in Methods. Markers
lactic dehydrogenase (mol wt 33,500) and myoglobin (17,500) were run at the
same time and molecular weights estimated.

vivo when concentrated by use of antisera (17).

DISCUSSION.

An RNA polymerase from yeast mitochondria has been isolated. The enzyme can be separated from the nuclear enzyme by DEAE-cellulose chromotography, eluting before the nuclear enzyme as two peaks. The enzyme is sensitive to rifampicin and insensitive to α-amanitin (Table I) as are the enzymes from *N. crassa* (6) and rat liver mitochondria (7). This is in contrast to other reports (12,13) showing yeast mitochondrial RNA polymerases insensitive to both rifampicin and α-amanitin. Upon further purification the enzyme's sensitivity to rifampicin increases (Fig. 4) being inhibited 95 % and 75 % by 10 μg/ml rifampicin and rifamycin SV respectively. The specific activities obtained with partially purified mitochondrial enzyme are 18-34 pmoles [^{3}H]UMP/μg protein/20 min which is low compared with a value of 2000 pmoles [^{3}H]UMP/mg protein/20 min (24) for the nuclear polymerase 2 and 906 pmoles [^{3}H]AMP/ μg protein/20 min for the *E. coli* RNA polymerase (25). This difference may be due to the labile nature of the mitochondrial enzyme.

The sensitivity of the mitochondrial enzyme to rifampicin can be demonstrated by the binding of [^{3}H]- rifampicin (Fig. 5) in both crude and partially purified extracts.

When analyzed by glycerol gradient centrifugation the mitochondrial RNA polymerase appears to have a molecular weight of about 200,000, but SDS gel electrophoresis shows that it consists of a single subunit of molecular weight 59-63,000. The formation of large aggregates having a small molecular weight subunit is similar to that found in *N. crassa* (6), rat liver (7) and *X. laevis* (10) mitochondrial polymerases. Also, similar is the enzymes' sensitivity to increasing concentrations of both potassium chloride and ammonium sulphate. After ammonium sulphate precipitation I was unable to detect an RNA polymerase activity sensitive to rifampicin.

The yeast mitochondrial RNA polymerase functions

well with mitochondrial DNA producing RNA up to 20S
in size whereas with native nuclear DNA little more
than 4S RNA is produced. Self-annealing of the transcription products indicates that some 70 % is transcribed from one DNA strand and of the remaining 30 %,
10 % is ribonuclease resistant prior to self-annealing. This contrasts with *N. crassa* mitochondrial DNA
and *E. coli* RNA polymerase products (19) which contain
some 35 % rapidly renaturing double-stranded regions.
Further the mitochondrial transcription products are
capable to directing an *E. coli* cell-free system to
produce mitochondrial membrane proteins (Fig. 9). These proteins upon analysis yield some 7 discrete sizes
of which three are similar to those membrane proteins
labelled *in vivo*. The remainder can be detected as minor components *in vivo* (17). Thus the rifampicin-sensitive mitochondrial enzyme is capable of producing
a functional messenger RNA which argues against the
presence of a sigma-like factor in yeast mitochondria.

REFERENCES.

1. D.J. South and H.R. Mahler, Nature 218, 1226(1968).
2. D.J.L. Luck and E.P. Reich, Proc.Nat.Acad.Sci. U.S.A. 52, 831 (1964).
3. D. Neubert and H. Helge, Biochem.Biophys.Res.Commun. 18, 600 (1965).
4. C. Saccone, M.N. Gadaleta and R. Gallerani, Eur. J.Biochem. 10, 61 (1969).
5. C. Saccone, M.N. Gadaleta and E. Quagliariello, Biochim.Biophys.Acta 138, 474 (1967).
6. H. Küntzel and K.P. Schäfer, Nature New Biol. 231, 265 (1971)
7. B.D. Reid and P. Parsons, Proc.Nat.Acad.Sci.U.S.A. 68, 2830 (1971).
8. A.E. Sippel and G.R. Hartman, Eur.J.Biochem. 16, 152 (1970).
9. F. Stirpe and L. Fiume, Biochem.J. 105, 779 (1967).
10. G.J. Wu and I.B. David, Biochemistry 11, 3589 (1972).
11. A.H. Scragg, Biochem.Biophys.Res.Commun. 45, 701 (1971).

12. M.J. Tsai, G. Michaelis and R.S. Criddle, Proc. Nat.Acad.Sci.U.S.A. 68, 473 (1971).
13. E. Wintersberger, Biochem.Biophys.Res.Commun. 48, 1287 (1972).
14. F.J. Bollum, J.Biol.Chem. 234, 2733 (1959).
15. C.W. Dingman and A.C. Peacock, Biochemistry 7, 659 (1968).
16. G.A. Bray, Anal.Biochem. 1, 279 (1960).
17. A.H. Scragg and D.Y. Thomas, in preparation.
18. D.Y. Thomas and D.H. Williamson, Nature New Biol. 233, 196 (1971).
19. K.P. Schäfer, G. Bugge, M. Grandi and K. Küntzel, Eur.J.Biochem. 21, 478 (1971).
20. H. Ponta, V. Ponta and E. Wintersberger, Eur.J. Biochem. 29, 110 (1972).
21. P. Borst, Ann.Rev.Biochem. 41, 333 (1972).
22. R.N. Bryan, M. Sugiura and M. Hayashi, Proc.Nat. Acad.Sci. U.S.A. 62, 483 (1969).
23. M. Schweiger, P. Herrick and R.L. Milette, J.Biol. Chem. 246, 6707 (1971).
24. S. Dezélée and A. Sentenac, Eur.J.Biochem. 34, 41 (1973).
26. R. Burgess, J.Biol.Chem. 244, 6160 (1969).
26. B. Alberts and G. Herrick, in: "Methods in Enzymology", S.P. Colowick and N.O. Kaplan Eds., Ac. Press New York, vol. 21, pt. 1, p. 198 (1971).

THE DNA-DEPENDENT RNA POLYMERASE FROM RAT LIVER MITOCHONDRIA

R. Gallerani and C. Saccone

Institute of Biological Chemistry, University of Bari, Bari, Italy.

The synthesis of RNA by rat liver mitochondria has been studied in our laboratory since 1965. It was demonstrated, using isolated organelles as the source of enzyme, that the process was clearly distinguishable from that occurring in nuclei and that the product of synthesis was RNA hybridizable with mitochondrial DNA (1-3). It was then deduced that the mitochondria contain a proper DNA-dependent RNA polymerase different from that present in the nuclei, possessing peculiar properties and able to use the mitochondrial DNA as template. The mitochondrial enzyme was, successively, solubilized (4) using different techniques and also, quite recently, purified (5). In the meantime the purification of the mitochondrial enzyme from three other sources has been reported: from *Neurospora crassa* (6), from yeast (7-8) and from *Xenopus* ovaries (9). The nature of these enzymes seems to be very similar in all cases with the exception of the yeast which, according to Criddle *et al.* (7), should possess more than one mitochondrial enzyme resembling more closely the nuclear polymerases.

A partial purification of the mitochondrial enzyme from rat liver has been reported by Reid and Parsons (10). In this paper we report our recent studies on the purified mitochondrial DNA-dependent RNA polymerase from rat liver. It is interesting to stress that the properties displayed by the purified enzyme are very similar to those found using the isolated organelles.

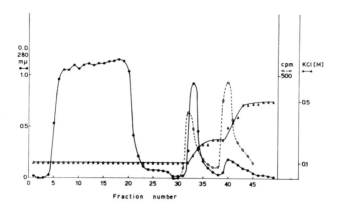

Fig. 1. Elution pattern of a (1x12 cm) DEAE-Sephadex A-25 column eluted by the stepwise technique. After DNAse treatment and ammonium sulphate fractionation (5) the proteins were adsorbed to the column at a flow rate of about 10 ml/h. The column was then washed by B_{2k} buffer and eluted by 10 ml of the same buffer containing KCl 0.25 M and extensively by B_{2k} buffer containing 0.5 M KCl. The volume of each fraction was 1.1 ml and the activity was determined by incubating 20 µl aliquots in the following incubation mixture: 53 mM Tris-HCl pH 7.4, 64 mM KCl, 3 mM $MgCl_2$, 1 mM $MnCl_2$, 0.1 mM of each ATP, CTP, GTP, UTP, [^3H]UTP at a specific activity of 1 Ci/mmole (Bio Schwarz, New York), 4 % glycerol, 0.2 mM DTT and $E.\ coli$ DNA 100 µg/ml. The KCl concentration was determined by conductivity measurements.

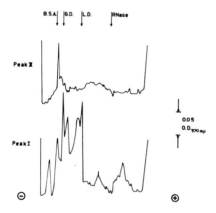

Fig. 2. SDS gel electrophoresis of peak I and peak II proteins after stepwise elution of the DEAE-Sephadex A-25 column. The proteins were concentrated by dialysis against PEG 30 % (12) and were put on (0.7 x 6.5 cm) gels in a maximal volume of 75 µl. All other experimental procedures are as described by Weber and Osborn (13). The gels were scanned in a Gilford 2400-S model Spectrophotometer at 570 nm.

RESULTS AND DISCUSSION.

Purification procedure and determination of the molecular weight of the mitochondrial enzyme. The purification procedure used to prepare the DNA-dependent RNA polymerase from rat-liver mitochondria is only slightly different from that already reported (5). The main difference concernes the elution of the DEAE-Sephadex column. In these experiments the stepwise technique has been used. The enzyme after the ammonium sulphate fractionation was adsorbed on a (1x12 cm) DEAE-Sephadex A-25 column equilibrated by B_{2k} buffer. After adsorption the column was washed by B_{2k} buffer and eluted by solutions of 0.25 and 0.5 M KCl in B_{2k} buffer. Fig. 1 shows the elution pattern of the DEAE-Sephadex column obtained using this particular technique. In these conditions two distinct peaks of activity appear to be eluted resp. at 0.25 and 0.5 M KCl. They have been characterized side by side for their sensitivity to inhibitors, divalent cations optimal concentration, DNA dependence and template specificity (manuscript in preparation). The common properties displayed by the two peaks (11) led us to conclude that both represent the same activity eluted at different salt concentrations.

The physical characterization of the two peaks seems to support this hypotesis. The SDS-polyacrylamide gel electrophoresis of peak I and peak II is shown in Fig. 2. The tracing of a stained gel after electrophoresis of the proteins coming from peak II contains only one band, whereas that of peak I contains more bands, one of them corresponding, in electrophoretical mobility, to that of peak II. On the basis of these experiments and because of the higher specific activity of peak II we believe that mitochondrial DNA-dependent RNA polymerase is present in the DEAE-Sephadex fractions of peak II in a virtually purified form although, the presence of minor impurities in this fraction cannot be excluded owing to the low protein concentration of the material analyzed on gel. The electrophoretic data can offer an explanation for the elu-

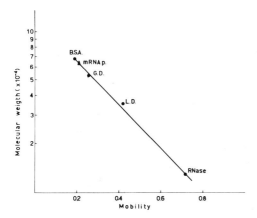

Fig. 3. Molecular weight determination of mitochondrial RNA polymerase (DEAE-Sephadex fraction peak-II). The standard proteins were: bovine serum albumin, glutamate dehydrogenase, lactate dehydrogenase and pancreatic RNAse. The amount of protein ranged from 3 to 4.2 µg.

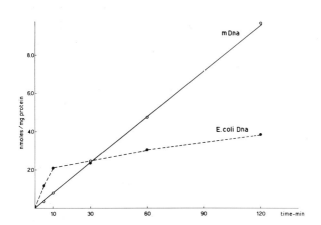

Fig. 4. Transcriptionkinetics of *E.coli* DNA and mitochondrial DNA by the mitochondrial RNA polymerase from rat liver. The protein of peak II was incubated at 30° C in 175 µl of a mixture having the composition described in the legend of Fig. 1 and the mtDNA or the *E.coli* DNA at a concentration of 100 µg/ml. Aliquots of 25 µl were taken at the different times and chilled in 1 ml of cold trichloroacetic acid containing 100 µg of bovine serum albumin as carrier. The samples were filtered through GF/C Whatmann glass fibre paper using cold 5 % T.C.A., dried and counted in a TRI-CARB Scintillator.

tion of the same enzyme (peak I and peak II) from the DEAE-Sephadex column at two different salt concentrations. The enzyme present in peak I is probably still linked to membrane proteins or to the template DNA which could modify its binding affinity.

The SDS gel electrophoresis results allow us to estimate the molecular weight of the mitochondrial DNA-dependent RNA polymerase. The molecular weight of the band commonly present in peak I and II has been determined using a semilogarithmic plot of the molecular weight of several standard proteins, against their electrophoretic mobilities. (Fig. 3). Using this technique it is possible to estimate a molecular weight value ranging from 62,000 to 64,000 daltons for the rat liver mitochondrial RNA polymerase. In all further experiments here reported the DEAE-S, peak II fraction has been used as source of enzyme.

Properties of the purified mitochondrial RNA polymerase. Table I summarizes some properties of the

TABLE I

PROPERTIES OF RAT LIVER MITOCHONDRIAL RNA POLYMERASE PURIFIED THROUGH DEAE-SE-PHADEX COLUMN.

The complete system contained all the components mentioned in the legend of Fig. 1, only the KCl was omitted. 100 % activity corresponded to 2.520 pmoles/mg protein/10 min.

Omission or addition	Activity %
Complete system	100
– DNA	15
– ATP, CTP, GTP	55
+ KCl 125 mM	52
+ Spermine 1.5 mM	85
+ DTT 1.5 mM	67

DNA-dependent RNA polymerase extracted from rat-liver mitochondria. The enzyme is almost fully DNA dependent. The 45 % dependence on the other ribonucleosides triphosphates shows that this enzyme displays also a homopolymer synthetase activity. In this regard we should like to stress that we have found a poly A polymerase activity associated with mitochondria. This activity is copurified with RNA polymerase at least until the ammonium sulphate step. We are now studying

the properties of this enzyme and its further purification. It is not clear at the moment if poly A-poly-
merase is a different specific enzyme or if mitochondrial RNA polymerase itself could be able, in appropriate conditions, to synthesize poly A. It must be
recalled that recently Ohasa and Tsugita (14) have
found that the α-subunit of the *E.coli* polymerase possesses such an activity.

The enzyme is strongly inhibited by salt. Already
at a concentration of KCl 0.1 M the inhibition is
about 50 %. Similar behaviour of other mitochondrial
RNA polymerases (9,10) seems to suggest that this is
a general property of the mitochondrial enzyme.

Sensitivity of inhibitors. Table II shows the sen-

TABLE II

SENSITIVITY OF THE RAT LIVER MITOCHONDRIAL RNA POLYMERASE TO DIFFERENT INHI-
BITORS.
All the experimental conditions are as in the legend of Fig. 1.

Experimental conditions	Inhibitor concentration		Inhibition %
Complete system	---		0
+ α-amanitin	0.2	μg/ml	0
+ Rifampicin	20	μg/ml	60
+ Cordycepin	60	μg/ml	0
+ Ethidium bromide	40	μg/ml	23
+ Cycloheximide	100	μg/ml	63

sitivity of the DNA-dependent RNA polymerase to different inhibitors. The RNA polymerase extracted from
rat liver mitochondria is insensitive to α-amanitin
and sensitive to rifampicin. These characteristics
confirm our previous results (15,16) obtained with
isolated organelles or solubilized enzyme. The rifampicin sensitivity seems to be a peculiar property of
organelles-polymerase although in some cases this is
still controversial. The major discrepancy regards the
yeast from which Scragg isolates a mitochondrial enzyme sensitive to rifampicin (8) whereas Criddle and
Wintersberger (7,17) are unable to show any sensitivity of the enzyme (5) to this inhibitor. The *Xenopus*
enzyme is, on the other hand, sensitive only to a derivative of rifampicin (9). It is interesting to no-

tice the *in vitro* sensitivity to cycloheximide. Some authors have already reported the *in vivo* effect by this drug on the nucleolar RNA synthesis whereas no inhibition seems to occur on the extranucleolar RNA synthesis (18). Recent experiments made *in vitro* seem to suggest that the cycloheximide is inhibiting the nucleolar RNA polymerase directly. Timberlake and Griffin (19) have demonstrated that this inhibition is strongly dependent on the procedure used to solubilize the enzyme. Using the detergent Na-deoxycholate, the same used to lyse mitochondria in our experiments, the nucleolar DNA-dependent RNA polymerase becomes extremely sensitive to cycloheximide. By using different techniques this sensitivity is lost. The hypothesis pur forward by the authors is that a separable factor is involved in the inhibition and the differences in solubilization techniques result in a differential loss of this factor. A similar situation can apply to the mitochondrial enzyme also in the case of other inhibitors, like the rifampicin. Different isolation procedures and incubation conditions could explain the discrepancies regarding the sensitivity of the mitochondrial enzyme toward this antibiotic.

Template specificity.   Table III shows the effecti-

TABLE III.

TEMPLATE SPECIFICITY OF THE MITOCHONDRIAL RNA POLYMERASE FROM RAT LIVER.
The mitochondrial and nuclear DNA from rat liver were extracted according to Leffler (20) and Borst (21) respectively. For experimental conditions see legend of Fig. 1.

Template used	Specific activity (pmoles/mg prot./10 min)
Poly d(A-T)	6,310
E. coli DNA native	2,750
" " " denatured	3,005
Mitochondrial DNA native	2,060
" " denatured	1,930
Nuclear DNA native	1,445
" " denatured	1,330
calf thymus DNA native	940
" " " denatured	1,380

veness of different templates during a 10 min incubation. The synthetic polymer poly d(A-T) seems to be

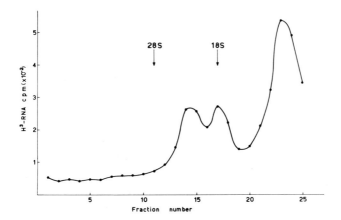

Fig. 5. Sedimentation pattern on sucrose gradients of the product made *in vitro*
by the mitochondrial RNA polymerase in the presence of homologous template. The
enzyme was incubated for 30' and 30° C under the same conditions as reported in
the legend of Fig. 4. The incubation was stopped by adding SDS to a final concentration of 1 %. After cold phenol extraction the newly synthesized RNA was
precipitated by ethanol adding cold cytoplasmic RNA from rat liver as carrier.
0.2 ml of the ethanol precipitate, dissolved in 2 x SSC was applied to a linear
5 % - 23 % sucrose gradient.

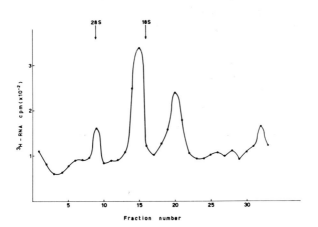

Fig. 6. SDS gel electrophoresis of the product made *in vitro* by the mitochondrial RNA polymerase on the homologous template. The transcription product, in
a maximal volume of 75 μl was applied to a 0.6 x 5 cm gel containing 2.7 % acrylamide. The gels and the solutions were prepared according to Loening (22). The
pattern of radioactivity was measured by cutting the gels in 1.5 mm slices,
digesting the slices in 0.3 ml of H_2O_2 at 60° C overnight and counting each fraction using Bray's solution.

the best template for the enzyme. Taking into account that the nucleotide dAMP is present in it with a frequency twice higher than in a natural template the difference in [3H]UTP incorporation between poly d(A-T) and the *E.coli* DNA is not very high indeed. What is really interesting is the different kinetics of incorporation using incubation times as long as 120 min in presence of *E.coli* DNA or mitochondrial DNA. As is shown in Fig. 4 the transcription of the *E.coli* DNA reaches a plateau already after 10-20 min, whereas the transcription of mitochondrial DNA goes on linearly for 120 min and probably longer. To explain the clear different specificity for these DNAs in relation to the incubation time one can assume that the binding of mitochondrial enzyme to the heterologous template could be irreversible. On the contrary just after the transcription of a first molecule of mitochondrial DNA the enzyme would easily have the possibility to come off from it an to start again the transcription of a new DNA molecule.

Transcription product. In order to study the transcription product made *in vitro* by the mitochondrial RNA polymerase in the presence of mitochondrial DNA, the enzyme was incubated for 30 min and the RNA synthesized was extracted by the cold phenol-SDS procedure (see legend of Fig. 5). The size of the transcription product was characterized both by sucrose gradient centrifugation and by SDS gel electrophoresis. The sensitivity to RNAse was also assayed and the results demonstrated that 25 % of the material transcribed is RNAse resistant. Probably a partially symmetric transcription is occurring under the conditions indicated. In Fig. 5 the sucrose gradient profile of the newly synthesized RNA using rat-liver cytoplasmic RNA as marker is shown. The synthesized product produces three distinct peaks, two of high molecular weight (about 22 and 18 S) and the third of small size (4 or 5 S). Fig. 6 shows the characterization of the *in vitro* product on SDS gel electrophoresis. Using this method the two high molecular weight RNA molecules display sedimentation coefficients of 19 S and 14 S respectively. These small differences regarding the "S" values found using the two different techniques, could be due to the calibration method used and/or to some intrinsic properties of mitochondrial RNA itself as already described in the case of mitochondrial riboso- mal RNA (23). The sedimentation coefficients found for high molecular weight RNAs clearly resemble that re- ported by many authors for mitochondrial ribosomal RNA of animal cells. This could suggest that transcription of ribosomal RNA of of its precursors occurs under the conditions reported above. However other studies now in progress in our laboratory are necessary to clari- fy the nature of the *in vitro* transcribed mitochon- drial RNA by the purified enzyme.

REFERENCES.

1. C. Saccone, M.N. Gadaleta and E. Quagliariello, Biochim.Biophys.Acta 138, 474 (1967).
2. C. Saccone, M.N. Gadaleta and R. Gallerani, Eur. J.Biochem. 10, 61 (1969).
3. C. Aaij, C. Saccone, P. Borst and M.N. Gadaleta, Biochim.Biophys.Acta 199, 373 (1970).
4. C. Saccone and M.N. Gadaleta, 8th Intern.Congress of Biochem. abstr. 185 (1970).
5. R. Gallerani, C. Saccone, P. Cantatore and M.N. Gadaleta, FEBS Letters 22, 37 (1972).
6. H. Küntzel and K.P. Schäfer, Nature New Biol. 231, 265 (1971).
7. M.J. Tsai, G. Michaelis and R.S. Criddle, Proc. Natl.Acad.Sci. USA 68, 473 (1971).
8. A.H. Scragg, Biochem.Biophys.Res.Commun. 45, 701 (1971).
9. G.J. Wu and I.B. Dawid, Biochem. 11, 3589 (1972).
10. B.D. Reid and P. Parsons, Proc.Nat.Acad.Sci. USA 68, 2830 (1971).
11. R. Gallerani, C. de Benedetto and C. Saccone, 8th FEBS Meeting, Amsterdam, abstr. 619 (1972).
12. A. Bollen, personal communication.
13. K. Weber and M. Osborn, J.Biol.Chem. 244, 4406 (1969).
14. S. Ohasa and A. Tsugita,Nature New Biol. 240, 35

(1972).
15. C. Saccone, R. Gallerani, M.N. Gadaleta and M. Greco, FEBS Letters 18, 339 (1971).
16. M.N. Gadaleta, M. Greco and C. Saccone, FEBS Letters 10, 54 (1970).
17. E. Wintersberger, Biochem.Biophys.Res.Commun. 48, 1287 (1972).
18. M. Muramatsu,N.Shimada and T. Higashinakagawa, J. Mol.Biol. 53, 91 (1970).
19. W.E. Timberlake, G. Hagen and D.H. Griffin, Biochem.Biophys.Res.Commun. 48, 823 (1972).
20. A.T. Leffler II, E. Creskoff, S.W. Luborsky, V. McFarland and P.T. Mors, J.Mol.Biol. 48, 455 (1970).
21. P. Borst, G.J.C.M. Ruttenberg and A.M. Kroon, Biochim.Biophys.Acta 149, 140 (1967).
22. U.E. Loening, Biochem.J. 102, 251 (1967).
23. M. Edelman, I.M. Verma and U.Z. Littauer, J.Mol. Biol. 49, 67 (1970).

EXPRESSION OF THE MITOCHONDRIAL GENOME IN WILD TYPE AND IN AN EXTRANUCLEAR MUTANT OF *NEUROSPORA CRASSA*

Hans Küntzel, Igbal Ali and Christian Blossey

Max-Planck-Institut für experimentelle Medizin, Abteilung Chemie, 34 Göttingen, Germany.

The only mitochondrial gene products which have been identified with certainty are ribosomal RNAs and transfer RNAs (1). In *Neurospora* as well as in other organisms it has been shown that each mtDNA circle contains one gene for each of the two rRNAs (2). The existence of mitochondrial structural genes coding for proteins is suggested by our observation that the *in vitro* transcription product of mtDNA directs the synthesis of a few distinct polypeptides in a cell-free system (3). However, little is known about the structure and function of mitochondrially coded proteins. The phenotypical properties of some extranuclear mutants of *Neurospora* (4) suggest the following rather complex mitochondrial gene functions which may be more or less independent from each other: 1. Control of growth rate; 2. control of the content of cytochromes c and $a+a_2$ and of; 3. cytochrome b in the inner mitochondrial membrane; 4. control of protoperithecial formation (female fertility). We have recently proposed another mitochondrial gene function, the synthesis of a regulatory protein which is exported to the nucleus where it controls genes for mitochondrial enzymes, in order to explain our finding that mitochondrial RNA polymerase and other enzymes can be induced by a selective block of mitochondrial transcription and translation (5,6).

To learn more about mitochondrial genes we have studied biochemically the stopper growth mutant "abn-1" of *Neurospora* (7). In contrast to the "leaky" extranuclear mutants of the "poky" type this mutant

71

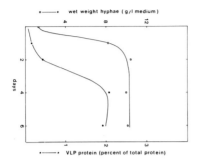

wet weight hyphae (g/l medium)

days

VLP protein (percent of total protein)

Fig. 1. Production of virus-like particles (VLP) during the growth period of abn-1.

TABLE I.

PULSE LABELLING OF PROTEINS FROM VLP, MITOCHONDRIA AND CYTOPLASM IN THE PRESENCE AND ABSENCE OF SELECTIVE INHIBITORS.

| | abn-1 | | wild type | |
	VLP	S-100	mitochondrial lipoprotein	S-100
Specific activity (cpm/mg):				
– CHI	3,167	8,370	67,920	73,230
+ CHI	396	156	66,740	1,490
+ CAP	852	6,874	–	–
Percent inhibition:				
+ CHI	87	98	1.5	98
+ CAP	73	18	–	–

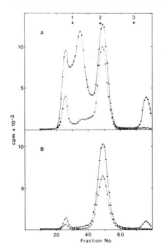

cpm x 10⁻³

Fraction No.

Fig. 2. Sephadex G-200 filtration in the presence of SDS of mitochondrial proteins labelled *in vivo* in the presence of cycloheximide with [^{14}C]leucine (•——•) and of VLP proteins labelled with [^{3}H]leucine (o——o). A: total proteins; B: proteins soluble in 90% methanol. The internal markers are (1) bovine serum albumin (MW 69,000) and (2) β-lactoglobulin (MW 18,500).

72

seems to have lost some mitochondrial gene functions, which is not only manifested by a lower density of mitochondrial DNA (8) and by a loss of control of the nuclear gene for mitochondrial RNA polymerase, but also by an irreversible alteration of the cytochrome composition of the inner mitochondrial membrane. However, instead of finding out which mitochondrial genes have been knocked out in this mutant we discovered a new mitochondrial cistron present not only in abn-1 but also in wild type.

The most prominent feature of abn-1 is the abundance of a particulate fraction not detectable in wild type. This fraction has a somewhat lower density (around 1.18 g/cm^3) than mitochondria and appears under the electron microscope in negative contrast (9) or in thin sections after OsO_4 fixation as polymorphic vesicles containing a dense nucleoid surrounded by a membrane envelope (10). The particles are composed of RNA, lipo- and glycoproteins and phospholipids; the RNA is single-stranded, sediments with 33S and after heat treatment with 7-9S, and has a base composition differing from mitochondrial and cytoplasmic rRNA (10). From their morphology and their chemical composition these particles resemble much more certain enveloped RNA viruses (for example RNA tumour viruses) than other cellular structures like mitochondria, and have therefore been designated "virus-like particles" (VLP) although they obviously do not infect wild type hyphae. However, it is conceivable that VLP from abn-1 belong to the well-known group of "infectious" agents which are transmitted only by cytoplasmic inheritance and not by infection, and that VLP are responsible for the suppressiveness of the abn-1 character in heteroplasmons (11). From Fig. 1 it is obvious that VLP are produced mainly in the beginning stationary phase, indicating that VLP formation is connected with slow growth and ageing, and perhaps only indirectly caused by the mitochondrial mutation.

We will discuss now several lines of evidence suggesting that VLP originate within mitochondria and are composed of mitochondrial gene products. In ear-

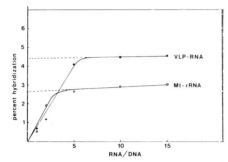

Fig. 3. Hybridization of [32p]labelled 33S RNA from VLP (o   o) and rRNA from wild type mitochondria (o   o) with mtDNA (B) in a S-30 from *E.coli*. The proteins labelled with [14C]leucine in the absence (o   o) and with [3H]leucine in the presence of exogenous RNA (o   o) were analyzed by Sephadex G-200 filtration in the presence of SDS.

TABLE II.

HYBRIDIZATION BETWEEN [32P]LABELLED VLP-RNA AND WILD TYPE mtDNA IN THE PRESENCE OF VARIOUS COMPETITOR RNA's (50-FOLD EXCESS).

Competitor	Percent hybridization
rRNA from purified mt ribosomes	97
total mtRNA from mid log cells	53
total mtRNA from stationary cells	18
no competitor	100

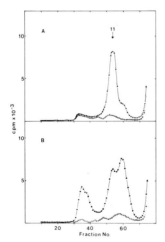

Fig. 4. *In vitro* translation of VLP-RNA (A) and of the *in vitro* transcription product of wild type mtDNA (B) in a S-30 from *E.coli*. The proteins labelled with [14C]leucine in the absence (o———o) and with [3H]leucine in the presence of exogenous RNA (o———o) were analyzed by Sephadex G-200 filtration in the presence of SDS.

74

lier stages of the growth cycle (between 1 and 3 days)
the majority of VLP are firmly associated with mitochondria and can be liberated only by selective lysis
of mitochondrial membranes (the VLP membrane is rather resistent against Triton X), whereas in the late
period most of the particles are found in the postmitochondrial supernatant. VLP contain cardiolipin, the
marker lipid of the inner mitochondrial membrane and
as major protein component a lipoprotein of MW 15000
(10) which is synthesized on CAP-sensitive mitochondrial ribosomes (Table I). This lipoprotein appears
to be homogenous in SDS-polyacrylamide gels (10) and
is indistinguishable from a major product of mitochondrial protein synthesis in wild type (Table I and Fig.
2) by at least two criteria, molecular weight and solubility in 90 % methanol (Fig. 2). The wild type mitochondrial proteins shown in Fig. 2 have been pulse-
labeled in the presence of cycloheximide; however,
virtually the same elution pattern of the methanol-
soluble proteins (Fig. 2B) is obtained if mitochondria
are labelled in the absence of cycloheximide indicating that the only mitochondrial protein extractable
with 90 % methanol is the intramitochondrially synthesized lipoprotein of MW 15000. We conclude that the
major protein component of VLP is synthesized on mitochondrial ribosomes; it is possibly identical with the
lipoprotein found in wild type to be located in the
inner mitochondrial membrane, although the identity
remains to be proven by immunological techniques. Fig.
3 shows that the RNA component of VLP hybridizes efficiently with mitochondrial DNA from wild type cells,
reaching a saturation plateau at 4.5 % of the genome.
Preliminary results have shown that VLP-RNA saturates
a considerably higher portion of mtDNA from abn-1, although we had difficulties in obtaining [^{3}H]labelled
mtDNA from abn-1 of sufficiently high specific activity to correct for filter losses.

 The competition data of Table II suggest that VLP-
RNA is coded by a mitochondrial cistron different from
the ribosomal cistron, and that this hitherto unknown
cistron is also transcribed in wild type mitochondria,

especially in older cells, although no significant
amounts of 33S RNA or VLP can be detected in wild type. The question arises whether VLP-RNA contains the
message for the "coat" protein (lipoprotein) of VLP.
To answer this question we have translated heated VLP-
RNA in a crude S-30 from *E.coli* and analyzed the product by Sephadex G-200 filtration in the presence of
SDS. Fig. 4A shows that VLP-RNA directs the synthesis
of a main polypeptide component of MW 11000. A component of the same MW is also formed in the same cell-
free system programmed with the *in vitro* transcription
product of mtDNA from wild type, together with polypeptides of MW 8000, 41000 and 47000 (Fig. 4B). This
result demonstrates that VLP-RNA has messenger activity and carries the information for at least one polypeptide. None of the *in vitro* translation products
listed in Table III have yet been identified with pro-

TABLE III.

ESTIMATED MW OF POLYPEPTIDES SYNTHESIZED *IN VITRO* IN A S-30 FROM *E.COLI* OR
FROM MITOCHONDRIA.

Messenger RNA:	Transcript of wild type mtDNA		VLP-RNA
S-30	mitochondria	*E.coli*	*E.coli*
MW of polypeptides:	47,000	47,000	
	41,000	41,000	
	15,000		
	11,000	11,000	11,000
	8,000	8,000	

teins synthesized *in vivo* on mitochondrial ribosomes.
However, we have to consider the attractive possibility that the MW 11000 species directed *in vitro* by
VLP-RNA or by wild type mtDNA is identical with the
apoprotein of the MW 15000 lipoprotein. Our results
are summarized in a tentative scheme shown in Fig. 5.
We have shown that mtDNA from *Neurospora* carries in
addition to the ribosomal cistron which is transcribed into a 32S precursor RNA (12) a cistron for 33S
RNA. This cistron is transcribed *in vivo* both in wild
type and in abn-1, but only in abn-1 the primary gene
product, a single-stranded 33S RNA, is incorporated
into VLP together with phospholipids and proteins. The
main protein component of VLP, a lipoprotein of MW
15000, is synthesized on mitochondrial ribosomes and
is possibly identical with a major product of mito-

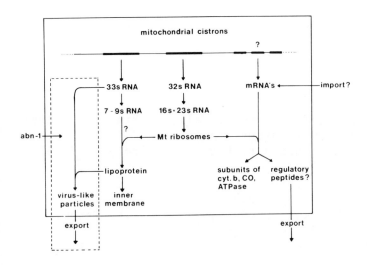

Fig. 5. Tentative presentation of mitochondrial cistrons and their products.

chondrial protein synthesis in wild type which is located in the inner mitochondrial membrane. From *in
vitro* translation experiments it is obvious that both
wild type mtDNA and one of its gene products, VLP-RNA,
carry the information for a polypeptide of MW 11000.
Although it remains to be shown that this product is
identical with the protein component of the lipoprotein it appears to us that a lipoprotein of MW 15000
is at the moment the most likely candidate for a true
mitochondrial gene product.

ACKNOWLEDGEMENTS.

One of us (I.A.) holds a fellowship of the Deutsche Akademische Austauschdienst. This work was supported by the Deutsche Forschungsgemeinschaft.

REFERENCES.

1. P. Borst, Ann.Rev.Biochem. 41, 333 (1972).
2. K.P. Schäfer and H. Küntzel, Biochem.Biophys.Res.
   Commun. 46, 1312 (1972).
3. H.C. Blossey and H. Küntzel, FEBS Letters 24, 335
   (1972).

4. H. Bertrand and T.H. Pittenger, Genetics 71, 521 (1972).
5. Z. Barath and H. Küntzel, Nature New Biol. 240, 195 (1972).
6. Z. Barath and H. Küntzel, Proc.Nat.Acad.Sci.U.S.A. 69, 1371 (1972).
7. L. Garnjobst, J.F. Wilson and E.L. Tatum, J.Cell Biol. 26, 413 (1965).
8. H. Küntzel, Z. Barath, I. Ali, J. Kind, H.H. Althaus and H.C. Blossey, in press.
9. R.W. Tuveson and J.F. Peterson, Virology 47, 527 (1972).
10. H. Küntzel, Z. Barath, I. Ali, J. Kind and H.H. Althaus, Proc.Nat.Acad.Sci.U.S.A. 70, 1574 (1973).
11. E.G. Diakumakos, L. Garnjobst and E.L. Tatum, J. Cell Biol. 26, 427 (1965).
12. Y. Kuriyama and D.J.L. Luck, J.Mol.Biol. 73, 425 (1973).

TRANSCRIPTION OF mtDNA BY MITOCHONDRIAL RNA POLYMERASE FROM *XENOPUS LAEVIS*

Igor B. Dawid and Guang-Jer Wu

Department of Embryology, Carnegie Institution of Washington, 115 West University Parkway, Baltimore, Maryland 21210, U.S.A.

RNA polymerase has been isolated from ovarian mitochondria of the frog *Xenopus laevis* (1). This enzyme is distinct from bacterial RNA polymerase and from the 3 known nuclear polymerases of *Xenopus*. *Xenopus* mtRNA polymerase is not inhibited by rifampicin or by α-amanitin, but is sensitive to inhibition by several rifamicin analogues. The enzyme consists of a single polypeptide subunit of a molecular weight of 46000, and forms large aggregates in low salt. These physical properties are similar to those of the *Neurospora* mtRNA polymerase described by Küntzel and Schäfer (2). We have studied the transcription of *Xenopus* mtDNA by homologous mtRNA polymerase and report here several properties of this system.

STRAND SELECTION BY mtRNA POLYMERASE.

Most of the stable products of *Xenopus* mtDNA *in vivo* are transcribed from the H-strand (3). In HeLa cells Aloni and Attardi (4) reported that both strands of mtDNA are transcribed, but again most stable products are H-strand transcripts. During *in vitro* transcription with *E.coli* RNA polymerase of mtDNA of the rat (5), and of *Xenopus* (6) the L-strand is preferred. Nuclear RNA polymerases I and II from *Xenopus* transcribe mtDNA essentially symmetrically, whether the template is native or denatured (Dawid and Roeder, unpublished). With homologous mtRNA polymerase and mtDNA-I (closed-circular DNA) both strands are transcribed and the H-strand is preferred (Table I). With

TABLE I.

STRAND SELECTION DURING *IN VITRO* TRANSCRIPTION OF mtDNA BY mtRNA POLYMERASE. The percentage of RNA homologous to the H-strand was calculated as (RNA hybridized to H-strand)/(RNA hybridized to H-strand plus RNA hybridized to L-strand). Between 30 and 70 % of total input RNA hybridized to DNA in these experiments.

Template	RNA homologous to H-strand percent $\pm$ S.D. (n)
mtDNA-I	72.4 $\pm$ 8.5 (5)
mtDNA-II	93.7 $\pm$ 2.3 (7)
Denatured mtDNA	20.5 $\pm$ 1.9 (4)

mtDNA-II (open or nicked DNA) as template H-strand preference was almost complete (Table I). This preference was not influenced by the degree of purification of the polymerase, nor by variations in the reaction conditions like the salt concentration, inclusion of manganese ions, length of incubation or reduction of the temperature from 30° to 20°. Strand preference was drastically altered by denaturation of the template (Table I). These results show that mtRNA polymerase recognizes the secondary structure of its template and interacts with it differently than do nuclear and bacterial polymerases.

The presence of mt-rRNA sequences in the *in vitro* transcript was demonstrated by competition experiments. The hybridization of *in vitro* synthesized RNA with mtDNA was competed with mt-rRNA to an extent of about 15 %. This corresponds closely to the fraction of the H-strand of mtDNA which is homologous to mt-rRNA (3). Therefore, the most likely interpretation is that mtRNA polymerase transcribes all sequences of the H-strand with about equal frequency.

THE ASSOCIATION OF NASCENT RNA WITH THE TEMPLATE.

When an *in vitro* transcription reaction of mtRNA polymerase was stopped by the addition of sodium dodecyl sulfate (SDS) and analysed on SDS-sucrose gradients the radioactivity (nascent RNA) traveled with the DNA (Fig. 1). With mtDNA-I as template most of the label sedimented at about 39S, and a smaller amount at 27S. These are the sedimentation coefficients of the closed-circular and open-circular components of mtDNA, respectively (7). With mtDNA-II as template

TABLE II.

RIBONUCLEASE RESISTANCE OF *IN VITRO* TRANSCRIPTION PRODUCT.

Template	Treatment	RNase resistant,* per cent
mtDNA-I	RNase	37
	RNase and DNase	10
mtDNA-II	RNase	35
	RNase and DNase	6

*Acid precipitable radioactive RNA after incubation with pancreatic RNase at 10 µg/ml in 0.1 M NaCl, 5 mM $MgCl_2$, 0.03 M Tris, pH 7.5, for 30 min at 25°, with or without DNase at 40 µg/ml.

most of the labelled RNA travelled at about 27S (Fig. 1b). These facts suggested that the nascent RNA was bound to DNA. This conclusion is confirmed by Fig. 1c: After heating (or after treatment with DNAse: experiment not shown) the *in vitro* synthesized RNA sedimented between 4 and 12S. The binding of RNA to the template is not mediated by the polymerase, since the same results were obtained when the samples were digested with pronase before sedimentation analysis. Our interpretation is further supported by the fact that treatment of a sample like the one shown in Fig. 1a with a very small amount of DNAse (10^{-4} µg/ml, 10 min at 25°) converted all the label into a form sedimenting at 27S without releasing substantial amounts of label into the 4 to 12S region. It is known that low concentrations of DNAse convert mtDNA-I into mt-DNA-II. Very little RNA is released from the complex with DNA during the synthesis reaction, even if the incubation is prolonged for 90 min.

RNA synthesized *in vitro* by mtRNA polymerase contains a fraction which is not sensitive to RNAse, unless the template is destroyed or the RNA/DNA complex exposed to conditions known to dissociate nucleic acid duplexes (Table II). The size of the region which is protected from RNAse was found to be quite heterogenous, ranging from about 50 to about 500 nucleotides. We tentatively conclude that a helical region is formed during the transcription process between one strand of the template and the nascent RNA.

CHAIN INITIATION BY mtRNA POLYMERASE.

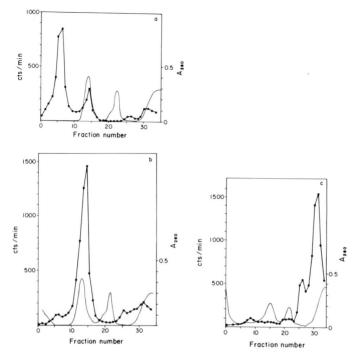

Fig. 1. Sedimentation profile of RNA synthesized *in vitro* by mtRNA polymerase. The reaction conditions were described by Wu and Dawid (1); incubation was for 30 min at 30°. Sedimentation in SDS-sucrose gradients was carried out as described previously (3). The thin line without points is the absorbance trace of *Xenopus* rRNA added as marker (nominal sedimentation coefficients of 18 and 28S). a: mtDNA-I used as template; b: mtDNA-II used as template; c: denaturation of product: a synthesis reaction as in Fig. 1a was mixed with SDS and EDTA and heated for 3 min at 90° before sedimentation analysis.

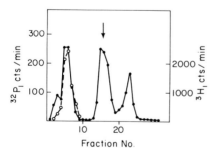

Fig. 2. Measurement of chain initiation. A mtRNA polymerase reaction with $[^3H]$-UTP and γ-$[^{32}P]$ATP was carried out, the RNA isolated by acid precipitation, and hydrolyzed in KOH for 1 h at 70°. The hydrolysate was fractionated on a DEAE-Sephadex column (*cf*. ref.8). The arrow shows the position of marker ATP. Solid circles, $[^{32}P]$; open circles, $[^3H]$.

TABLE III.
RNA CHAIN INITIATION BY mtRNA POLYMERASE.
The apparent chain length was calculated as the molar ratio ($[^3H]$UTP x 3):$[^{32}P]$-NTP incorporated. The $[^{32}P]$incorporation was calculated from label recovered as pppNp (see text and Fig. 2).

γ-$[^{32}P]$nucleotide triphosphate	internal nucleotides per 5'-terminus ("chain length")
ATP	830*
GTP	1770
CTP	>3200
UTP	>6000

*Average of 3 determinations.

To study this problem we use γ-$[^{32}P]$labelled nucleoside triphosphates. The label from this position is retained at the 5'-end of nascent RNA chains but is eliminated in all internal positions. Due to the relatively low activity of the mtRNA polymerase and the large amounts of labelled precursors required (*e.g.*, $[^{32}P]$ATP) we could not eliminate entirely a background of $[^{32}P]$label by washing the product with TCA. Therefore, we hydrolyzed the *in vitro* synthesized RNA with alkali and separated the nucleotides on DEAE-Sephadex. Contaminating precursor ("background") appears as $[^{32}P]$ATP, or as its hydrolysis product, inorganic phosphate. The label derived from the 5'-end of an RNA chain is converted during hydrolysis to adenosine tetraphosphate, pppAp. Fig. 2 shows a chromatogram in which $[^{32}P]$label occurs in an early peak, probably inorganic phosphate, and in a peak at the position of ATP. In addition, there is a later peak of radioactivity at the position expected for pppAp. The $[^3H]$label in the chromatogram is UMP, arising from the internally incorporated $[^3H]$UTP. In this way the level of $[^{32}P]$incorporation into the 5'-end of RNA chains was determined with all 4 nucleoside triphosphates. Table III shows that ATP is the most frequent initiator of chains in this system, followed by GTP. Pyrimidine nucleotides initiate very rarely, if at all. This result is similar to that obtained with other types of RNA polymerases (see ref. 9).

The fact that two nucleotides occur at the ends of RNA chains synthesized *in vitro* shows only that at

least two sites on mtDNA are involved in initiation.
We asked whether initiation occurred at a few specific sites or at many sites. Fot this purpose RNA was
synthesized with γ-[^{32}P]ATP of GTP, the product digested with pancreatic RNAse and the oligonucleotides
separated on DEAE-Sephadex. Several oligonucleotide
peaks containing [^{32}P]label were found in either case,
showing that mtRNA polymerase initiates on mtDNA on
at least 6, and probably more sites. This multiplicity of initiation sites make it unlikely that *in vitro*
RNA synthesis by mtRNA polymerase mimics the specific
mode of RNA synthesis *in vivo*. Nevertheless, the enzyme shows considerable selectivity in its interaction with the mtDNA, which is presumably its template
in vivo. The extent and manner in which this *in vitro*
selectivity relates to the situation *in vivo* remains
to be determined.

REFERENCES.

1. G.-J. Wu and I.B. Dawid, Biochemistry 11, 3589 (1972).
2. H. Küntzel and K.P. Schäfer, Nature New Biol. 231, 265 (1971).
3. I.B. Dawid, J.Mol.Biol. 63, 201 (1972).
4. Y. Aloni & G. Attardi, J.Mol.Biol. 55, 251 (1971).
5. H.F. Tabak and P. Borst, Biochim.Biophys. Acta 217, 356 (1970).
6. I.B. Dawid, Develop.Biol. 29, 139 (1972).
7. I.B. Dawid and D.R. Wolstenholme, J.Mol.Biol. 28, 233 (1967).
8. G.M. Tener, Methods Enzymol. XIIA, 398 (1967).
9. P.H. von Hippel and J.D. McGhee, Ann.Rev.Biochem. 41, 231 (1972).

EFFECT OF RIFAMPICIN ON MACROMOLECULAR SYNTHESIS IN RAT-LIVER MITOCHONDRIA

Maria N. Gadaleta and Cecilia Saccone

Institute of Biological Chemistry, University of Bari, Bari, Italy.

Although the sensitivity of mitochondrial DNA-dependent RNA polymerase to rifampicin is still controversial in several organisms, we have clearly demonstrated that the rat-liver mitochondrial enzyme, both in isolated organelles and in purified form, is sensitive to the drug like bacterial polymerase (1). Using this inhibitor we studied the stability of *in vitro* synthesized RNA. The stability of newly synthesized RNA molecules was determined by adding rifampicin to mitochondria that had been allowed to incorporate $[^3H]$-UTP for 30 min. Fig. 1 shows that, after the addition of 50 µg/ml of rifampicin, about 50 % of acid insoluble radioactivity became acid soluble within 60 min incubation. From the degradation kinetics this unstable mitochondrial RNA, probably mRNA, appears to decay with a half life of about 3 min. Literature data in this connection are quite variable. Wintersberger (2) studying yeast mitochondrial RNA degradation after a block of the synthesis with actinomycin D found that RNA synthesized *in vitro* in a 10 min incubation, turned over rapidly, whereas RNA synthesized *in vitro* in 60 min was stable. Gamble and McCluer (3) report that in rat-heart mitochondria, 60 min after the addition of 150 µg rifampicin/ml only 10 % of the RNA, synthesized *in vitro* during a 15 min incubation, was degraded. Grant and Poulter (4) using also rifampicin found that in *Physarum polycephalum* 100 % of mitochondrial RNA, synthesized *in vitro* during a 30 min incubation, was degraded with a half life of less than 5 min.
Given that mitochondrial rat liver RNA polymerase

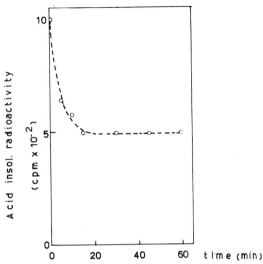

Fig. 1. Kinetics of decay of labeled *in vitro* synthesized mitochondrial RNA in the presence of rifampicin. Rifampicin (10 mg/ml in 70 % (v/v) ethanol), at the final concentration of 50 µg/ml was added to 2 mg of mitochondrial protein which had been incubated for 30 min in 1 ml of RNA synthesis assay mixture containing: 44 mM Tris-HCl pH 7.4, 3 mM MgCl₂, 3 mM MnCl₂, 50 mM KCl, 40 µg pyruvatekynase, 100 µM ATP, GTP, CTP and 50 µM [³H]UTP (s.a. 10³ µC/µmole). At the indicated times aliquots of 0.1 ml were withdrawn and denatured in 10 % trichloroacetic acid.

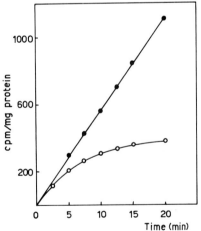

Fig. 2. Effect of rifampicin on *in vitro* mitochondrial protein synthesis. 2.5 mg of mitochondrial protein, after standing 30 min at 0° C, were incubated at 37° C in 1 ml of a medium containing: 50 mM sucrose, 50 mM Tris-HCl, 20 mM KCl, 30 mM NH₄Cl, 6 mM MgCl₂, 1 mM EDTA, 20 mM potassium phosphate buffer, 3 mM ADP, 30 mM sodium succinate, 100 µg/ml cycloheximide, 50 µg/ml of a synthetic amino-acid mixture (10) minus leucine and 95 µM [¹⁴C]leucine (s.a. 10⁶ µC/µmole). Heat stable, acid insoluble radioactivity was measured. ●——●: control; 0——0: protein synthesis in the presence of 100 µg/ml of rifampicin.

is sensitive to rifampicin, this drug must interfere with all the processes depending on the concomitant transcription of mitochondrial DNA such as the synthesis of proteins directed by mitochondrial mRNA. In Fig. 2 the time course of [^{14}C]leucine incorporation in the presence and in the absence of rifampicin is reported. Between 7 and 15 min after the addition of the antibiotic the decline in the synthetic capacity was exponential, following first order kinetics. Assuming that this decline only depends on the block of mDNA transcription we calculated from this experiment a half-life of mRNA of 4.8 min. Not very different results we obtained using ethidium bromide in place of rifampicin.

It has been reported recently by us (5) that in rat-liver mitochondria also DNA synthesis is affected by rifampicin. The results of Table I show that the

TABLE I.

EFFECT OF RIFAMPICIN AND CHLORAMPHENICOL ON [^{3}H]dATP INCORPORATION IN RAT LIVER MITOCHONDRIA.

2.5 mg of mitochondrial protein in 0.25 M sucrose plus 2 mM EDTA, preincubated for 10 min at 37° C were added to 0.5 ml of a reaction mixture containing: 50 mM Tris-HCl buffer, 20 mM sodium phosphate, 20 mM sodium succinate, 4 mM KCl, 0.5 mM ATP, 7 mM MgCl$_2$, 60 mM sucrose, 4 μM [^{3}H]dATP (s.a. 800 μC/μmole) and 15 μM of the other deoxyribonucleoside triphosphates. Incubation was carried out at 37° C. At the times indicated samples with and without the inhibitor were stopped by adding 10 % trichloroacetic acid. For other experimental details, see ref. 5.

Additions	Incubation time (min)	Inhibition (%)
Rifampicin (50 μg/ml)	10	30
	30	45
	60	60
Chloramphenicol (90 μg/ml)	30	10

inhibition of [^{3}H]dATP incorporation by rifampicin increases with time reaching 60 % inhibition after 60 min incubation. The drug concentration used was 10 μg/ mg protein. The effect of rifampicin on DNA synthesis is not secondary to its inhibition of mitochondrial protein synthesis since the reaction is practically unaffected by chloramphenicol. Both *in vitro* and *in vivo* a synergism between RNA and DNA polymerase has recently been observed in bacteria (6) as well as in higher organisms (7), suggesting that transcription can be coupled to duplication. In the light of these

studies we think that the effect of rifampicin on DNA
synthesis may be secondary to its inhibition of the
transcription process. This would indicate that the
synergism between DNA and RNA polymerase is probably a
general phenomen occurring also in mitochondria. Recent data of Kock (8) and of Miyaki *et al*. (9) who have found some alkali and RNAse sensitive regions, respectively in human mitochondrial DNA and in the DNA of
rat ascites hepatoma cells, seem to confirm this hypothesis. In conclusion the data reported above show
that in rat liver mitochondria rifampicin affects all
processes of macromolecular synthesis probably acting
directly only at the level of the transcriptional process.

REFERENCES.

1. R. Gallerani and C. Saccone, this volume.
2. E. Wintersberger, in:"Regul. of Metab. Processes in
   Mitochondria", J.M. Tager, S. Papa, E. Quagliariello and E.C. Slater,Eds., Elsevier, Amsterdam, 439
   (1966).
3. I.G. Gamble and R.H. McCluer, J.Mol.Biol. 53, 557
   (1970).
4. W.D. Grant and R.T.M. Poulter, J.Mol.Biol. 73, 439
   (1973).
5. C. Saccone and M.N. Gadaleta, FEBS Letters 34, 106
   (1973).
6. J.D. Karkas, Proc.Nat.Acad.Sci. USA 69, 2288 (1972).
7. G. Stavrianopoulos, J.D. Karkas, E. Chargaff, Proc.
   Nat.Acad.Sci USA 68, 2207 (1971).
8. J. Kock, Eur.J.Biochem. 33, 98 (1973).
9. M. Miyaki, K. Koide and Tetsuo Ono, Biochim.Biophys.
   Res.Commun. 50, 252 (1973).

CHARACTERIZATION OF YEAST GRANDE AND PETITE MITOCHONDRIAL DNA BY HYBRIDIZATION AND PHYSICAL TECHNIQUES

Murray Rabinowitz, James Casey, Paul Gordon, Joseph Locker, Huey-Juang Hsu and Godfrey S. Getz

Departments of Medicine, Biochemistry, Biology and Pathology, The University of Chicago and the Franklin McLean Memorial Research Institute (operated by The University of Chicago for the United States Atomic Energy Commission; formerly the Argonne Cancer Research Hospital), Chicago, Illinois 60637, U.S.A.

Yeast mtDNA differs from the mtDNA of higher organisms in that intact DNA molecules have never been isolated, although 25 μ circles have been visualized by Hollenberg *et al.* (1) after osmotic shock of mitochondria. Despite this experimental difficulty, the availability of cytoplasmic petite and mitochondrial antibiotic resistant mutants, which result respectively in massive and fine changes in the mtDNA, makes the study of yeast extremely advantageous. In collaboration with the laboratories of Dr. P. Slonimski at Gif-sur-Yvette, France, we have tried to characterize the structure and function of grande and petite mtDNA using a combination of molecular hybridization, physical, electron microscopic and genetic techniques. Some of these studies have been reviewed recently (2). It has been concluded that the mtDNA of cytoplasmic petite mutants may contain large deletions ranging from 25 to more than 99 percent of the wild-type mtDNA sequences. Different parts of the mitochondrial genome may be retained in the different petite strains, as indicated by the retention or loss of mitochondrial genetic markers for antibiotic resistance, as well as by the variable capacity of mtDNAs from the same group of petites to hybridize with mitochondrial ribosomal or transfer RNAs. In some cases petites may continue

to transcribe mitochondrial ribosomal RNA and tRNAs
capable of accepting their cognate amino acids. In addition to deletion, petite mtDNA may contain base sequence changes and repetition of DNA segments. Long
tandem repeats of the mitochondrial genome may be present in petites, and are particularly apparent in
strains having mtDNA of low kinetic complexity. Mitochondrial tRNA and ribosomal RNA hybridizations have
also been used as molecular markers, which combined
with genetic analysis provides the opportunity to map
mitochondrial ribosomal and tRNA cistrons. With the
help of genetic markers, selective enrichment of different gene segments has been accomplished and physical characterization of these segments has been carried out.

In this presentation we will summarize some of the
more recent investigations from our laboratory relating to the structural and physical characterization
of petite mtDNA. The succeeding paper from Dr. Slonimski's laboratory will emphasize the characterization of mtDNA molecules as well as the purification
of different gene segments obtained from a series of
genetically defined stable petite strains. We will also present experiments which bears on the question
of whether grande mtDNA transcribes a full complement
of mitochondrial tRNAs.

tRNA-DNA hybridization has been used in the attempt
to define the tRNA transcripts of yeast mtDNA. All 14
of the yeast mt-tRNAs studies to date hybridize with
grande mtDNA: fmet (3), leu, val (4), ileu, ala, gly,
phe (5), tyr, asp, glu, lys, his. pro, ser (6). The
hybridization system is similar to that used by Weiss
et al. (7) to study T4 tRNAs, and by Nass and Buck (8)
for rat liver mt-tRNA. Mt-tRNA was acylated *in vitro*
with individual amino acids of high specific radioactivity. Hybridization is carried out in formamide at
33°, pH 5.2, to minimize deacylation. The hybridization of several mitochondrial tRNAs not previously reported by us is illustrated in Fig. 1. Hybridizations
are specific, as indicated by sharp melting curves of
the hybrids having appropriate Tm's, and by competition

studies that show unlabelled mt-tRNA, but not unlabelled cytoplasmic tRNA, or *E.coli* tRNA, to compete in
the hybridization. Hybridization studies with total
mt-tRNA labelled with [^{32}P], by Borst *et al.* (9) and
by our group (10), indicate the presence of 20 or more
tRNA cistrons in yeast mtDNA. It is probable that
yeast mtDNA specifies a full complement of mt-tRNAs.
Thus, yeast mtDNA probably differs from mtDNA of higher organisms, which appear to code for only 12 tRNAs
(11).

Mt-tRNA-DNA hybridizations were also used as molecular markers to analyze the mtDNA of cytoplasmic petite mutants. We have studied (12) the mtDNA of a
spontaneously derived cytoplasmic petite mutant (R1-6)
isolated and characterized in Dr. Slonimski's laboratory many years ago. The hybridization of grande [^{3}H]-
leucyl mt-tRNA to grande and petite mtDNA is shown in
Fig. 2A. Not only was hybridization with petite mtDNA
evident, but at saturation, the level of hybridization
was twice as great with petite as with grande mtDNA.
In contrast, grande [^{3}H]valyl mt-tRNA hybridized only
to a very small extent with petite mtDNA (Fig. 2B).
The cistron for leucyl tRNA is thus retained, but cistrons for valyl-tRNA are eliminated or greatly changed in the petite mtDNA of this strain. We have also
established that a functional tRNA is transcribed from
the mt-leucyl tRNA cistron in this petite. The tRNA
isolated from petite mitochondria and aminoacylated
in vitro with [^{3}H]leucine hybridizes with both grande
and petite mtDNAs (12).

The petite strain (R1-6) used initially proved to
be heterogeneous (as suggested to us by Dr. Slonimski),
since it had never been subcloned after initial isolation. Genetic and molecular heterogeneity of newly
isolated clones of mitochondrial mutants have been
shown by Slonimski's (13) and Linnane's (14) groups.

The hybridization of grande leucyl mt-tRNA to the
mtDNA of four subclones of strain R1-6 are shown in
Fig. 3 (12). Hybridization was obtained with mtDNAs
of all subclones, but to markedly different extents.
Petite R1-6/6 mtDNA hybridized to the same level as

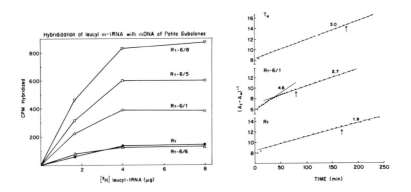

Fig. 1. Hybridization of mt-tRNA with grande mtDNA (6). Each filter contained 20 µg mtDNA. Hybridization conditions were as described previously (4).

Fig. 2. Hybridization of [³H]mt-leucyl tRNA (A) and [³H]valyl tRNA (B) to grande (R1) and petite (R1-6) mtDNA. Each filter contained 5 µg mtDNA. (Adapted from ref. 12).

Fig. 3. Hybridization of [³H]mt-leucyl tRNA to mtDNA (10 µg) of grande (ρ⁺) and four petite (ρ⁻) subclones (R1-6/1, R1-6/5, R1-6/6 and R1-6/8). (Adapted from ref. 12).

Fig. 4. Second-order renaturation plot (18) for T4 DNA, grande (R1 ρ⁺) and petite (R1-6/1 ρ⁻). (Adapted from refs. 15-17).

grande mtDNA, petite subclone R1-6/1 hybridized to 2 to 3 times this level, and petite subclone R1-6/8 to 6 times this level. Not only is heterogeneity apparent, but hybridization levels are markedly enhanced in many of the subclones, an observation that suggests the possibility of reiteration of segments of mtDNA that contain the cistron for leucyl tRNA. An alternate explanation could be deletion of different amounts of mtDNA in the different petite subclones, with retention of the leucyl tRNA cistron.

To test whether deletion could account for the high leucyl tRNA hybridization saturation levels, we examined these levels in relation to the genetic complexity of the DNAs measured by renaturation kinetic analysis (15-17). With deletion alone, one would expect a good correlation between renaturation rate and tRNA hybridization levels.

Renaturation kinetic analysis of mtDNA of petite subclone R1-6/1, which had a 2.5-fold enhancement of leucyl-tRNA hybridization, is compared with wild type mtDNA and T4 DNA in Fig. 4. The procedure and second-order rate plots were carried out according to the method of Wetmur and Davidson (18). The mtDNA was sheared to 5.4S. Of particular note is the presence of a fast renaturing fraction in the petite mtDNA, representing approximately 40 % of the DNA. The second-order rate constant, K_2, of the fast component is about 2.5 times greater than the K_2 of wild-type mtDNA, whereas the K_2 of the slow component is only about 1.5 times greater than that of wild-type mtDNA. Rapid renaturation of a fraction of the petite mtDNA indicates that copies of these sequences are more numerous than copies of sequences in the slow fraction. This repeated DNA does not represent heterogeneity of cell types in the subclone, since the same results were obtained after further subcloning.

Fast renaturing fractions were also observed in three other petite subclones (16,17), as seen in Fig. 5. The rapidly renaturing fractions represent different proportions of the mtDNA of the three strains. The K_2 of the fast fractions were 2- to 8-fold greater

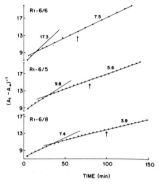
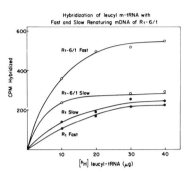

Fig. 5. Renaturation of mtDNA from petite subclones R1-6/5 (ρ 5), R1-6/6 (ρ 6) and R1-6/8 (ρ 8) (16, 17).

Fig. 6. Hybridization of [³H]mt-leucyl tRNA to "fast" and "slow" renaturing fractions of mtDNA (10 µg) of petite (R1-6/1) and grande (R1) yeast strains (16,17). Fractions were isolated by hydroxyapatite chromatography following controlled renaturation.

TABLE I

COMPARISON OF LEUCYL tRNA HYBRIDIZATION LEVELS, SECOND-ORDER RENATURATION RATES, AND SEQUENCE HOMOLOGY ESTIMATED BY DNA-DNA HYBRIDIZATION IN A SERIES OF PETITE SUBCLONES.
(From refs. 17 and 20).

Strain	Leucyl tRNA Hybridization (Relative to Grande)	Renaturation Rate (K_2 Relative to Grande)		Percent Homology to Grande mtDNA by DNA-DNA Hybridization in Solution
		Fast Fraction	Slow Fraction	
R1 (grande)	1	1	1	100
R1-6/1	2.5	2.3	1.3	70
R1-6/5	4	4.9	2.8	54
R1-6/6	1	8.7	3.7	74
R1-6/8	6	3.7	1.9	64

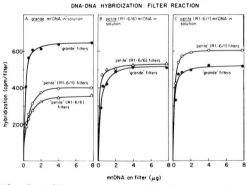

Fig. 7. Filter DNA-DNA hybridization analysis of petite (R1-6/1, R1-6/6) and grande (R1) mtDNA. A: [³H]grande mtDNA in solution; B: [³H]petite R1-6/6 mtDNA in solution; C: [³H]petite R1-6/1 mtDNA in solution. (From refs. 19-21).

94

than that of grande mtDNA. It is apparent that the high leucyl tRNA hybridization in some subclones could be accounted for if the leucyl tRNA cistron were located in the fast renaturing fraction, but not if it were in the slowly fraction.

Direct evidence for localization of leucyl tRNA cistrons in the fast fraction was obtained by partial separation of fast and slow renaturing fractions, and by hybridization to leucyl tRNA (16,17). Conditions were chosen at which renaturation of the fast fraction was 70 % complete, but renaturation of the slow fraction has proceeded to only 20 %. Renatured and denatured DNA were separated on hydroxyapatite columns. As seen in Fig. 6, leucyl tRNA hybridizes selectively to the enriched fast fraction of petite mtDNA. If a correction is made for the cross contamination between fast and slow fractions, the leucyl tRNA cistron can be localized exclusively in the fast fraction. Control studies with grande mtDNA do not show preferential hybridization with early renaturing DNA. We can conclude that the mtDNAs of the petite strains have repetitions of the portion of the mitochondrial genome carrying the leucyl tRNA cistron. It must be established, however, whether slow and fast renaturing fractions are on a single mtDNA molecule with reiteration of some DNA segments, ot whether there are stable heterogenous populations of mtDNAs within single petite cells.

Further evidence supporting either selective reiteration of some DNA segments, or heterogeneity of the intracellular molecular population of petite mtDNA is derived from the lack of correlation between leucyl tRNA hybridization levels and the kinetic complexity or DNA-DNA homology (19,29) (Table I). For example, strain R1-6/6 mtDNA renatures fastest, but shows no increase in leucyl tRNA saturation levels, whereas strain R1-6/8, with a 6-fold increase in hybridization levels, shows only a 40-50 % decrease in homology with grande mtDNA. Furthermore, in experiments not presented here, R1-6/6 and R1-6/8 mtDNAs could not be distinguished from one another by DNA-

DNA hybridization, even though they differed by a factor of 6 with respect to leucyl-tRNA hybridization (19,20).

DNA-DNA hybridization was also employed to characterize additional changes in the petite mtDNA. Hybridization both in solution (20) and to DNA immobilized on filters (20,21) was used. The hybridization of sheared tritiated grande mtDNA in solution with filter-bound mtDNA of the two petite subclones is seen in Fig. 7A. Conditions were chosen so that renaturation of DNA in solution was negligible and hybridization to filter-bound DNA was complete. Hybridization was carried out with individual filters, and the content of DNA on each filter was progressively increased. The depressed hybridization of the two petite mtDNAs suggests at least a 30-50 % alteration in these DNAs.

In one of the petite subclones (R1-6/6), deletion appears to be the predominant change (20) (Fig. 7B). Tritiated R1-6/6 mtDNA hybridized to the same level with grande mtDNA as it did to homologous DNA. Thus, all sequences in petite R1-6/6 mtDNA are represented in the grande mtDNA, as expected for a deletion mechanism.

Very different results were observed with petite subclone R1-6/ 1 (21) (Fig. 7C). In many experiments, a 10-20 % greater (p < 0.001) hybridization was consistantly observed for petite to petite than for petite to grande mtDNA (21). The results appear to be the consequence of changed sequence as well as deletion.

Hybridization studies carried out in solution with analyses on hydroxyapatite columns gave similar results (19,20). The hybridization of [^{3}H]petite DNA was driven by the homologous sequences of unlabelled grande mtDNA added in 100-200-fold excess (Table II). As with filter hybridization, about 20-30 % of petite R1-6/1 mtDNA did not hybridize with grande mtDNA under stringent hybridization conditions.

Further evidence for base sequence change in petite mtDNA is derived from hydroxyapatite thermal elu-

TABLE II

DNA-DNA HYBRIDIZATION IN SOLUTION.
Homology between grande (R1) and petite (R1-6/1, R1-6/6) mtDNA, estimated from hybridization in solution with analysis on hydroxyapatite chromatography. Renaturation of [^{3}H]petite or grande mtDNA was driven by the addition of 100-200-fold excess of [^{32}P]grande or petite mtDNA. Renaturation of excess DNA was carried to 85-90 % completion. Under these conditions, control [^{3}H]mtDNA alone renatured to less than 5 %. (From refs. 19 and 20).

Strain	Petite Sequences Homologous to Grande mtDNA	Grande Sequences Homologous to Petite mtDNA
	Percent	
R1	100	--
R1-6/1	70	71
R1-6/6	74	96

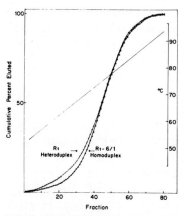

Fig. 8. Hydroxyapatite thermal elution chromatography of R1-6/1 petite-petite homoduplex, and a petite (R1-6/1)-grande (R1) heteroduplex. Renaturation reactions were carried out as described in Table II. Melting profiles of the hybrids were obtained by elution from hydroxyapatite columns by raising the temperature 12°/h using a Haake linear temperature programmer (19,20).

tion chromatography of the petite grande heterohybrid (Fig. 8). The sharp melting curves with little change in Tm indicate excellent base pairing of the hetero- hybrids. However, all heterohybrids showed a signifi- cant decrease in thermal stability of the high AT DNA melting at lower temperatures.

It may then be concluded that the mtDNA of one of the petite subclones contained sequences that are suf- ficiently changed so as not to hybridize with grande mtDNA under moderately stringent conditions, but that all the petite strains studied contained some base se- quence changes. The nature or function of these chang- ed sequences is unclear, but it is possible that they represent reiteration of small segments of grande mt- DNA, which results in the formation of sequences that do not hybridize with grande mtDNA.

In collaboration with Dr. H. Fukuhara, the molecu- lar hybridization procedure has also been used to de- termine the representation of tRNA cistrons within the mtDNA of a series of petite mutants constructed in the Gif laboratory where they have been characterized with respect to genetic markers and extent of deletion. At- tempts to map the tRNA genes in relation to the gene- tic markers are now in progress (22). In these stu- dies we have noted extremely high levels of hybridi- zation of several tRNAs with the mtDNA of some strains. The increased hybridization levels may be explained in part by a deletion mechanism. However, the widely differing increments of hybridization for histidyl, leucyl, tyrosyl, alanyl, and isoleucyl tRNA and for ribosomal RNA in strain D21 indicate either indepen- dent amplification of different segments of the DNA, or heterogeneous populations of mtDNA molecules in this petite subclone.

It is also apparent that petite mtDNAs having low kinetic complexity may be composed of long tandem re- peats of the entire mitochondrial genome. Recent elec- tron-micrographic studies of the petite mtDNA by Locker *et al.* (23), in collaboration with the Gif- sur-Yvette group (2,24), reveal the presence of cir- cular and long linear molecules. On contrast to grande

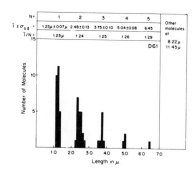

Fig. 9. Frequency distribution of circular mtDNA molecules from petite strain D61, measured from electron micrographs. A uniform polymeric distribution is seen with a monomer size of 1.2 μ. In this strain, about 20 % of DNA molecules were circular and many circles were supercoiled. (From ref. 23).

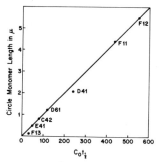

Fig. 10. Comparison of monomer size of circular molecules and renaturation rates of mtDNA expressed as Cot 1/2 ($A_{260 \text{ nm}}$ sec) (26) in 7 petite strains. Only strain F13, which has an extremely high A+T content, diverges significantly from the linear relationship between these parameters. (From ref. 23).

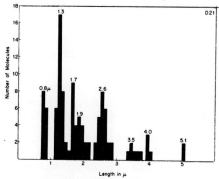

Fig. 11. Frequency distribution of circular mtDNA molecules from strain D21. Two or more polymeric series of circular molecules must be present to explain these data. (From ref. 23).

mtDNA, where only rare circular molecules are isolated, and where none are 26 μ in length (the assumed genome size), considerable numbers of circular molecules were observed in the eight petite strains studied so far.

Data for one of the ethidium bromide induced genetically defined strains are shown in Fig. 9. The frequency distribution of the circular molecules indicates that all are multimers of molecules with a monomeric size of 1.2 μ. Michel and Fukuhara's kinetic analysis (25) of this DNA would indicate a kinetic complexity approximately 1/25 that of wild type mtDNA. Thus, the monomer size is exactly that predicted by renaturation kinetics. In seven of the eight strains studied in this group, a single or predominantly single polymeric series of circles was seen, with a unique monomer size in each of the petite strains.

A comparison of basic monomer size and the Cot 1/2 in the seven petite strains is seen in Fig. 10. There is excellent correlation between the two measurements in these strains, where monomer size varies between 0.1 and 3.7 μ. The data extrapolates to 25 μ for the Cot 1/2 or K_2 of grande mtDNA.

In one strain, the size distribution of circular molecules does not follow the simple pattern expected for a polymeric series derived from a single population of monomeric units (Fig. 11). These data are best explained by assuming the presence of two or more polymeric series derived from different monomer units.

In all petite strains, circular and linear molecules many times larger than monomer size were observed. A remarkable finding in some of the petite mtDNAs was the presence of a considerable amount of self-renaturing DNA (27). Hairpin loops and a variety of double-stranded configurations were observed after formimide spreading of denatured DNA (Fig. 12). The renatured segments have a uniform length and are interspersed with small single-stranded regions. The length of self-renatured DNA segments is identical with the monomeric circle size in this petite strain (23). These observations indicate the presence of reverse repeats

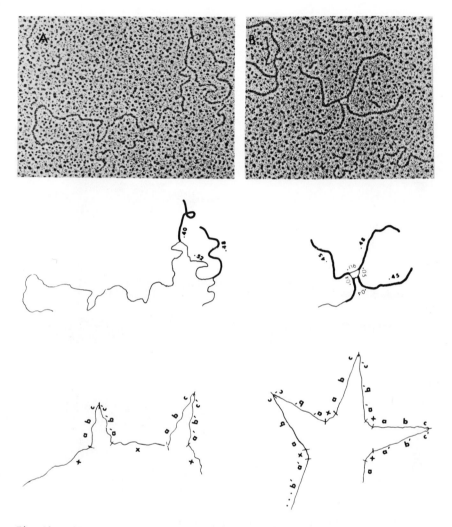

Fig. 12. Self-renaturing molecules of mtDNA from petite strain E41. The postulated explanations for the molecular configurations are given in the diagrams below. A B C and A' B' C' represent complementary DNA sequences. In (A), reverse repeats are nonalternating. In (B), the reverse repeats alternate with small regions (X) of non-homology. Eighty to ninety percent of the mtDNA of this strain self-renatures, and the predominant molecular configuration indicates alternating reverse repeats; a regular spacing of renatured segments of approximately 0.45 μ was present. (From ref. 27).

101

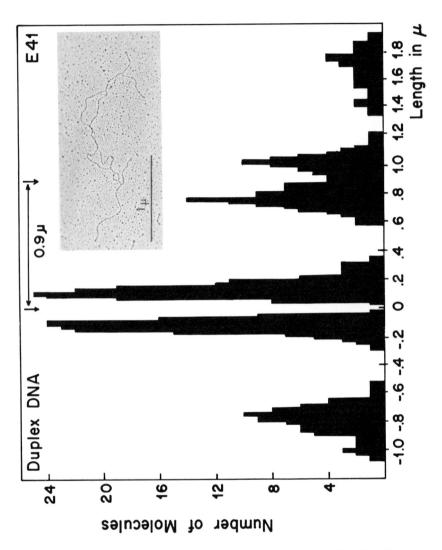

Fig. 13. Denaturation mapping of mtDNA of petite strain E41. The distribution of double-stranded segments is indicated in the figure, normalized at 0μ, to the center of a small loop. A typical denatured molecule is seen at the top, right. The spacing of a unit repeat, which consists of a large, and a small denatured region, is 0.9 μ. This unit probably represents two mtDNA genomes (each composed of 1/2 of a large and 1/2 of a small denatured loop) connected as a reverse repeat. (From ref. 27).

of the mitochondrial genome.

Denaturation mapping of the mtDNA of the same strain (Fig. 13) indicated a regular unit spacing that is twice as large as that observed for circular mono- mers and for self-renaturing segments (27). The unit spacing, which included a small and large denaturation loop (Fig. 13), probably represents two copies of the mitochondrial genome connected head to head, that is, as a reverse repeat. The amount of self-renaturing DNA observed in the different petite strains varied be- tween 5 and 90 % and did not correlate with kinetic complexity. Monomeric units thus appear to be connec- ted head to tail, or head to head, to form circular and perhaps linear molecules of different sizes.

In summary, data have been presented which indica- te that most, if not all mt-tRNAs are transcribed from mtDNA. Mt-tRNA-DNA hybridization and DNA-DNA hybridi- zation have been used to characterize the changes in petite mtDNA. Although the predominant change in peti- te mtDNA appears to be deletion, changed sequences in some petite strains also appear to exist. The possibi- lity of repetition of some segments of mtDNA, or of heterogeneity of the mtDNA population within a cell, has also been raised. Genetic and hybridization me- thods have been combined in an attempt to map mt-tRNA cistrons. Finally, electron microscopy has shown that several petite strains contain polymeric series of circular molecules; the monomeric units are of a size expected from the second-order renaturation rates. So- me strains, however, may be heterogenous with respect to monomer size. Self-renaturing regions indicative of reverse repeats show that, in many cases, the monomers are connected head to head as well as head to tail.

ACKNOWLEDGEMENT.

These studies were supported in part by USPHS grants HL09172, HL04442, and GM18558 from the National Institutes of Health, and the Kathryn Tobin Grant for Cancer Research from the American Cancer Society.

REFERENCES.

1. C.P. Hollenberg, P. Borst and E.F.J. van Bruggen, Biochim.Biophys.Acta 209, 1 (1970).
2. G. Faye, H. Fukuhara, C. Grandchamp, J. Lazowska, F. Michel, J. Casey, G.S. Getz, J. Locker, M. Rabinowitz, M. Bolotin-Fukuhara, P. Coen, J. Deutsch, B. Dujon, P. Netter and P.P. Slonimski, Biochimie, in press.
3. A. Halbreich and M. Rabinowitz, Proc.Nat.Acad.Sci. U.S.A. 68, 294 (1971).
4. J. Casey, M. Cohen, M. Rabinowitz, H. Fukuhara and G.S. Getz, J.Mol.Biol. 63, 431 (1972).
5. M. Cohen and M. Rabinowitz, Biochim.Biophys.Acta 281, 192 (1972).
6. J. Casey, H. Hsu, M. Rabinowitz, G.S. Getz and H. Fukuhara, manuscript in preparation.
7. S.B. Weiss, W. T. Hsu, J.W. Foft and N.H. Scherberg, Proc.Nat.Acad.Sci.U.S.A. 61, 114 (1968).
8. M.M.K. Nass and C.A. Buck, J.Mol.Biol. 54, 187 (1970).
9. L. Reijnders and P. Borst, Biochem.Biophys.Res. Commun. 47, 126 (1972).
10. A. Halbreich and M. Rabinowitz, unpublished observation.
11. M. Wu, N. Davidson, G. Attardi and Y. Aloni, J. Mol.Biol. 71, 81 (1972).
12. M. Cohen, J. Casey, M. Rabinowitz and G.S. Getz, J.Mol.Biol. 63, 441 (1972).
13. M. Bolotin, P. Coen, J. Deutsch, B. Dujon, P. Netter, E. Petrochile and P. Slonimski, Bull.Inst. Pasteur 69, 218 (1971).
14. P. Nagley and A.W. Linnane, J.Mol.Biol. 66, 181 (1972).
15. M. Fauman and M. Rabinowitz, FEBS Letters 28, 317 (1972).
16. J.W. Casey, P. Gordon and M. Rabinowitz, J.Cell Biol. 55, 35a (1972).
17. J.W. Casey, P. Gordon and M. Rabinowitz, manuscript in preparation.
18. J.G. Wetmur and N. Davidson, J.Mol.Biol. 21, 34 (1968).

19. P. Gordon, J. Casey and M. Rabinowitz, Fed.Proc.
    32, 663 Abs (1973).
20. P. Gordon, J. Casey and M. Rabinowitz, manuscript
    in preparation.
21. P. Gordon and M. Rabinowitz, Biochemistry 12, 116
    (1973).
22. J. Casey, H. Hsu, M. Rabinowitz and H. Fukuhara,
    manuscript in preparation.
23. L. Locker, M. Rabinowitz and G.S. Getz, manus-
    cript in preparation.
24. C. Grandchamp *et al.* manucript in preparation.
25. F. Michel *et al.* manuscript in preparation.
26. R.J. Britten and D.E. Kohne, Science 161, 529
    (1968).
27. J. Locker, M. Rabinowitz and G.S. Getz, manus-
    cript in preparation.

ELONGATION FACTORS FOR ORGANELLAR PROTEIN SYNTHESIS IN *CHLORELLA VULGARIS*

Orio Ciferri, Orsola Tiboni, Gabriella Lazar[*], James van Etten[**]

Institutes of Plant Physiology and Genetics, University of Pavia, Pavia, Italy.

It is commonly accepted that mitochondria and chloroplasts contain autonomous protein synthesizing systems which resemble those of prokaryotic organisms and hence are distinct from the system present in the cytoplasm. There is a paucity of evidence, however, on the source of the genetic information and the site of synthesis of the translation factors involved in organellar protein synthesis. Present evidence, which has been obtained only from the fungi *Saccharomyces cerevisiae* and *Neurospora crassa*, suggests that some of the mitochondrial translation factors are coded in the nucleus and synthesized in the cytoplasm (1-4). Similar data about the origin of the translation factors in chloroplasts are lacking. Also since chloroplast and mitochondrial protein synthesizing systems are similar, it is possible that a plant contains one set of translation factors which catalyzes protein synthesis on both organellar ribosomes or that each possesses a unique set of translation factors.

The present report describes the separation of two species of elongation factors-G (EF-G) of the prokaryotic type from the green alga *Chlorella vulgaris*; one factor is probably of mitochondrial origin and the other of chloroplast origin. Additional data suggest

---

[*]Permanent address: Biological Research Center, Hungarian Academy of Sciences, Szeged, Hungary.
[**]Permanent address: Department of Plant Pathology, University of Nebraska,Lincoln, Nebraska, U.S.A.

that the genetic information for mitochondrial elongation factors is translated on cytoplasmic ribosomes.

MATERIALS AND METHODS.

Chlorella vulgaris wild type and the apochlorotic mutant CM-20 were grown autotrophically, mixotrophically, or heterotrophically in the dark or light as described previously (5). Usually inhibitors were added 24-48 h prior to harvesting when the cell concentration had reached 1 to 5 x 10^6 cells/ml. The cells were harvested and washed as described previously (5).

In vivo amino acid incorporation into mitochondria.
The procedure for measuring *in vivo* amino acid incorporation into mitochondria was modified from that reported by Schatz and Saltzgaber (6). Fifty ml (1 x 10^7 cells/ml) samples of *C.vulgaris* CM-20, which had been grown in the presence or the absence of inhibitor, were washed twice with growth medium (in appropriate cases containing the same inhibitor) and the cells were then incubated for 30-60 min in the presence of the same inhibitor plus or minus 100 µg/ml of cycloheximide. After incubating the cells for 20 min with a mixture of 12 [^{14}C]amino acids an excess of unlabeled amino acids was added and the cells incubated for an additional 15 min. The cells were then harvested, washed twice with buffer A (0.25 M mannitol; 0.02 M Tris, pH 7.4; 0.003 M EDTA; 0.1 % albumin) before subjecting them to two passages through a French Press at approximate 10-12,000 psi. The extracts were centrifuged twice at 2,000 xg for 10 min to remove rapidly sedimenting material and the mitochondria were pelleted by centrifuging at 20,000 xg for 30 min and then washed two times with buffer A. The final pellet was resuspended in a small amount of buffer A and aliquots assayed for cytochrome c oxidase activity by the procedure of Smith (7). The remaining material was pelleted and solubilized in a small amount of 3.3 % (w/v) deoxycholate. After diluting with 0.01 M Tris, pH 7.8 containing 0.1 M KCl, aliquots were precipitated with trichloroacetic acid, the precipitate collected on filter papers, and the radioactivity determined.

Enzyme extracts. The cells were either homogenized by grinding in a cold mortar and pestle (Table IV)
and the enzyme extracts prepared as previsously reported (5) or approximately 1.5 - 3 mg wet weight of
cells per 5 ml of buffer B (0.02 M Tris, pH 7.8; 0.05
M magnesium acetate; 0.02 M 2-mercaptoethanol; 0.01 M
KCl; 40 µg/ml spermine; 10^{-4} M reduced glutathione)
were homogenized by two passages through a French
Press at 10 - 12,000 psi (Table III). After removal
of rapidly sedimenting material and the ribosomes by
centrifugation, the supernatant was treated with pro-
tamine sulfate (0.17 mg/ml of extract) for 15 min,
centrifuged and fractionated with ammonium sulphate.
The proteins which precipitated at 30 - 70 % ammonium
sulphate were dissolved in a small volume of buffer C
(0.02 M Tris, pH 7.8; 0.005 M magnesium acetate; 0.02
M 2-mercapthoathanol; 0.01 M KCl; and 10^{-4} M reduced
glutathione) and dialyzed (5). The presence of elongation factors was assayed with ribosomes from *E.coli*
(70S) or cytoplasmic ribosomes from *S.cerevisiae* (80S)
by measuring poly U directed polyphenylalanine synthesis in a standard assay mixture (8) containing 17
pmoles of *E.coli* [^{14}C]phenylalanyl-tRNA and either
125 µg of *E.coli* or *S.cerevisiae* ribosomes. The data
are expressed as pmoles of [^{14}C]phenylalanine incorporated in 15 min at 30º per mg of enzyme protein.

The separation of two EF-G's, which function with
70S ribosomes, was accomplished by chromatography on
DEAE-Sephadex columns as described elsewhere (5).

RESULTS AND DISCUSSION.

Separation of EF-G$_{mit}$ and EF-G$_{chl}$. The existence
of two protein fractions which exibit EF-G activity
on *E.coli* ribosomes in autotrophically-grown *C.vulga-
ris* is shown in Fig. 1A. The following observations
suggest that the first peak is probably of chloroplast
origin and the second peak is of mitochondrial origin.
Only one active fraction corresponding to the second
peak of activity is observed when the parental strain
is grown heterotrophically in the dark (Fig. 1B) or
from an apochlorotic mutant of *C.culgaris* (Fig. 1D). In

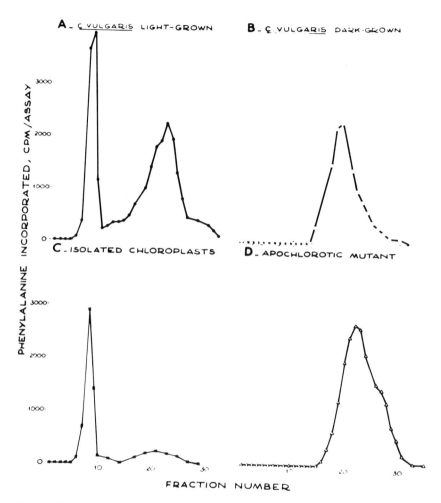

Fig. 1. Separation of two protein factors from *C. vulgaris* which exhibit EF-G activity on *E. coli* ribosomes. (A) wild type cells grown in the light, (B) wild type cells grown in the dark, (C) chloroplasts isolated from light-grown cells, (D) apochlorotic mutant CM-20. Experimental details reported in ref. 5.

addition, essentially only the first peak is obtained from extracts prepared from isolated chloroplasts (Fig. 1C).

Further confirmation that the two EF-G's are probably unique proteins is supported by their differential response to fusidic acid, an antibiotic reported to differentially affect the activity of bacterial EF-G and mitochondrial EF-G (9,10). The EF-G$_{chl}$ is more

TABLE I

EFFECT OF FUSIDIC ACID ON POLYPHENYLALANINE SYNTHESIS BY *E.COLI* RIBOSOMES IN THE
PRESENCE OF CHLOROPLAST AND MITOCHONDRIAL EF-G's FROM *C.VULGARIS* AND EF-G FROM
E.COLI.
Assay conditions as previously reported (5) except that the following amounts (µg
protein/assay) of EF-G's were used: *E.coli* 9; mitochondria from *C.vulgaris* 8;
chloroplasts from *C.vulgaris* 11. EF-T from *E.coli* (32 µg of protein/assay) was
present in all reaction mixtures. -: not tested.

Source of EF-G	Fusidic acid concentration (M)					
	5×10^{-7}	1×10^{-6}	1×10^{-5}	5×10^{-5}	5×10^{-4}	5×10^{-3}
	(Percent inhibition)					
E.coli	-	0	0	43	81	94
C.vulgaris mitochondria	-	0	0	0	0	70
C.vulgaris chloroplasts	0	36	63	76	93	100

sensitive to fusidic acid than is EF-G$_{mit}$ or *E.coli*
EF-G (Table I). These results, therefore, would suggest that the two organelles contain unique proteins
with EF-G activity. Furthermore, the results with fusidic acid suggest that the amount of EF-G$_{chl}$ and EF-
G$_{mit}$ in crude enzyme extracts from *C.vulgaris* and possibly other photosynthetic eukaryotes may be assayed
rapidly by their differential response to the drug.

*Site of synthesis of mitochondrial elongation fac-
tors.* Presently all of the evidence that mitochondrial elongation factors are synthesized in the cytoplasm has been obtained from studies on *S.cerevisiae*
or *N.crassa* . Therefore, experiments patterned from
thoss conducted on *N.crassa* were attempted on the
apochlorotic strain of *C.vulgaris* in which no EF-G$_{chl}$
was detected by chromatography (Fig. 1). The effect
of six inhibitors on growth, *in vivo* amino acid incorporation into mitochondria, and cytochrome *c* oxidase activity is reported in Table II. Although nali-
dixic acid and rifampicin inhibited cell growth, they
did not selectively inhibit mitochondrial amino acid
incorporation. In contrast, erythromycin, chloramphenicol, lincomycin and ethidium bromide inhibited *in
vivo* amino acid incorporation into mitochondria. In
addition, cytochrome *c* oxidase activity was depressed
in erythromycin- and chloramphenicol-treated cells;
this enzyme is synthesized, at least in part, in the
mitochondrion (for review, see ref. 11). Therefore, we

111

TABLE II

EFFECT OF SEVERAL INHIBITORS ON GROWTH, *IN VIVO* AMINO ACID INCORPORATION INTO MITOCHONDRIA AND CYTOCHROME *C* OXIDASE ACTIVITY OF THE APOCHLOROTIC MUTANT OF *C. VULGARIS*.
The experimental conditions are described in the materials and methods.

Inhibitor	Concentration µg/ml	Cell conc. when antibiotic was added, cells/ml	% inhibition of growth after		S.A. of mitochondrial protein, cpm/mg of protein		% inhibition of cytochrome *c* oxidase activity
			24h	48h	−	+cycloheximide	
Nalidixic	0		0		18,574	1210	0
acid	250	1.2×10^6	38		16,210	1760	0
	500		41		17,160	1400	0
Rifampicin	0		0		36,890	1800	0
	250	1.1×10^6	37		37,500	1950	0
	500		42		38,310	2080	0
Ethidium	0		0	0	32,777	2132	−
bromide	50	3.5×10^5	16	46	2,534	1136	−
Lincomycin	0		0	0	25,350	1610	0
	400	1.2×10^6	29	35	21,587	1354	0
Erythromycin	0		0		30,998	1739	0
	1000	9×10^5	29		11,267	1370	14
	2000		50		4,993	810	26
Chloramphenicol	0		0	0	7,491	1047	0
	4000	1.74×10^6	22	48	3,913	307	86
	6000		23	52	3,090	266	76

chose to investigate the effect of chloramphenicol and erythromycin on the synthesis of mitochondrial elongation factors. Exposure of the cells to high concentrations of either antibiotic for 48 h had no effect on the specific activities of the cytoplasmic elongation factors and possibly a stimulatory effect on the specific activities of the mitochondrial elongation factors (Table III). This was more pronounced with the chloramphenicol-treated cells than with the erythro-

TABLE III

PRESENCE OF ELONGATION FACTORS OF THE PROKARYOTIC AND EUKARYOTIC TYPE IN THE APOCHLOROTIC MUTANT OF *CHLORELLA VULGARIS* GROWN IN THE PRESENCE OR ABSENCE OF ERYTHROMYCIN OR CHLORAMPHENICOL FOR 48h.
Antibiotics added when the cell concentration was 5×10^5 cells/ml. Specific activity expressed as pmoles of phenylalanine incorporated/mg enzyme protein/ 15 min.

Antibiotic/Concentration	Specific activity of elongation factors	
	Yeast ribosomes	*E. coli* ribosomes
Erythromycin/−	42.5	3.7
1 mg/ml	47.4	4.2
2 mg/ml	46.3	4.0
Chloramphenicol/−	143.3	4.8
4 mg/ml	146.4	9.8
6 mg/ml	132.7	10.2

mycin-treated cells. In separate experiments the cells were exposed to 50 µg/ml of cycloheximide, an inhibitor of cytoplasmic protein synthesis, for periods up to 72 h. If the mitochondrial elongation factors were synthesized on the mitochondrial ribosomes one might observe a preferential increase in the specific activity of mitochondrial elongation factors in contrast to the cytoplasmic elongation factors. However, the ratio of these two sets of elongation factors was constant over the time period studied. These latter experiments revealed that cytoplasmic and mitochondrial elongation factors were very stable. Even after inhibiting the growth of the cells for 72 h with cycloheximide or cycloheximide plus chloramphenicol there was no decrease in the specific activity of either set of elongation factors or in the total amount of extractable protein as compared to cells not exposed to the drug.

Therefore, the data obtained with the inhibitors suggest that mitochondrial elongation factors from *C.vulgaris* are synthesized on the cytoplasmic ribosomes. Such results are in agreement with the conclusions previously reported for *N.crassa* and *S.cerevisiae* (1-4).

Site of synthesis of chloroplast elongation factors. The parent strain of *C.vulgaris* was grown in the dark to a concentration of approximately 5 to 10 x 10^6 cells/ml and then exposed to light for 24 h. Chloramphenicol (4 mg/ml) or cycloheximide (100 µg/ml) was added to some cultures 1 h before exposure to the light. Assaying the crude enzyme fractions for elongation factor activity on *E.coli* ribosomes revealed that the enzyme preparations from cells exposed to continuous light or 24 h of light were more sensitive to low concentrations of fusidic acid then cells grown continuously in the dark or cells exposed to the light in the presence of either antibiotic (Table IV). If the sensitivity to low concentrations of fusidic acid accurately reflects the concentrations of chloroplast elongation factors, these data suggest that the synthesis of activity of these factors is dependent on

TABLE IV

EFFECT OF FUSIDIC ACID ON POLYPHENYLALANINE SYNTHESIS BY *E.COLI* RIBOSOMES IN THE
PRESENCE OF CRUDE ENZYME PREPARATIONS FROM WILD TYPE *C.VULGARIS* CELLS
The cells were grown in the dark as reported in the text except for experiment 1
in which the cells were cultured in the light. The cells were harvested when they
reached a concentration of 1-2 x 10^7 cells/ml. This usually required 4 days.

Conditions	Fusidic acid concentration, M			
	1×10^{-5}	5×10^{-5}	5×10^{-4}	5×10^{-3}
		(Percent inhibition)		
1. Cells grown continuously in the light.	43	68	80	100
2. Cells grown in the dark, exposed to light for 24h.	38	55	65	100
3. Cells grown continuously in the dark.	12	20	19	100
4. 4mg/ml chloramphenicol added 1h before 24h of light.	12	17	17	100
5. 100 μg/ml cycloheximide added 1h before 24h of light.	0	13	7	100

the presence of light and possibly of a functional
chloroplast. Therefore, the control mechanism for the
synthesis of chloroplast elongation factors appears
to be different from that operating in mitochondria of
N. crassa and *S. cerevisiae*.

ACKNOWLEDGEMENT.

This work was supported by a grant from the Laboratorio di Genetica Biochimica ed Evoluzionistica of the Consiglio Nazionale delle Ricerche.

REFERENCES.

1. A.M. Scragg, FEBS Letters 17, 111 (1971).
2. B. Parisi and R. Cella, FEBS Letters 14, 209 (1971).
3. D. Richter, Biochemistry 10, 4422 (1971).
4. Z. Barath and H. Küntzel, Proc.Nat.Acad.Sci. USA
69, 1371 (1972).
5. O. Ciferri and O. Tiboni,(submitted for publication).
6. G. Schatz and J. Saltzgaber, Biochem.Biophys.Res.
Commun. 37, 996 (1969).

7. L. Smith, in:"Methods of Biochemical Analysis", D. Glick, Ed., vol.2, Interscience Publ. Inc. N.Y., 427 (1955).
8. B. Parisi, G. Milanesi, J. van Etten, A. Perani and O. Ciferri, J.Mol.Biol. 28, 295 (1967).
9. M. Grandi, A. Helms and H. Küntzel, Biochem.Biophys.Res.Commun. 44, 864 (1971).
10. O. Tiboni and O. Ciferri, FEBS Letters 19, 174 (1971).
11. R. Sager, in:"Cytoplasmic Genes and Organelles", Academic Press, N.Y. (1972).

SYNTHESIS AND PROCESSING OF MITOCHONDRIAL RIBOSOMAL RNA IN WILD TYPE AND POKY STRAIN OF *NEUROSPORA CRASSA*

Yoshiaki Kuriyama and David J.L. Luck

Laboratory of Cell Biology, The Rockefeller University, New York, N.Y. 10021, U.S.A.

In mitochondria of *Neurospora crassa*, ribosomal RNA's are first transcribed from mitochondrial DNA as a high molecular weight procursor which is cleaved to give both large and small subunit RNA's (1). This mechanism, which resembles the one described for eukaryotic cytoplasmic ribosomal RNA's (2,3), could ensure the coordinate production of large and small ribosomal subunits. It is known however that in mitochondria of the poky mutant of *Neurospora*, ribosomal subunits are not present in equal numbers and there is a striking deficienct of small ribosomal subunits (4). The results presented here, focus the problem in poky to the events in maturation and cleavage of the mitochondrial ribosomal RNA precursor. First, we will review briefly the evidence for ribosomal RNA production from a high molecular weight precursor. We will then present new evidence concerning the role of RNA methylation in the precursor maturation process and demonstrate that this sequence is altered in poky mitochondria.

At the beginning it is important to emphasize that all of these studies have been made with mitochondrial fractions obtained from exponentially growing cultures and purified by flotation in sucrose gradients. We have shown that these preparations are free of detectable contamination by cytoplasmic ribosomes, which considerably simplifies analysis of our results.

BIOSYNTHESIS OF MITOCHONDRIAL RIBOSOMAL RNA's.

We began our studies of ribosomal RNA synthesis in mitochondria with pulse-labeling experiments. The re-

117

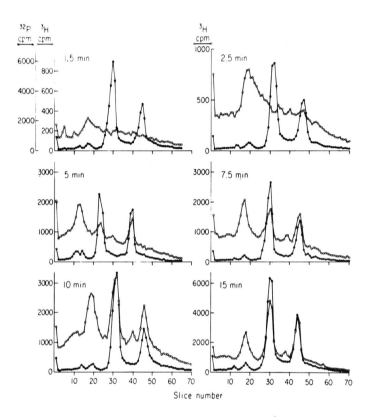

Fig. 1. Incorporation of 5[^{3}H]uracil and [^{32}P]P$_i$ into high molecular weight mitochondrial RNA's. Cells were grown for 14.5 h in the presence of [^{32}P]phosphoric acid (0.1 mCi/150 ml) in medium containing a low concentration of phosphate (KH$_2$PO$_4$ 200 μg/ml) and 5[^{3}H]uracil (0.8 mCi/150 ml) was added. After various times of labeling, the cells were cooled rapidly by pouring cultures into 4 vol of ice-cold 0.25 M sucrose (2 mM tricine, pH 7.6) and combined with the same amount of cells grown in media without radioactive material. The cells were collected by filtration and ground with sand. Mitochondria were prepared by flotation, the RNA was extracted by the diethyl pyrocarbonate method, and ~0.5 OD$_{258}$/μl was analyzed on 2.4 % gels containing 0.8 % agarose, by electrophoresis for 4 h at 130 V at 6° C. The gels were sliced and processed for counting. 0 - 0: [^{3}H]counts/min; ● - ●: [^{32}P]counts/min.

118

sults of analytical experiments are shown in Fig. 1. Cells were continuously labeled with [^{32}P] from the beginning of culture to obtain a mass marker of RNA. After 14 hours of preculture, the cells were pulse-labeled for varying times with [^{3}H]uracil. The closed circles show the [^{32}P]-labeled stable RNA components: 19S, 25S, 32S RNA. To the left of 32S a small peak is due to DNA. Only this peak disappeared with DNAse treatment. The open circles show the RNA pulse-labeled with tritium uracil. It is apparent that at 1.5 and 2.5 min, the major peak of tritium radioactivity coincides with a high molecular weight RNA at 32S position. This RNA is present only as a small peak in the [^{32}P]-labeled stable RNA. At later time points, tritium label appears in the region of 19S and 25S RNA's. In addition, there are at least two other pulse-labeled components. One is visible after 2.5 min and at all later time points as a small tritium peak between 19S and 25S. We refer to this peak as the P19S. The other peak appears as a small shoulder to the left of 25S and this is called P25S. These results are summarized in Fig. 2.

Here it can be seen that the 32S component is the most rapidly labeled and reached a plateau at 10 min. The 19S and 25S RNA's begin to accumulate tritium radioactivity only after a few minutes lag. During the early period of pulse, the 19S RNA is labeled faster than the 25S. This observation is highly reproducible. These results suggested the possibility of a precursor product relationship between 32S and ribosomal RNA's, a hypothesis which was further tested by pulse-chase experiments. A summary of such an experiment is shown in Fig. 3. The loss of radioactivity from 32S could be accounted for by the increases in the radioactivity of the ribosomal RNA's. As a final test of a precursor role for the 32S RNA, competition for its hybridization to mitochondrial DNA by 19S and 25S RNA's was studied.

[^{32}P]-labeled 32S RNA and varying concentrations of cold 19S and/or 25S were incubated with a mitochondrial DNA-fixed filter. The results are shown in Fig.

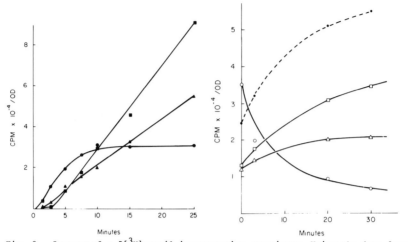

Fig. 2. Summary of a 5[³H]uracil incorporation experiment. Using the data from
Fig. 1 and additional time points not included there, total [³H]radioactivity
in each RNA species was derived by summation of the radioactivity in the gel fractions containing that species (5 to 6 gel fractions around each peak). The data
from different gels were made comparable by normalizing with respect to the OD_{258}
of total RNA loaded on the gels. ● - ●: 32S RNA; ■ - ■ : 25S RNA; ▲ - ▲: 19S RNA.

Fig. 3. Pulse-chase experiments using 5[³H]uracil. Cells were grown for 14 h at
25° C in the presence of [³²P]phosphoric acid (0.1 mCi/150 ml, 1 mg KH_2PO_4/ml),
then shifted to 7° C and incubated for 60 min. The culture was pulse-labeled with
[³H]uracil (0.8 mCi/150 ml) at 7° C for 2 h, filtered rapidly and washed with 50
ml of cold chase medium, and finally resuspended and incubated in a chase medium
at 25° C. After various time intervals, cells were chilled quickly in ice-cold
0.25 M sucrose and RNA was prepared from purified mitochondria. Analysis of RNA
on gel electrophoresis (2.1 % gels), and radioisotope counting were as described
in Fig. 1. A summary of these results is plotted in a manner similar to Fig. 2.
0 - 0: 32S RNA; □ - □ : 25S RNA; Δ - Δ: 19S RNA; ● - ●: 25S RNA + 19S RNA.

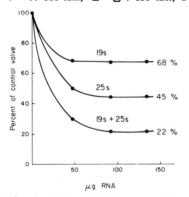

Fig. 4. RNA-DNA hybrid competition experiments with 32S RNA. Millipore filters
containing about 5 μg of [³H]labeled mitochondrial DNA were incubated with 36,000
counts/min of [³²P]labeled 32S RNA and unlabeled 19S RNA, 25S RNA and 19S + 25S
rRNA in 2 ml of 5 x SSC at 65° C for 15 h. The filter membranes were washed and
treated with ribonuclease. 5 % of input 32S RNA (1800 counts/min) was hybridized
with mitochondrial DNA in the absence of added mitochondrial rRNA's, and this value was designated as 100 %. The data were corrected for nonspecific background,
which was between 140 and 160 counts/min (about 8 % of hybrid values).

120

4. From these plateau levels we can conclude that 32 % and 55 % of 32S RNA are homologous to the sequence of 19S and 25S RNA respectively. When 19S and 25S RNA are added together in excess, the loss of 32S hybridization ot mitochondrial-DNA is 78 %. This value is close to the sum of the results obtained with 19S and 25S singly. It indicates that the 32S precursor contains about 20% excess nucleotides over what is represented by mature ribosomal RNA's.

These results can be summarized in the model shown in Fig. 5. We have assigned molecular sizes, but the difficulty in obtaining absolute molecular weights for mitochondrial rRNA's is well known (5). The low G+C content of these molecules and their unusual secondary structure introduce uncertainty in interpreting electrophoretic and sedimentation data. So we tentatively made use of the molecular weights of mitochondrial ribosomal RNA's electrophoretically determined by Neupert *et al.* (6). The molecular weights of 32S, P25S, and P19S were determined from electrophoretic mobilities in 2.1 % and 2.4 % gels. The molecular size ratios for 32S to 25S and 32S to 19S are 0.53 and 0.3, respectively. These ratios are in good agreement with the RNA-DNA hybridization competition results which suggest that 25S and 19S RNA's comprise 55 % and 32 % respectively of the precursor molecules. The model also includes P19S and P25S as possible maturation intermediates but a role for these components as ribosomal RNA precursors is merely an assumption.

RIBOSOMAL RNA SYNTHESIS IN MITOCHONDRIA OF "POKY".

Rifkin and Luck (4), and Neupert *et al.* (7) found that poky mitochondria were deficient in the amount of ribosomal small subunits and suggested that the poky phenotype was related to this deficiency. Presumably it would seriously reduce the rate of protein synthesis in the mitochondria. We undertook experiments to examine the course of ribosomal RNA synthesis and processing in poky mitochondria. In the first experiment, cells were pulse labeled for 30 min with [3H]uracil.

From the results shown in Fig. 6 it is apparent

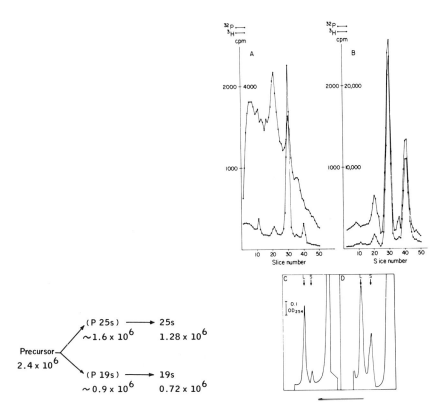

Fig. 5. Scheme of *Neurospora crassa* mitochondrial ribosomal RNA synthesis. The molecular size estimates were made as described in the text.

Fig. 6. Incorporation of 5[^{3}H]uracil and [^{32}P]P$_i$ into high molecular weight mitochondrial RNA's of poky and wild type cells. Cells were grown for 14 h and 24 h for wild type and poky strains respectively, in the presence of [^{32}P]phosphoric acid (0.2 mCi/150 ml) in media containing a low concentration of phosphate (KH$_2$PO$_4$ 1 mg/ml). The cells were labeled with 5[^{3}H]uracil (0.8 mCi/150 ml) for 25 min and 30 min for wild type and poky respectively and then cooled rapidly by pouring cultures into 4 vol of ice-cold 0.25 M sucrose (2 mM tricine-KOH, pH 7.6). Mitochondria were prepared by flotation, the RNA was extracted from one half of the mitochondrial preparation by the diethyl pyrocarbonate method. A 50 μl sample containing 0.5 OD$_{258}$ was analyzed on 2.1 % gels containing 0.8 % agarose, by electrophoresis for 3 h at 120 V at 6° C. The gels were sliced and processed for counting. The other half of the mitochondrial preparation was used for the analysis of ribosomal profiles. Preparation of total ribosomal pellet from mitochondrial lysate and analytical centrifugation of ribosomal subunits in the sucrose gradient were carried out. The gradient was analyzed by recording the absorbance at 254 nm using an Isco density gradient fractionator.

122

that the mass ratio of stable 19S to 25S RNA (labeled with $[^{32}P]$) in poky is much lower than in wild type. This deficiency of small ribosomal RNA is reflected by the patterns of ribosomal subunits shown in the lower half of Fig. 6. The distributions of RNA pulse-labeled with tritium uracil are also different. In poky there is a large accumulation of radioactivity to the left of 32S. It appears to be in the form of very large molecules heterogeneous in size. Since the electrophoretic mobility of these molecules is not altered by heating to 90° C in running buffer or by electrophoresis in nonaqueous formamide gels, it is not likely that the low mobility is the result of aggregation. Compared to wild type the relative content of 32S RNA appears to be higher in poky. Furthermore a comparison of tritium radioactivity of 19S and 25S RNA's in the two kinds of mitochondria reveals a significant deficiency of 19S in poky. There appears to be an unusual accumulation of radioactivity in a shoulder just to the right of 25S RNA. The position of this shoulder in poky corresponds to a small peak in wild type and as already mentioned, it was designated P19S, a hypothetical precursor intermediate for 19S.

These findings point to a possible alteration in the processing of 32S ribosomal precursor RNA of poky mitochondria. Because of the evidence that in the processing of eukaryote ribosomal RNA's extensive methylation is required for a normal cleavage of the precursor RNA (8,9,10), studies of mitochondrial RNA methylation were undertaken.

METHYLATION IN MITOCHONDRIAL RIBOSOMAL RNA SYNTHESIS.

The kinetics of $[^{3}H]$radioactivity incorporated into 32S, 25S and 19S RNA from $[^{3}H]-CH_3$ methionine in wild type mitochondria are plotted in Fig. 7. Cells were pulse-labeled with $[^{3}H]-CH_3$ methionine for varying times, mitochondria were prepared, the RNA extracted, and analyzed by gel electrophoresis. The content of tritium label in 32S, 25S and 19S were summed for each time point, normalized for total RNA content, and plotted against time. The 32S component is the

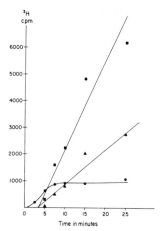

Fig. 7. Kinetics of incorporation of $[^3H]CH_3$ methionine into high molecular weight RNA's of wild type mitochondria. Cells were labeled with $[^3H]CH_3$ methionine (2 mCi/ 150 ml) for varying periods, RNA was extracted from the purified mitochondria, and analyzed by gel electrophoresis. $[^3H]$radioactivity in each RNA species was derived by summation of the radioactivity in the gel fractions containing that species (5 to 6 gel slices around each peak). The data from different gels were made compara- ble by normalizing with respect to the OD_{258} of total RNA loaded on the gels.

most rapidly labeled and reaches a plateau at 7.5 min. The ribosomal 19S and 25S RNA begin to accumulate [^{3}H] radioacti ity after a few minutes lag. Thus methylation appears to be an early event, occurring on 32S RNA.

For quantitative study of the content of RNA methyl groups, and in order to minimize the randomization of the label of a [^{3}H]-CH$_3$ methionine we undertook all subsequent studies using methionineless strains. A methionineless poky was made by a cross with meth-3. To make certain that through long labeling periods the label of [^{3}H]-CH$_3$ methionine was not incorporated into the carbon skeleton of adenine and guanine in the RNA, the [^{3}H] radioactivity of 5S RNA from cytoplasmic ribosomes was studied. This molecule is totally unmethylated (11). Our experiments indicated that under the conditions used, randomization of the label was negligibly small. To obtain a measure of mass RNA in these experiments all culture media contained [^{32}P] phosphoric acid along with [^{3}H]-CH$_3$ methionine.

The methyl content of the doubly labeled RNA's was analyzed by two methods. First, RNA extracted either from mitochondria or from cytoplasmic ribosomes in the postmitochondrial supernatant was resolved into large and small subunit RNA's on gel electrophoresis. Second, RNA's extracted from mitochondrial or cytoplasmic ribosomal subunits were hydrolyzed with alkali and the hydrolysate were applied to DEAE cellulose in 7M urea (12) to separate mononucleotide and alkaline-resistant di- or trinucleotides. Since the methylation at 2'-OH ribose in RNA makes the adjacent phosphodiester bond resistant to alkaline hydrolysis, the latter method allows us to estimate the methyl content at base and 2'-O ribose of the RNA's.

These results are summarized in Table I. The numbers marked with an asterisk are sums obtained from DEAE cellulose 7M urea chromotography. As noted, mitochondrial 25S, 19S and 4S RNA's though methylated, show a lower content of methyl groups than their cytoplasmic counterparts. The cytoplasmic rRNA are much richer in 2'-OH ribosome methylation than the mito-

TABLE I.

THE EXTENT OF METHYLATION IN VARIOUS RNA SPECIES OF WILD TYPE AND POKY CELLS.
The percent of methylated nucleotides was calculated from the [^{3}H] to [^{32}P] ratios
of gel analyses and from the specific radioactivity of [^{32}P] and [^{3}H] of the culture media used. To minimize the labeling of the purine rings from [^{3}H]CH$_3$ methionine, adenosine and guanosine were added at a concentration of 10^{-4} M to the growth
medium during the labeling period and this ring labeling was shown to be negligibly
small by checking the [^{3}H] radioactivity of cytoplasmic 5S RNA, because 5S RNA is
known to have no methylated bases (11). In the column marked total, data were obtained from total RNA extracts after fractionation by electrophoresis. The figures
between brackets represent totals obtained from analysis of base and 2'-0 methylation described below. To estimate 2'-0 ribose and base methylation was extracted from fractionated ribosomal subunits with phenolcresol-SDS. Mono-, di- and
trinucleotides of an alkaline hydrolysate were separated by DEAE-cellulose chromotography (12). The percent of mathylated bases of nucleotides was calculated
from the ratios of [^{3}H] to [^{32}P] in mononucleotides using as a standard the ratios of [^{3}H] to [^{32}P] of dinucleotides of cytoplasmic rRNA's. The percent of methylated ribose of nucleotides was calculated from [^{32}P] radioactivity in dinucleotides, trinucleotides compared with total [^{32}P] counts. In some cases where
[^{32}P] radioactivity in dinucleotides was very low (poky mitochondrial rRNA), the
percent was calculated from [^{3}H] radioactivity in the dinucleotide fraction of
poky mitochondrial RNA compared with the ratios of [^{3}H] to [^{32}P] in the dinucleotides of the poky cytoplasmic rRNA.

		Methyl Groups per 100 Nucleotides		
Mitochondrial RNA's		Total	Base Methylation	2'-0 Methylation
M-3	32S	1.15		
	25S	1.2 (*1.43)	1.2	0.23
	19S	1.10 (*1.43)	1.2	0.21
	4S	2.92		
Poky	25S	0.83 (*0.93)	0.8	0.13
	19S	0.65 (*0.70)	0.63	0.07
	4S	2.41		
Cytoplasmic RNA's				
M-3	28S	2.15 (*2.3)	1.2	1.1
	18S	2.37 (*2.6)	1.4	1.2
	4S	7.3		
Poky	28S	2.00 (*2.3)	1.1	1.2
	18S	2.20 (*2.3)	1.1	1.2
	4S	7.7		

chondrial counterparts, while base methylation is more
or less the same between cytoplasmic and mitochondrial
rRNA.

The comparison of 32S, 25S and 19S in control mitochondria show that these components are methylated to
more or less the same extent. This result, in combination with the previous kinetic data, suggest that 32S

RNA is a major site for methylation of rRNA. When one compares the poky results with those of control cells, rRNA's of poky mitochondria are clearly less methylated than their counterparts in control cells. On the other hand, no differences were found in the methylation of mitochondrial 4S, cytoplasmic 18S and 28S RNA. Thus the undermethylation is rather specific for rRNA's of poky mitochondria and is not restricted to one category: Base and 2'-O sugar methylation are equally diminished. The level of 32S methylation in poky could not be calculated, because of the high background of methylation contributed by heterogeneous high molecular weight RNA.

We knew from the work of Sebald and coworkers (13) that some protein synthesis occurs in poky mitochondria and from the work of Lambowitz and coworkers (14) that poky mitochondria contain about 30 % of the cytochrome *c* oxidase activity of wild type. Therefore we were interested to know whether all ribosomal RNA's were equally undermethylated, especially those of ribosomes actively engaged in protein synthesis. We attempted to isolate polysomes from poky mitochondria using detergent lysis and centrifugation through dense sucrose. Our efforts were unsuccesful, each preparation contained a great excess of large ribosomal subunits. We therefore turned to a new approach which will be described briefly before giving the results concerning methylation. This approach was based on the recent finding by Chua *et al.* (15) that in *Chlamydomonas* active chloroplast ribosomes are attached to thylakoid membranes.

We prepared submitochondrial vesicles by sonication in a low salt buffer and separated them from the supernatant (S25) by centrifugation at 25,000x*g*. The majority of ribosomes were sedimented with membranes, and most of them could be released from the membrane by incubation with puromycin in high salt buffer. Sedimentation analyses of the ribosomal subunits present in the total mitochondria, the supernatant, and the membrane fraction are shown in Fig. 8.

The upper panels of Fig. 8 show the results with

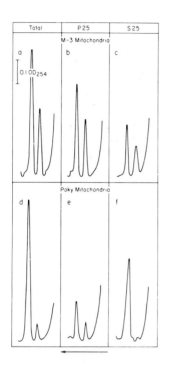

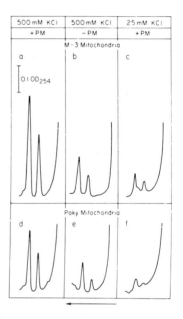

Fig. 8. Sucrose density gradient analysis of ribosomal subunits in S25, P25, and total mitochondria of wild type and poky cells. Purified mitochondria (20 culture flasks) from chloramphenicol treated cells were suspended in 10 ml of LKMTD. One half of the suspension was sonicated in a Branson sonifier and then centrifuged at 25,000 x g for 10 min to obtain S25 and P25. The concentration of KCl in S25 was adjusted to 0.5 M by adding solid KCl. After washing with LKMTD, P25 was suspended in the same volume of $HKMTD_{10}$. The other half of the mitochondrial suspension was lysed in 1 % nonidet P-40 and the lysate was layered over 1.8 M sucrose in LKMTD to obtain a total ribosomal pellet by centrifuging at 55,000 rpm in the A321 rotor of the IEC centrifuge, model B60 for 18 h. The total ribosome pellet was suspended in the equal volume of $HKMTD_{10}$. Solution of S25, P25 and ribosome pellet were treated with 1 mM puromycin at 35° C for 10 min and then the reaction mixture of P25 was again centrifuged at 25,000 x g for 10 min to remove membrane fragments. The same aliquots of these reaction mixtures (0.3 ml - 0.5 ml) were layered onto 5-20 % sucrose gradients in $HKMTD_{25}$ and centrifuged at 39,000 rpm for 3 h in a rotor SB283 of the IEC centrifuge. The gradients were analyzed by monitoring OD_{254} with an ISCO gradient fractionator. The same protocol was followed with poky mitochondria except that volumes of the total ribosome suspension and S25 were 1.5 fold that of P25.

Fig. 9. Effect of KCl concentrations and puromycin on the release of ribosomes from P25 of wild type and poky cells. P25 was prepared from chloramphenicol treated wild type or poky cells as described in Fig. 1. The low salt (25 mM KCl) reactions were carried out similarly in LKMTD. The reaction mixtures (about 15 and 25 mg protein/ml of P25 from wild type and poky mitochondria) were centrifuged at 25,000 x g for 10 min and aliquots of the supernatant from each reaction mixture were layered onto 5-20 % linear sucrose gradients in $HKMTD_{25}$. After centrifugation at 39,000 rpm for 3 h at 18° C in a SB283 rotor of the IEC centrifuge, the gradients were analyzed as described in Fig. 1.

128

wild type mitochondria, and the lower panels, the
results with poky mitochondria. A considerable number
of ribosomal subunits can be released from membrane
fractions of wild type mitochondria by high salt and
puromycin treatment. There is a striking deficiency
of small ribosomal subunits in the total mitochondrial
population of poky, as already mentioned. The mass ratio of large and small subunits here is about 10:1. In
striking contrast, from the membrane fraction of poky
mitochondria, large and small subunits are released in
the mass ratio of 2:1. This is the ratio expected for
ribosomal monomers. Excessive amounts of large subunits are also observed in the supernatant. We should
mention that the results obtained here are from cultures treated with chloramphenicol 1 h before harvest.
The purpose of this treatment is to lower the frequency of peptide chain termination in polysomes associated with the mitochondrial membrane. To quantitate the
chloramphenicol effect we determined the OD_{254} of ribosomal subunits in S25, those released from P25 by
puromycin high salt, and those remaining in the P25
pellet which required detergent lysis for release.

Table II shows the relative amount of ribosomes in
various subfractions of mitochondria; + CAP indicates
that the cells were treated with chloramphenicol 1 h

TABLE II.

DISTRIBUTION OF RIBOSOMES IN VARIOUS FRACTIONS DERIVED FROM MITOCHONDRIA OF
CONTROL AND POKY STRAIN TREATED WITH OR WITHOUT CHLORAMPHENICOL.
Experiments were carried out as described in Fig. 1. + CAP: cells were treated with chloramphenicol for 1 h before harvest. P25 was treated with high
salt and puromycin and then centrifuged at 25,000 rpm for 10 min. Released
subunits were analyzed in the supernatant and the membrane pellet was lysed
in 1 % nonided P-40. The lysate was layered over 1.8 M sucrose in LKMTD and
centrifuged at 55,000 rpm for 18 h to obtain a ribosome pellet. The ribosome
pellet was processed as described in Fig. 1 for analysis of subunits. Distribution (%) was obtained from analyses of ribosomal subunits on the 2-20 %
sucrose gradient at 254 nm.

	Wild Type		Poky			
	+ CAP	− CAP	+ CAP		− CAP	
			L	S	L	S
	(%)	(%)	(%)	(%)	(%)	(%)
Supernatant (S-25)	23	38	50	10	54	22
High Salt + PM (P-25)	48	37	11	54	8	42
Pellet	28	25	39	35	38	36

before harvest. Although the effect is quantitatively small, chloramphenicol has the effect of shifting free ribosomes to the membrane bound fraction. This has already been reported with chloroplast ribosomes of *Chlamydomonas* (15).

In eukaryotic cells (16,17) and chloroplasts of *Chlamydomonas* (15), ribosomes are attached to the membrane mainly by two factors; salt-sensitive bonds between the large ribosomal subunit and the membrane and an indirect linkage via the nascent peptide chain. The following experiments were carried out to explore whether disruption of the association of ribosomes with the membrane in mitochondria required simultaneous attacks on both kinds of bonds. P25 fractions of wild type and poky mitochondria were treated with a low (25 mM) or a high (500 mM) KCl concentration in the presence or absence of puromycin. The results are shown in Fig. 9. The upper panels show the results of wild type mitochondria and the lower panels the results with poky mitochondria. Low salt treatment with or without puromycin releases only a small amount of ribosomes from the membrane. Maximum release of the bound ribosomes was obtained only by the combination of high salt and puromycin. These results are entirely comparable to those obtained under similar conditions with rat hepatocyte and chloroplast of *Chlamydomonas* (15,16,17).

Finally, we have studied thin sections through P25 and P25 after puromycin high salt treatment by electron microscopy. These studies reveal that P25 is represented by a collection of vesicles heterogeneous in size. When vesicles are cut through their centers, the content for the most part is clear; however tangential grazing sections of vesicle surfaces are studded with ribosomes. The ribosomes frequently are grouped in small cluster or linear arrays. After puromycin treatment, the preparations are – as expected – almost devoid of ribosomes. It is interesting that when ribosomes are found to be present, they are usually trapped inside of vesicles or in poorly disrupted mitochondria.

Finally to return to poky. We made use of this separation procedure for free and bound ribosomes, and
using the $[^3H]-CH_3$ methionine and $[^{32}P]$ double labeling method, we measured the methylation of the functional ribosomal subunits released from the membrane
of poky mitochondria and the subunits in the supernatant. We found that 25S RNA and 19S RNA in subunits
released from P25, contained 0.76 and 0.63 methyl
groups per 100 nucleotides. For the large subunit RNA
in S25, the value was 0.76. In short, there was no
difference in methylation between the small fraction
of active ribosomes associated with membrane, free
subunit in the supernatant, and those of the total mitochondria. These data show that mitochondrial ribosomal subunits containing undermethylated RNA's may
function in protein synthesis.

In summary, the methylation and cleavage of 32S ribosomal precursor in poky mitochondria is altered. Apparently there is a small accumulation of 32S RNA and
potential 19S molecules may accumulate in the form of
a precursor (P19S). Most of these 19S precursors are
degraded before they can be assembled into small subunits. Both 25S and 19S RNA's are undermethylated,
but undermethylated RNA's can be assembled into ribosomes which associate with the membrane and are apparently active in protein synthesis. The true nature of
the defect in poky and its relationship to cytoplasmic inheritance remain to be solved. Since ribosomal
RNA methylation and ribosomal assembly are closely in-
terrelated it is not yet possible to emphasize one or
the other as the primary mechanism for the deficiency
of small ribosomal subunits in poky. It is intriguing
that poky, alone among cytoplasmic mutants of *Neuros-*
pora, is defective in the post-transcriptional modification of cytochrome *c* (18), known to be trimethyla-
tion of ε amino group of lysine residue 72. It is also possible that the unusual elution profiles of a
few poky mitochondrial tRNA's from BD cellulose is a
result of undermethylation (19). Thus one focus for
future experiments should be the regulation of S-ade-
nosyl methionine content in poky mitochondria.

ACKNOWLEDGEMENT.

This research was supported in part by a National Science Foundation grant GB 27509X2.

REFERENCES.

1. Y. Kuriyama and D.J.L. Luck, J.Mol.Biol. 73, 425 (1973).
2. J.E. Darnell Jr., Bacteriological Reviews 32, 262 (1968).
3. B.E.H. Maden, Prog.Biophys. and Mol.Biol. 22, 127 (1971).
4. M.R. Rifkin and D.J.L. Luck, Proc.Nat.Acad.Sci. USA 68, 287 (1971).
5. P. Borst and L.A. Grivell, FEBS Letters 13, 73 (1971).
6. W. Neupert, W. Sebald, A.J. Schwor, A. Pfaller and Th. Bücher, Eur.J.Biochem. 10, 585 (1969).
7. W. Neupert, P. Massinger and P. Pfaller, in:"Autonomy and Biogenesis of Mitochondria and Chloroplasts", N.K. Boardman, A.W. Linnane and R.M. Smillie, Eds.,North Holland Publ.Co./American Elsevier Publ. Co., 339 (1971).
8. M.H. Vaughan Jr., R. Soeiro, J.R. Warner and J.E. Darnell Jr., Proc.Nat.Acad.Sci. USA 58, 1527 (1967).
9. M.C. Liau, J.B. Hunt, D.W. Smith and R.B. Hurlbert, Cancer Res. 33, 323 (1973).
10. M.C. Liau, D.W. Smith and R.B. Hurlbert, J.Cell. Biol. 55 , 154a (1972).
11. G.G. Brownlee and F. Sangor, J.Mol.Biol. 23, 337 (1967).
12. R.V. Tomlinson and G.M. Tener, Biochem. 2, 697 (1963).
13. W. Sebald, Th. Bücher, B. Olbrich and F. Kaudewitz, FEBS Letters 1, 235 (1968).
14. A.M.Lambowitz, E.W. Smith and C.W. Slayman, J. Biol.Chem. 247, 4850 (1972).
15. N. Chua, G. Blobel, D. Siekevitz and G.E. Palade, Proc.Nat.Acad.Sci. USA 70, 1554 (1973).
16. G. Blobel and D.D. Sabatini, Proc.Nat.Acad.Sci.

USA 68, 390 (1971).

17. M.R. Adelman, D.D. Sabatini and G. Blobel, J.Cell Biol. 56, 206 (1973).

18. W.A. Scott and H.K. Mitchell, Biochemistry 8, 4282 (1969).

19. R.M. Bramble and D.O. Woodward, Nature New Biol. 238, 188 (1972).

BIOGENESIS OF MITOCHONDRIA FROM *ARTEMIA SALINA* CYSTS AND THE TRANSCRIPTION *IN VITRO* OF THE DNA

H. Schmitt, H. Grossfeld, J.S. Beckmann and U.Z. Littauer

Department of Biochemistry, The Weizmann Institute of Science, Rehovot, Israel.

MITOCHONDRIAL BIOGENESIS.

Recently we demonstrated that the developing cyst of *Artemia salina* is an excellent model system for studying the biogenesis of animal mitochondria (1). The relative ease in obtaining purified mitochondria and the availability of large quantities of the cysts from commercial sources are added advantages in working with this system.

The crustacean, *Artemia salina* (brine shrimp), can reproduce in two ways: either the fertilized egg in the female ovisac directly evolves into a nauplius larva or during the dry season development is halted at the stage of the gastrula, the egg encysts to be released in this form. These cysts may be desiccated by natural drying, or osmotically by high environmental salinity after which they enter a state of dormancy called cryptobiosis. When the gastrula is rehydrated it will undergo further differentiation to the prenauplius stage. This latter process takes place without cell division in about 15 h. Further incubation of the encysted prenauplii induces hatching to give rise to a free-swimming nauplius larva. At this stage cell division is resumed (2,3).

We have found that hydration of the cysts induces marked biochemical and morphological changes in the mitochondria (1). The oxygen consumption of the cysts was followed after different times of incubation at 30° C. The undeveloped cysts have a Q_{O_2} about one third of that found for cysts incubated for 1 h at 30°

C. After hatching at the nauplius stage, a further two-fold increase in Q_{O2} occurs yielding a final oxygen consumption equivalent to 70 µmoles O_2 /h/gram dry cysts.

We have also examined the fate of cytochrome *c* oxidase, a typical mitochondrial enzyme localized in the inner membrane of the organelle. It was shown that the increase in cytochrome *c* oxidase activity roughly parallels the increase in the respiratory rate. Using differential spectrophotometry the same conclusion is reached. Differential spectra of cytochromes were taken from a suspension of mitochondria isolated from cysts induced for various periods of time. The peak of cytochrome *c* oxidase at 605 nm attains a maximum value after incubation of the cysts for 1 h. On the other hand, the content of cytochrome *b* (α band at 564 nm) and cytochrome *c* (α band at 553 nm) continues to increase until a plateau is attained around the seventh hour of incubation. These results exclude the possibility that the rise in the respiratory rate results primarily from an increase in the level of the respiratory substrates (at least till the nauplius stage). It was also noted that the ratio of cytochrome *c* to cytochrome *b* increases during cyst incubation (1).

The morphology of mitochondria was investigated by means of electron microscopy. It was found that during the development of the cysts a striking change in the mitochondrial morphology takes place. The mitochondria from unincubated cysts lack cristae and are poorly stained. After 1 h of incubation the mitochondria seem to show a somewhat more stained matrix and perhaps some cristae start to develop. More pronounced changes are observed in the mitochondria from cysts which were incubated for 7 h at 30°. These mitochondria have definite cristae and a stained matrix while mitochondria isolated from nauplii possess even more cristae (1). It is interesting to note that Soslau and Nass (4) found that treatment of L cells with ethidium bromide leads to progressive mitochondrial changes resulting, finally, in mitochondria which are morphologically very similar to those observed in the unincubated *Arte-*

mia cysts.

What is the role of protein synthesis in the biogenesis of mitochondria ? The evolution of protein
synthesizing activity in isolated whole mitochondria
was examined. These experiments revealed a 30 min lag
before increased protein synthetic activity is observed. As the main part of the respiratory capability of
the cyst is gained during the first hour of incubation
at 30°, it may be concluded that the initial rise in
respiration is most probably due to assembly of precursor molecules present in the mitochondria. This is
in contrast to the biogenesis of yeast mitochondria
which requires protein synthesis *de novo*.

The results presented above show the developing
cyst of *Artemia salina* to be an excellent system for
studying the biogenesis of animal mitochondria. It
seems likely that in these cysts mitochondrial biogenesis proceeds in two stages. The first stage is characterized by a rapid increase in the respiratory capability of the mitochondria and perhaps some morphological changes; while in the second stage there is
an increase in the protein synthesizing capacity of
the mitochondria as well as striking changes in mitochondrial morphology leading to the formation of cristae (1).

MATURATION OF RIBOSOMAL RNA.

Artemia mitochondrial rRNA is by far smaller in size than mitochondrial rRNA from ascomycetes. On electrophoresis in polyacrylamide gels it is resolved into two components with an apparent molecular weight
of 0.35×10^6 (L) and 0.50×10^6 (H). A different pattern, however, was obtained for mitochondrial rRNA
from developing cysts. Hydration of the cysts induces
the appearance of a lighter mitochondrial rRNA compo-
nent (LL) with an apparent molecular weight of 0.28×10^6. As the differentiation of the cysts progresses
there is a simultaneous increase in the amount of the
LL peak, untill almost none of the original L peak is
evident. We still do not know whether the mitochondrial L peak is converted during the differentiation
process to the LL peak or the appearance of the LL

peak only represents newly synthesized RNA. The nature of the difference in electrophoretic mobility is also obscure. It may be that the L peak is indeed larger than the LL peak. An alternative hypothesis would be the post-transcriptional modification of the L peak. These questions are difficult to solve because of the impermeability of the cysts to nucleic acid precursors. It should be noted that the three rRNA components (H, L and LL) were shown to be of ribosomal origin since they were found to be associated with the ribosomal fraction isolated from purified mitochondria. It would be of interest to examine the protein synthesizing capacity of mitochondrial ribosomes isolated from unincubated cysts (containing H and L rRNA) and compare their properties to mitochondrial ribosomes isolated from the developing cysts (containing H and LL rRNA).

PHYSICAL PROPERTIES OF MITOCHONDRIAL rRNA.

Observations relating to conformation (5-10), size (11,12) nucleotide composition (*cf*. 13) and nucleotide distribution (14), have revealed that mitochondrial rRNA from ascomycetes differs markedly not only from its homologous cytoplasmic rRNA but from bacterial-type rRNA as well. The unique conformational properties of mitochondrial rRNA were first suspected from their anomalous electrophoretic mobilities on polyacrylamide gels. Reports from this and other laboratories have shown that the electrophoretic mobility of mitochondrial rRNA is lower than would be expected from sedimentation behaviour in sucrose gradients. Furthermore, the electrophoretic mobility relative to cytoplasmic or *E. coli* rRNA is much more sensitive to changes in ionic strength, temperature and gel concentration. These anomalous electrophoretic features were explained by the observation that mitochondrial rRNAs have a less compact and less stably ordered structure as compared to cytoplasmic and bacterial rRNAs (5, 7, 8, 15, 9, 10). For example, mitochondrial rRNA from fungi showed a greater percentage change in relative absorbance at 260 nm than 280 nm, while cytoplasmic and *E.coli* rRNA samples exhibited quite the opposite pattern. Thermal denaturation mid-points also occurred at considerably lower temperatures for mitochondrial rRNA (7,10). Additionally, circular dichroism and optical absorbance properties with formaldehyde-treated RNA indicated that, despite its lower stability, the ordered structure of mitochondrial rRNA, like all other rRNAs, results from base pairing in helical regions as well as from base stacking (16, 17).

Another distinctive property of mitochondrial rRNAs from ascomycetes is their extremely low G+C content 26-38 %). Furthermore, the G+C content in the ordered regions of the rRNA as calculated from the hyperchromicity ratios at 260 and 280 nm was also found to be low and reflects the bulk G+C content of the rRNA. Thus, the less stably ordered structure and the peculiar electrophoretic mobility of mitochondrial rRNA may be linked to the low G+C content in the helical regions of the mitochondrial rRNA (7). However, other factors must be involved as well. Thus, the G+C content for the heavy and light component of *Neurospora crassa* rRNA is 34 and 36.5 % respectively, yet its electrophoretic mobility is much more sensitive to changes in ionic strength than that of *Aspergillus* mitochondrial rRNA which has a lower G+C content of 30.5 adn 31.5 % respectively (10).

The G+C content of *Artemia* mitochondrial rRNA is 43 %, considerably higher than the G+C content of fungal or yeast mitochondrial rRNAs. We wondered whether the electrophoretic mobility of this rRNA would also be sensitive to changes in the ionic strength. Table I

TABLE I

ELECTROPHORETIC BEHAVIOR OF *ARTEMIA* rRNA.
Electrophoresis was carried out on 2.0 % polyacrylamide gels. The composition of the buffers is given in ref. 10.

Buffer	Temperature	Cytoplasmic		Mitochondrial		
		L	H	LL	L	H
		Apparent Molecular Weight x 10^6				
Low	20°	0.66	1.30	0.34	0.41	0.71
E	20°	0.66	1.40	0.29	0.39	0.62
Mg^{++}	20°	0.62	1.30	0.28	0.36	0.56
E	4°	0.62	1.40	0.29	0.35	0.50

illustrates the apparent molecular weights of mitochondrial and cytoplasmic rRNA measured against an *E. coli* rRNA standard. It is evident that in low salt buffer, the relative mobilities of the mitochondrial rRNA, especially for the H component, are severely retarded, while the mobilities of the cytoplasmic rRNA are much less affected. The mobility of mitochondrial rRNA is also more sensitive to changes in temperatures than cytoplasmic rRNA. A similar conclusion was reached by Dawid and Chase, for mitochondrial rRNA from *Xenopus laevis* (18)

The melting behaviour of rRNA isolated from the mitochondria and the cytoplasm of *Artemia salina* cysts was recorded. It was found that mitochondrial rRNA has a much lower Tm value both at 260 and 280 nm than its corresponding cytoplasmic rRNA. The relative low stability of the mitochondrial structure was even better relvealed when the thermal denaturation was carried out at pH 4.0. At this pH the cytidine residues in the polynucleotide chains are already ionized and hence the ordered structure is more labile. When the melting profile is followed at 260 nm, mitochondrial rRNA shows only a progressive noncooperative increase in absorbance which is probably due to unstacking of the bases. On the other hand, the melting of cytoplasmic rRNA still retains some degree of cooperativeness. When the measurements were made at 280 nm, both rRNA samples showed cooperative melting but the hyperchromic change was by far more pronounced with cytoplasmic rRNA. We concluded that mitochondrial rRNAs from both animals and ascomycetes mitochondria are distinguished by unusual electrophoretic properties and a labile secondary structure. This lability is not solely due to a low G+C content and probably indicates a relatively low number of base pairs in the rRNA helices.

TRANSCRIPTION *IN VITRO* OF MITOCHONDRIAL DNA WITH *E. coli* RNA POLYMERASE.

Artemia mitochondrial DNA has a density of 1.697 as determined by CsCl density gradient centrifugation. The density of mitochondrial DNA is slightly higher

than that of the nuclear DNA which is 1.694. These values correspond to a G+C content of 37.8 % and 34.7 %
for mitochondrial and nuclear DNA respectively.

The *Artemia* mitochondrial DNA, like mitochondrial
DNA from other animals, has a twisted circular structure of 5.1 microns corresponding to about 10^7 daltons. Using the circular mitochondrial DNA from *Artemia*, we attempted to transcribe it with purified *E.
coli* RNA polymerase. Particular attention was paid to
the mechanism of the transcription reaction, the size
of the RNA product and the number of promoter sites
recognized by the heterologous enzyme.

In order to transcribe the closed circular DNA
faithfully, it is essential that the preparation of *E.
coli* RNA polymerase used should be free of any contaminating nuclease. The DNA was incubated with the
enzyme, treated with SDS and separated by electrophoresis in the presence of ethidium bromide according
to the procedure of Aaij and Borst (19). This procedure separates closed-circular DNA from open-circular
DNA. By this technique it was shown that the enzyme
preparation did not introduce any nicks into the closed-circular DNA preparation.

E. coli RNA polymerase is composed of several subu-
nits: α_2, β, β' and σ. After chromatography on phos-
phocellulose columns, the σ subunit can be separated
from the core enzyme (α_2 β β'). It has been shown in
several laboratories that the σ factor is required for
initiation of RNA synthesis at the correct site on the
DNA template. Without the σ factor, RNA polymerase
will initiate polymerization along the DNA molecule in
a ramdom fashion. Less is known on the role of σ factor in the transcription of closed-circular DNA. Table
II shows that addition of σ factor, to a reaction mixture containing *Artemia* mitochondrial DNA and core enzyme, greatly stimulated the transcription reaction.

The antibiotic rifampicin is believed to block the
initiation step of DNA transcription by *E. coli* RNA polymerase. If the enzyme is incubated with DNA prior
to the addition of rifampicin and the nucleoside triphosphates, on subsequent incubation it will catalyze

TABLE II

EFFECT OF σ FACTOR ON TRANSCRIPTION OF MITOCHONDRIAL DNA
Reaction mixtures of 0.06 ml consisted of 50 mM Tris-HCl, pH 7.5; 50 mM KCl; 10 mM MgCl$_2$; 1 mM DTT; 10 μg/ml mitochondrial DNA; 70 μg/ml of core enzyme isolated from E.coli MRE-600 RNA polymerase and when required 75 μg/ml of σ factor. Reaction mixtures were incubated for 4 min at 37°, then supplemented with 0.5 mM ATP, GTP, CTP and 11 μM [^{3}H]UTP (1 Ci/mM) and further incubated for 10 min at 37° with or without 4 μg/ml rifampicin. Aliquots of 10 μl were applied on Whatman 3 MM filters discs, immersed in 10 % trichloroacetic acid containing 20 mM pyrophosphate, washed, dried and counted.

Enzyme fraction	Minus Rifampicin	Plus Rifampicin
	cpm	
Core enzyme	4,400	340
Core enzyme + σ factor	13,240	2,170

mainly one round of transcription before the reaction is inhibited (20). We have observed the same phenomenon with mitochondrial DNA from Artemia. Moreover the limited transcription in the presence of rifampicin is dependent on the concentration of σ factor in the reaction mixture, thus suggesting that in this case as well σ is involved in the initiation reaction and does recognize specific promoter sites (Table II).

The size of the transcription product was analyzed by electrophoresis on polyacrylamide gels. In the absence of σ factor, the core enzyme alone yields a transcript with a molecular weight of 3×10^6. In the presence of σ factor the mass of the transcript is 1.5×10^6 daltons and there is no radioactivity in the 4S and 5S regions, nor is there any material in the zone of the mitochondrial rRNA. Addition of the termination factor ρ or KCl to a final concentration of 0.2 M had no effect on the size of the product. To exclude the possible participation of weak promoter sites in the transcription reaction rifampicin was included in the reaction mixture. In its presence the product has the same molecular weight as in its absence. The large size of our product may indicate that in vivo, a precursor rRNA is synthesized which is further cleaved by a specific endonuclease to yield the mature r-RNA.

The resistance ot RNAse digestion of the product synthesized in vitro was examined before and after heating at 70°. Table III shows that upon self-annealing, only 10 % of the product is resistant to nuclea-

TABLE III

COMPLEMENTARITY OF THE RNA TRANSCRIPTS.
The RNA transcribed *in vitro* was treated with DNAse (10 µg/ml) for 15 min at 37°.
Aliquots of 50 µl (41,000 cpm) were incubated for 2 h at 0° (control) or for 2 h
at 70°, with or without 2.5 µg of mitochondrial rRNA or *E.coli* rRNA in 2 x SSC.
To each tube RNAse T$_1$ and pancreatic RNAse were added to a final concentration
of 10 units/ml and 13 µg/ml respectively and incubation was continued for one h
at 37°. From each tube 40 µl were applied to Whatmann 3 MM filter paper discs
and treated as in Table II. The numbers given in the table represent average values of duplicate experiments.

Experimental conditions	No addition	plus mit.rRNA	plus *E.coli* RNA
		% RNAse resistance	
Control	12	16	12
Heating at 70°	22	70	20
Self annealing (net percentage)	10	54	8

se digestion. Therefore, the product is not significantly self-complementary. On the other hand when the
synthesized [^{3}H]RNA was heated in the presence of mitochondrial rRNA, about 50 % of the labelled product
is rendered RNAse resistant. This would mean that
about 50 % of the sequences made *in vitro* are specifically complementary to mitochondrial rRNA and are
transcribed on the anti-rDNA strand.

Because, the RNA made *in vitro* contained high
amounts of anti-rRNA sequences, stepwise hybridization-competition experiments were performed with denatured mitochondrial DNA, immobilized on nitrocellulose filters. In the first step increasing amounts of
unlabelled mitochondrial rRNA were hybridized to the
DNA filters and excess unhybridized RNA was removed
by washing. In the second step, labelled RNA made *in
vitro* was hybridized to the different DNA filters. The
results (Table IV) indicate that unlabelled mitochondrial RNA competes to 65 % with the RNA synthesized
in vitro in the hybridization reaction. It is believed that this stepwise annealing method monitors only
true RNA-DNA hybrids, and is devoid of interferences
resulting from formation of RNA-RNA duplexes. However,
when unlabelled mitochondrial rRNA and the product
made *in vitro* were hybridized simulteneously, the same percentage of composition was observed (Table IV).

143

TABLE IV

COMPETITION-HYBRIDIZATION OF THE TRANSCRIPT MADE *IN VITRO* WITH MITOCHONDRIAL rRNA.
Hybridization was performed in 0.3 ml of 50 % formamide solution (21) with 0.05 µg
of labeled *in vitro* RNA and 0.5 µg of mitochondrial DNA, immobilized on nitrocellulose filters. Hybridization reactions were run in duplicates for 48 h at 37° with
an efficiency of 9 % (% input RNA found in the hybrid). The value of 100 % hybridization is equivalent to 10,500 cpm. Blank values were subtracted from each sample
and averaged to 15 % of the hybridized material.

Quantity of competitor mit. rRNA added	Step wise hybridization	Simultaneous hybridization
µg	% competition	
0.25	45	33
2.5	65	65

From the results of the self-annealing experiments
(Table III), we would have expected the simultaneous
method to give a higher percentage of competition than
that observed with the stepwise procedure. We presume
that the prehybridized RNA can be detached from the
DNA, by complementary RNA sequences and washed from
the filters thus resulting in an apparent increase in
the extent of competition. We conclude, therefore, that
RNA-RNA annealing may occur during the stepwise hybridization. Thus results from such experiments should be
regarded with caution.

From the above experiments, it is difficult to assess the percentage of RNA sequences transcribed on
each of the two DNA stands (in these experiments we
have used denatured mitochondrial DNA, since alkaline
CsCl gradients failed to separate the two DNA strands).
However, we may tentatively conclude that the differences between the values obtained by the hybridization-competition and the self annealing experiments
(65% - 50% = 15%) are due to transcription of RNA sequences homologous to the mitochondrial rRNA, and that
most of the transcription on ribosomal genes occurs on
the anti-rDNA strand. Therefore, it seems that transcription *in vitro* of mitochondrial DNA with *E. coli* RNA
polymerase partially occurs on both DNA strands in opposite directions. However, it seems that the synthesis does not proceed over the entire length of the r-
DNA so that the transcripts do not overlap significantly.

ACKNOWLEDGEMENTS.

We thank Dr.L.G. Silvestri of Gruppo Lepetit S.p.A.

for rifampicin and Dr. V. Daniel and Mr. Y. Tichauer for *E.coli* RNA polymerase. This work was supported in part by U.S. Public Health Service Agreement No. 45514 and the Robert Bosch Stiftung GMBH. H.S. has been supported by an EMBO fellowship.

REFERENCES.

1. H. Schmitt, H. Grossfeld and U.Z. Littauer, J.Cell Biol. in press (1973).
2. Y.H. Nakanishi, T. Iwasaki, T. Ogigaki and H. Kato, Annot.Zool.Jap. 35, 223 (1962).
3. F.J. Finamore and J.C. Clegg, in:"The Cell Cycle", G.M. Padilla Ed., Academic Press Inc. New York, p. 249 (1969).
4. G. Soslau and M.M.K. Nass, J.Cell Biol. 51, 514 (1971).
5. I.M. Verma, M. Edelman and U.Z. Littauer, Israel J.Chem. 7, 118p (1969).
6. M. Edelman, I.M. Verma and U.Z. Littauer, Israel J.Chem. 7, 119p (1969).
7. M. Edelman, I.M. Verma and U.Z. Littauer, J.Mol. Biol. 49, 67 (1970).
8. R. Herzog, M. Edelman, E. Galun, I.M. Verma and U.Z. Littauer, Israel J.Chem. 8, 123p (1970).
9. L.A. Grivell, L. Reijnders and P. Borst, Eur.J. Biochem. 19, 64 (1971).
10. M. Edelman, I.M. Verma, R. Herzog, E. Galun and U.Z. Littauer, Eur.J.Biochem. 19, 372 (1971).
11. I.M. Verma, M. Edelman, M. Herzberg and U.Z. Littauer, J.Mol.Biol. 52, 137 (1970).
12. L. Reijnders, P. Sloof and P. Borst, Eur.J.Biochem. 35, 266 (1973).
13. P. Borst and L.A. Grivell, FEBS Letters 13, 73 (1971).
14. I.M. Verma, M. Edelman and U.Z. Littauer, Eur.J. Biochem. 19, 124 (1971).
15. I.T. Forrester, P. Nagley and A.W. Linnane, FEBS Letters 11, 59 (1970).
16. I.M. Verma, C.M. Kay and U.Z. Littauer, FEBS Letters 12, 317 (1971).
17. M. Edelman, I.M. Verma, D. Saya and U.Z. Littauer,

Biochem.Biophys.Res.Commun. 42, 208 (1971).

18. I.B. Dawid and J.W. Chase, J.Mol.Biol. 63, 217 (1972).

19. C. Aaij and P. Borst, Biochim.Biophys.Acta 269, 192 (1972).

20. A.E. Sippel and G.R. Hartmann, Eur.J.Biochem. 16, 152 (1970).

21. R.A. Weinberg, S.O. Warnaar and E. Winocour, J. Virol. 10, 193 (1972).

STRUCTURE OF PETITE mtDNA; PROSPECTS FOR MITOCHONDRIAL GENE ISOLATION, MAPPING AND SEQUENCING

P. Borst

Section for Medical Enzymology, Laboratory of Biochemistry, University of Amsterdam, Eerste Constantijn Huygensstraat 20, Amsterdam, The Netherlands.

One of the most attractive features of the mitochondrial genetic system is its simplicity. In the long run we should be able to get a full inventory of all genes on mtDNA and by analysis of intercistronic regions and regulatory substances this should lead to a complete description of the regulation of mitochondrial biogenesis. Since a large part of the mitochondrial biogenesis is controlled by the nucleus, an analysis of the regulation of the mitochondrial genetic system may eventually also yield information about nuclear control mechanisms in eukaryotic cells, control mechanisms that are intimately linked with some of the more interesting problems in biology like differentiation, dedifferentiation, ageing, etc.

In the final analysis detailed insight into the regulation of mitochondrial biogenesis will require knowledge of the base sequences of large parts of mtDNA. In this paper I want to briefly consider the possibility of using cytoplasmic petite mutants of yeast as an aid in obtaining specific segments of the mitochondrial genome in high yield, suitable for sequence analysis. Rather than present theoretical schemes to obtain interesting mutants, I shall discuss some of the facts now available about one mutant that we have studied in detail.

EARLY EXPERIMENTS ON mtDNA IN PETITE MUTANTS; MUTANT RD1A

In 1966 it was independently shown by Mounolou *et al.*(1) and Carnevali *et al.*(2) that in some cytoplasmic petite mutants normal mtDNA is replaced by an equivalent (or near-equivalent) amount of an abnormal DNA, which may differ considerably from wild-type DNA in buoyant density. Further analysis of some of these mutants by Bernardi *et al.*(3) and Mehrotra and Mahler (4) showed that the change in buoyant density was due to a change in base composition, some of these mutant DNAs containing less than 5 mole percent GC.

We were interested to find out how such mutants arise and whether they represent nonsense DNA or the amplification of a very AT-rich segment of the wild-type genome. After drastic ethidium bromide treatment and recloning (see ref. 5), Hollenberg in my laboratory obtained mutant RD1A in which the wild-type mtDNA was replaced by an equivalent amount of a low-density DNA. Early experiments (5-8) with this low-density DNA established three points:

1. After denaturation the DNA rapidly renatured on cooling. The melting temperature of the renatured DNA was within 1° C of that of the native DNA and this showed that the renatured structure was a perfect duplex, essentially without mismatch.

2. Quantitative renaturation experiments proved that renaturation followed perfect second-order kinetics. This excluded the possibility that our mutant mtDNA consisted of repeated runs of alternating poly $[d(A-T) \cdot d(A-T)]$ as found in crab satellite DNA. Furthermore, these experiments allowed a calculation of the complexity of RD1A mtDNA and this turned out to be less than 0.3% of the wild-type mtDNA.

3. Renaturation of mixtures of wild-type and mutant mtDNA clearly showed that the mutant mtDNA, despite its unusual character, was homologous to wild-type mtDNA and that it must, therefore, represent an amplification of part of the wild-type genome.

MORE RECENT EXPERIMENTS ON RD1A mtDNA

Our conclusion that RD1A mtDNA is a highly repeated DNA in which all repeating units are identical in sequence is confirmed by more recent experiments, summarized with the older data in Table I. RD1A mtDNA

TABLE I

PROPERTIES OF mtDNA FROM THE YEAST PETITE MUTANT RD1A

1. Base composition

Duplex DNA :	3.0% GC	(wild-type 17%)
Heavy strand:	3.0% G,	<0.1% O
Light strand:	<0.1% G,	3.0% O

2. Renaturation experiments

Renaturation follows second-order kinetics; complexity <0.3% of wild-type mtDNA.
T_m renaturated DNA - T_m native DNA.

3. Pyrimidine tract analysis

>95% of ^{32}P in 5 pyramidine tracts: T_1, T_2, T_3, T_6 and TTTCCTT.

4. Hybridization experiments

RD1A mtDNA hybridizes to a plateau of 0.3% with wild-type mtDNA.

contains only 3 mole percent GC and after separation of the complementary strands in alkaline CsCl, all the G is found in the H-strand and all the C in the L-strand (9). The pyrimidine-tract analysis (10) gives an extremely simple picture confirming the conclusion that sequence repetition is perfect. The sequence of the T_5C_2 tract has recently been determined (9) by the analysis of RNA transcripts made *in vitro*. Finally, the homology between RD1A mtDNA and wild-type mtDNA has been confirmed by a variety of hybridization experiments (11).

These results fit nicely into the fragmentation-amplification model shown in Fig. 1. Addition of ethidium leads to fragmentation of the wild-type genome and if the ethidium is removed in time, suitable genome fragments will resume replication and be amplified to restore a full complement of mtDNA (see ref. 12 for extensive discussion). More sophisticated variants of this model have been worked out by Slonimski and co-workers (personal communication) to account for the results on marker retention and RNA-DNA hybridiza-

ORIGIN OF ALTERED mtDNA IN YEAST PETITE (SPECULATIVE)

a b c d e f g h i j k l m n o p q r s t u v w x y z

↓ Fragmentation after
acridine intercalation

a b c + d e f g h i j k l m n o p + q + r s t + u v w z y z

↓ Compensatory amplification
and clone selection

d e f g h i j k l m n o p d e f g h i j k l m n o p

and

q q q q q q q q q q q q q q q q q q q q q q q q q q q q

etc.

Fig. 1. Origin of altered mtDNA in yeast petite (speculative).

IS THE RD1A mtDNA REPEATING
SEQUENCE STABLE ?

After 70 generations:

1. T_M renatured DNA =
   T_M native DNA
2. Fingerprint of pancreatic RNAse digest of
   complementary RNA has
   remained the same.

Fig. 2. Is the RD1A mtDNA repeating sequence stable ?

ORIGIN OF GROSSLY ALTERED mtDNA IN YEAST PETITES

a̲ b c d e f g h i j k l m n o p

a̲ a̲ a̲ a̲ a̲ a̲ a̲ a̲ a̲ a̲ a̲ a̲ a̲ a̲ a̲ a̲ a̲ a̲ a̲ a̲   a̲ b b b b b b b b̲ b b b b b b b b b

and   and

a̲ b a̲ b a̲ b a̲ b a̲ b a̲ b a̲ b a̲ b a̲ b a̲ b   a̲ c c c c c c c c c c c c c c c c c

and   and

a̲ b c a̲ b c a̲ b c a̲ b c a̲ b c a̲ b c   a̲ d d d d d d d d d d d d d d d d d d

etc   etc

Replicator is part of repeating unit   Replicator is not part of the repeat-
                                         ing unit

Fig. 3. Origin of grossly altered mtDNA in yeast petites.

tion plateaus in petite mtDNAs.

This model predicts that sections of mtDNA can be highly purified in the form of petite mtDNAs. To judge the potential of this unique purification-by-nature process for sequence studies, a number of questions must be answered.

ARE HIGHLY-REPETITIVE PETITE mtDNAs STABLE ?

Since all highly-repeated nuclear DNAs are loaded with mutations of various sorts (13), it could be that replication of repetitious DNA is particularly error-prone. Especially runs of T or alternating AT could lead to slippage errors. We have, therefore, checked after an additional 70 generations whether our mutant DNA had changed, judged from the criteria given in Fig. 2. A 1% random sequence change would result in a decrease of the T_m of renatured DNA by 1.5° C (see ref. 11). A uniform sequence alteration would, albeit with less sensitivity, show up in the fingerprint of complementary RNA. Neither of these changes were found (9) and this shows that the sequence is stable enough for sequence experiments. (By accident we have found that the mtDNA is lost when the cultures are grown at low temperature. Whether this is due to a block in mtDNA synthesis in this mutant at low temperature is being checked.)

IS THE REPLICATOR GENE PART OF THE REPEATING UNIT ?

10 years ago Jacob and Brenner (14) presented a simple model for the control of DNA synthesis. According to this replicon model every independently repli-cating DNA molecule contains the structural gene for an initiator. When the diffusible initiator acts on a second gene - the replicator gene - replication starts. For mtDNA it seems likely that the structural gene for the initiator is on nuclear DNA, but a replica-tor should be present on every DNA. One can now en-visage the two alternative possibilities shown in Fig. 3. Either the replicator a is part of the repeat-ing unit or it is not. If it is, we should only be able to get highly-repeated mutant DNA of the

region spanning a and this would severely limit the scope of gene purification by mutant selection. An obvious way to check this is to compare the base sequence of different low-density petites. Drs Mahler and Tecce were kind enough to send us the low-density petite mutants first described from their laboratory. In our hands (see ref. 11) the mtDNA of the Mahler mutant D310 is very similar in density and T_m to RD1A mtDNA. The T_m of the Tecce mutant DM1 is 1.5-2.0° C higher and we think that its GC content may have been underestimated by Bernardi et $al.$ (3). This would agree with the recent reports by Carnevali and co-workers (15,16) that this mtDNA is heterogeneous in density after fragmentation and still hybridizes with a mitochondrial tRNA.

The possible base sequence homology between these three mtDNAs was checked both by DNA-DNA hybridization experiments and by RNA-DNA hybridization experiments with complementary RNA made in $vitro$ on all three DNAs. In both cases cross-hybridization was undetectable, $i.e.$ <1% (ref. 11). Since the size of the repeating unit of RD1A is probably less than 150 nucleotides (9), this implies that no cross-homology between the repeating units of the three petite mtDNAs exists. The conclusion is that the replicator is not part of the repeated sequence. It is possible, therefore, that potentially any part of the DNA could end up as repeated sequence in a petite mutant, although some parts may have a better chance to do so than others.

It may seem surprising that the replicator gene does not show up at all in these experiments because is could be there, even if it is not part of the repeating unit. The reason for this is probably that it is such a small part of the total DNA. Lysis of RD1A mitochondria by osmotic shock and analysis of the extruded DNA by the protein monolayer technique shows that this DNA is in the same size range as wild-type mtDNA, if intact (17). Therefore, a few hundred repeating units are hooked up to one replicator gene and even if this gene would be as large as 75 nucleotides, it would still represent only 0.1% of the total DNA.

This is too close to the background of our hybridization experiments to be detectable.

IS THERE NONSENSE DNA IN RD1A mtDNA ?

This question can be rephrased as:"Are there new sequences in RD1A mtDNA that do not hybridize with wild-type mtDNA ?" This question was answered by hybridizing one of the complementary strands of RD1A mtDNA with very high amounts of denatured wild-type mtDNA. Table II shows that all RD1A mtDNA is converted

TABLE II.

PROPORTION OF THE RD1A mtDNA SEQUENCE PRESENT IN WILD-TYPE mtDNA
^{32}P-labelled RD1A L-strand DNA was hybridized in solution with a large excess of denatured wild-type yeast mtDNA. The proportion of ^{32}P converted into duplex DNA was measured with S_1 nuclease. See ref. 11 for experimental details.

DNA preparation	Input	Resistance to S_1 nuclease ^{32}P (cpm)	% of input
1	12 ng	128	94
2	33 ng	110	105

into a duplex form, resistant to S_1 nuclease (a nuclease specific for single-stranded DNA). This shows that no new sequences are present in RD1A mtDNA as far as we can detect with this type of experiment.

IS AMPLIFICATION ERROR-FREE ?

This can be determined by making a heteroduplex between RD1A and wild-type mtDNA and determining its melting profile on hydrocyl-apatite. Whereas native RD1A mtDNA and renatured RD1A mtDNA melt exactly at the same temperature, the heteroduplex of RD1A and wild-type mtDNA melts about 6° C lower (11). This unfortunate difference in T_m could be explained in two ways, illustrated in Fig. 4. The original mutagenic event could have led to copying errors leading to point mutation in the RD1A sequence. The alternative is that

Mis-matching due to point-mutations

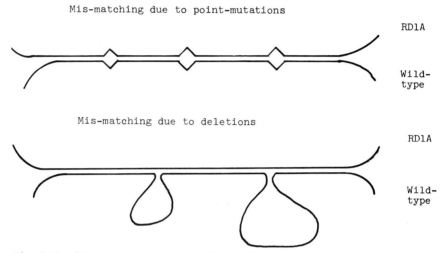

RD1A

Wild-
type

Mis-matching due to deletions

RD1A

Wild-
type

Fig. 4. Possible structure of the heteroduplex of RD1A and wild-type mtDNA.

in RD1A, AT-rich stretches have been fused which are far apart in the wild-type genome. We cannot decide between these alternatives, but it is clear that both would have serious consequences for any attempt at sequence analysis of mtDNA by means of repetitive petite DNAs.

CONCLUDING REMARKS

In summary then, our analysis of the highly repeated mtDNA of the cytoplasmic mutant RD1A has answered a number of questions:

1. Repetition is perfect.
2. The repeated sequence is stable.
3. The repeated sequence is an amplified segment of the wild-type genome.
4. The repeated sequence does not appear to contain the mitochondrial replicator gene and it may, therefore, potentially come from any part of the genome.
5. The repeated sequence does not contain detectable amounts of newly-generated nonsense sequences.
6. The heteroduplex of RD1A and wild-type mtDNA is significantly mismatched. This severely limits

the utility of this mutant DNA as a tool for the sequence analysis of wild-type DNA and it will be of importance to verify this point for other highly-repetitive petite DNAs.

ACKNOWLEDGEMENTS

The investigations were supported (in part) by a grant from the Netherlands Foundation for Chemical Research (S.O.N.) with financial aid from The Netherlands Organization for the Advancement of Pure Research (Z.W.O.).

REFERENCES

1. J.C. Mounolou, H. Jacob and P.P. Slonimski, Biochem.Biophys.Res.Commun. 24, 218 (1966).
2. F. Carnevali, G. Piperno and G. Tecce, Atti Accad. Nazl.Lincei, Rend.Classe Sci.Fis.Mat.Nat. 41, 194 (1966).
3. G. Bernardi, F. Carnevali, A. Nicolaieff, G. Piperno and G. Tecce, J.Mol.Biol. 37, 493 (1968).
4. B.D. Mehrotra and H.R. Mahler, Arch.Biochem.Biophys. 128, 685 (1968).
5. C.P. Hollenberg, P. Borst and E.F.J. van Bruggen, Biochim.Biophys.Acta 277, 35 (1972).
6. P. Borst in: "Autonomy and Biogenesis of Mitochondria and Chloroplasts",N.K. Boardman, A.W. Linnane and R.M. Smillie Eds., North Holland, Amsterdam, 260 (1971).
7. C.P. Hollenberg, Ph.D. Thesis, Mitochondriaal DNA in wildtype bakkersgist en enige cytoplasmatische petite-mutanten, Mondeel-Offsetdrukkerij, Amsterdam (1971).
8. C.P. Hollenberg, P. Borst, R.A. Flavell, C.F. van Kreijl, E.F.J. van Bruggen and A.C. Arnberg, Biochim.Biophys.Acta 277, 44 (1972).
9. J.N.M. Mol and P. Borst, unpublished experiments.
10. C.F. van Kreijl, P. Borst, R.A. Flavell and C.P. Hollenberg, Biochim.Biophys.Acta 277, 61 (1972).
11. J.P.M. Sanders, R.A. Flavell, P. Borst and J.N.M. Mol, Biochim.Biophys.Acta 312, 441 (1973).
12. P. Borst, Ann.Rev.Biochem. 41, 333 (1972).

13. P.M.B. Walker, Progr.Biophys.Mol.Biol. <u>23</u>, 147 (1971).
14. F. Jacob and S. Brenner, Compt.Rend. <u>256</u>, 298 (1963).
15. F. Carnevali and L. Leoni, Biochem.Biophys.Res. Commun. <u>47</u>, 1322 (1972).
16. F. Carnevali, C. Falcone, L. Frontali, L. Leoni, G. Macino and C. Palleschi, Biochem.Biophys.Res. Commun. <u>51</u>, 651 (1973).
17. R.A. Flavell, G.S.P. Groot and J.N.M. Mol, unpublished experiments.

Abbrevations: H-strand, heavy strand; L-strand, light strand; T_m, midpoint of the thermal transition of DNA.

THE PURIFICATION OF MITOCHONDRIAL GENES USING PETITE MUTANTS OF YEAST

Phillip Nagley, Elliot B. Gingold and Anthony W. Linnane

Department of Biochemistry. Monash University, Clayton, Victoria, 3168, Australia.

Studies on the mitochondrial genomes of cytoplasmic petite mutants of *Saccharomyces cerevisiae* have been in progress in our laboratory for several years (reviewed in ref. 1). In this communication we present further evidence for, and discuss the concept that petite mutants provide a unique system for the purification of selected regions of mitochondrial DNA (mtDNA; see ref's. 2,3).

GENERAL STATEMENTS ON GENE PURIFICATION IN PETITE MUTANTS

A summary of the genetic and biochemical studies leading to this idea will be briefly presented. Gingold *et al.* (4) showed that petites which were derived from a grande parent strain carrying a mitochondrial gene specifying resistance of mitochondrial protein synthesis to erythromycin (ρ^+ER^r) either retained this gene (ρ^-ER^r) or lost this gene (ρ^-ER^o). This finding showed that different petites retained or lost different regions of mtDNA. A study of the inheritance of the ER locus during vegetative growth of petite cells shows that there was considerable heterogeneity in freshly isolated petite clones (4,5); however after subcloning some ρ^-ER^r clones, a high degree of stability could be attained (6,7). Similar types of studies on the inheritance of suppressiveness values from mother to daughter cells showed freshly isolated petite clones to be rather unstable, whilst after subcloning relatively faithful transmission of a particular suppressiveness value during vegetative growth often occurred (6,8). These genetic studies were extended to a biochemical level by Nagley & Linnane (2) who investigated the petite clones derived by ethidium bromide (EthBr) treatment of a ρ^+ER^r strain. It was shown that many of the petite cells induced lacked mt-DNA (ρ^0) and were of zero suppressiveness (*cf.* ref's. 9,10). Those clones which retained mtDNA could be very heterogenous in composition, containing cells of many different suppressiveness values, as well as a number of ρ^0 cells. These results were interpreted by Nagley & Linnane (2) to indicate that a considerable molecular heterogeneity exists in freshly isolated EthBr-induced petite clones, with some cells carrying a wide range of the mtDNA fragments produced by EthBr action on the ρ^+ mtDNA (11-13). On subcloning such unstable petites the achievement of genetic stability results from the simplification of the range of mtDNA fragments within the one population of cells termed a clone, so that only a portion of the genetic information originally present in the grande parent is retained.

One of the most important features of the mitochondrial genomes of stable petite clones which retain mt-DNA (termed ρ^-) is that the total quantity of mtDNA is very similar to that of the grande parent strain (2). The presence of a homogeneous population of base sequences representing a small region of the grande mitochondrial genome in one petite clone, means that a considerable amplification of this base sequence must occur in order to maintain the required cellular level of mtDNA. This is the essential idea of gene purification.

A diagrammatic representation of the concepts described above is presented in Fig. 1. The various stages in the purification of one specific sequence in mtDNA, for example the ER locus (represented as the black bar), are drawn. The scheme starts with a hetorogeneous population of fragmentary base sequences (upper left), through possible early subcloning steps leading to reduction in the heterogeneity (upper right)

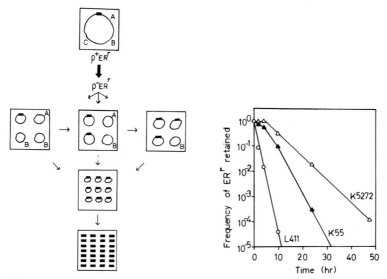

Fig. 1. Schematic representation of the purification of a particular mitochondrial gene in petite mutants of *Saccharomyces cerevisiae*. In the top panel is represented one of the fifty (see ref's 1 and 2) mitochondrial genomes in a haploid ρ^+ER^r strain. Each of the fifty genomes is probably 50×10^6 daltons in size (14, 15); one genome is drawn here to show the ER^r gene (i.e. that gene containing the ER locus, present as the ER^r allele) as the black bar; also drawn are several other regions of mtDNA designated A, B, C, The five lower panels represent possible genomes in ρ^-ER^r cells derived from the ρ^+ER^r parent; each petite genome in one panel contains the equivalent of 50×10^6 daltons of mtDNA. The total mtDNA content of ρ^-cells is similar to that of the grande parent (2). The upper trio of ρ^-ER panels represents early stages in gene purification as often occurs in freshly isolated petite clones which can be rather heterogeneous (see text).

Purification of the ER^r gene proceeds from left to right and from top to bottom, and is envisaged to occur by a process of deletion (by spontaneous or induced mutation) of some sequences in mtDNA and amplification of the residual sequences to maintain the total cellular level of mtDNA. Note also in the step from the upper left to the upper centre panels a reduction in the degree of purification of the region B of mtDNA is shown to occur (*cf*. Fig. 3, the subcloning step K5 to K52). The sequences of mtDNA in the petite cells have been drawn as separate circles for convenience and clarity. As emphasized in the text, the physical size of mtDNA and its structural organization in different ρ^-cells is not known with certainty. This diagram intends only to indicate possible assays of various fragmentary sequences of mtDNA in ρ^-cells.

Fig. 2. Kinetics of elimination of ER loci from a ρ^+ER^r strain L411 and two derived ρ^-ER^r clones K55 and K5272. Cells were grown in a supplemented yeast extract-salts medium containing glucose (5 %) in the presence of EthBr (20 μg/ml), subculturing where necessary to keep cells in the logarithmic phase. Samples were withdrawn at intervals, cells were washed twice with water and then crossed with strain L2200 (ρ^+ERs) to determine the frequency of ρ^+ER^r zygotic clones exactly as described in ref. 3. This frequency is a measure of the level of ER^r genes in the haploid cultures. Control cultures of each strain growing in the absence of EthBr showed the same cellular growth as the EthBr treated samples and did not show any detectable reduction in the level of ER^r genes.

and down through stages of obtaining mitochondrial genomes of simpler complexity but with corresponding increases in the degree of purification obtained. It
should be noted that Fig. 1 does not purport to indicate the physical size of mtDNA molecules in different ρ^- clones, but rather the size of the unique mt-
DNA sequence retained and the extent of its reiteration. Whilst a general description of the physical size of mtDNA in ρ^- cells *in vivo* is not yet settled,
there is no doubt that mtDNA from different ρ^- cells
shows considerably reduced genetic complxity compared
with grande mtDNA as indicated by the faster renaturation rates of many ρ^- mtDNA's (16-18) as well as by
sequence homology studies (19-21).

EVIDENCE FOR THE PURIFICATION OF ERYTHROMYCIN RESIS-
TANCE GENES IN PETITE CLONES USING ETHIDIUM BROMIDE
TARGET ANALYSIS.

We sought to obtain evidence for the amplification
of ER loci in ρ^-ER^r cells by applying a target analys
sis procedure with EthBr as a specific mutagen of mt-
DNA. In work described in detail elsewhere (3) we
showed that EthBr acts on the mtDNA of ρ^- cells in the
same manner as previously described for ρ^+ cells by
Goldring *et al.* (11) and Perlman & Mahler (13). EthBr
action in ρ^+ as well as ρ^- cells leads to inhibition
of mtDNA replication, fragmentation of mtDNA, and the
production of ρ^0 cells, often in high frequency. Cons
sidering the cheme for gene purification in Fig. 1, an
important prediction would be that in ρ^- cells containing highly amplified ER loci (present as ρ^-ER^r) it
would be harder to eliminate ER loci from such cells
compared with ρ^+ER^r cells, because of the greater number of targets in the ρ^-ER^r cells. In addition different ρ^-ER^r clones might be distinguished by their relative amplifications (purity) of ER loci as judged by
the relative ease of conversion of ρ^-ER^r cells to the
ER^0 state by EthBr.

The kinetics of elimination by EthBr (20 µg/ml) of
ER loci from two ρ^-ER^r clones (K 5272) and from the
parent ρ^+ER^r strain (L411) are presented in Fig. 2.

The data show that the two petite clones are very much
less sensitive than the grande strain and also differ
from each other in the sensitivity of the ER loci to
elimination by EthBr. We have reported elsewhere that
it is important to correct the genetic data for sus-
ceptibility of the mtDNA itself towards EthBr action
in ρ^- and ρ^+ cells (3). Petites in general are less
susceptible than the grande parent. However in these
petites derived from L411 of low and moderate suppressiveness a very similar response of mtDNA towards
EthBr was observed as judged by studies on the EthBr
induced reductions in mtDNA levels of K5272 and K55.
The important result is that after correction for the
general effects of EthBr on mtDNA, it was found by
Nagley *et al.* (3) that there are marked differences
between K5272, K55 and L411 in regard to the retention
of ER loci. It is concluded from these analyses that
the relative extents of reiteration of ER loci in the
clones are in the order K5272 (ρ^-ER^r)> K55 (ρ^-ER^r)>
L411 (ρ^+ER^r).

It is interesting and instructive to consider the
steps of subcloning by which clones K5272 and K55 were
derived from L411 (Fig. 3). The purification of ER loci was monitored by measuring the retention of ρ^-ER^r
cells after a fixed ethidium bromide treatment (20 µg/
ml for 8 generations of growth). Both K55 and K5272
are descendants of clone K5, a spontaneous petite isolate of L411 (*cf.* Saunders *et al.*, ref. 6). Whilst the
grande parent strain L411 (ρ^+ER^r) yields more than
90 % of the ρ^+ zygotic clones to be of the type ρ^+ER^r,
after crossing L411 with a reference strain L2200
(ρ^+ER^s), clone K5 exhibits only 57 % ρ^+ER^r zygotic
clones out of total ρ^+ zygotes, after a cross with
L2200. Thus K5 is heterogenous for ρ^-ER^r and ER^o
cells. After treatment of K5 with 20 µg EthBr/ml for
eight generations of growth, there is already a strong
retention of ER loci (2 %) as compared with L411 (<
0.05 %).

Clones K51, K52 and K55 were selected for study
from amongst subclones of K5. Clones K51 and K55 are
similar to one another in their properties, and show

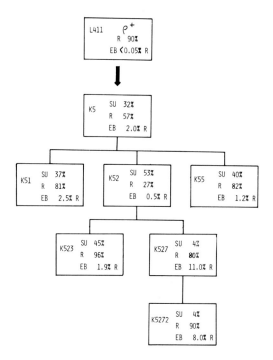

Fig. 3. Analysis of the properties of ρ^-ER^r clones and subclones monitoring the amplification of ER loci by an EthBr target analysis procedure. K5 is a spontaneously arising petite derived from L411; the properties of this clone and several generations of subclones were determined as follows. Each clone was crossed with L2200 (ρ^+ER^s) to measure the suppressiveness (SU) and the content of ρ^-ER^r cells (namely the level of ER^r genes) exactly as described in ref. 3. In addition, each clone was grown for eight generations in the presence of EthBr (20 µg/ml) and after washing, the treated cells were crossed with L2200 to measure the level of ER^r genes retained (expressed as a percentage of the R value for that particular clone). As described in the text, an increased level of retention of ER^r genes after EthBr treatment is taken to imply a higher degree of amplification of ER loci in the (untreated) clone in question.

162

a greater homogeneity for ρ^-ER^r cells than does K5, but there is no notable further purification of ER loci as judged by the results of EthBr treatment (namely retentions of ρ^-ER^r cells after EthBr treatment of 2.5 % and 1.2 % for K51 and K55, respectively). Clone K52 by contrast is less homogenous than K5 (% ρ^+ER^r zygotes is only 27 %) and has also suffered a significant fall in the retention of ER loci in EthBr (to 0.5 %). Subcloning of K52 gave the most interesting results. The derived petite K523 resembles K51 and K55 in most respects, but during the evolution of a sister subclone K527 profound changes took place in the mitochondrial genome. The suppressiveness changed from 53 % in K52 to 4 % in K527, but with no loss in ER loci. Moreover the test of EthBr action on K527 shows that considerable further amplification of ER loci has taken place in this clone (gene retention after treatment with EthBr is 11 %). Clone K5272 is a subclone of K527 and shows little further change from K527 except that a higher proportion of ρ^-ER^r cells (90 %) appears to be present.

Thus in considering the history of clones K55 and K5272 it can be concluded from the above data that little significant purification of ER loci was seen in K55 over than obtained during the initial mutagenic events leading to the primary petite clone K5: however K5272 is considerably more purified for ER loci. The critical event in the history of K5272 took place during the subcloning of K52 to give K527. The importance of this event is manifested by a large change in suppressiveness together with a 10-fold increase in the retention of ER loci after EthBr treatment.

OTHER EVIDENCE FOR THE PURIFICATION OF SPECIFIC BASE
SEQUENCES IN mtDNA IN PETITE MUTANTS.

The studies of Rabinowitz and colleagues on the retention or loss of specific mitochondrial tRNA genes in petite mutants have yielded results very similar to our own studies on the retention of ER loci in petite mutants. Cohen *et al.* (22) showed that while the mitochondrial leucyl-tRNA gene was retained in all subclones studied of a particular spontaneous petite isolate, the mitochondrial valyl-tRNA was retained in some subclones but lost from others. More importantly, the level of hybridization at saturation with leucyl-tRNA of some of the petite mtDNA's were considerably elevated (up to 6-fold in one clone) above that of grande parent mtDNA. This finding indicates some reiterative purification of leucyl-tRNA genes in such petite clones. This type of analysis of petite mtDNA has recently been extended to mitochondrial genes specifying a number of other tRNA species; the same general patterns were observed, with loss from some petites, retention and possibly some purification in others (23).

A different type of base sequence in yeast mtDNA has been studied by Borst and collegues, who studied a petite clone RDIA containing mtDNA of 4-6 % G+C (17). RDIA mtDNA was found to be homologous to 0.3-0.5 % of grande mtDNA (19) and consists of a perfectly repeating sequence of the order of 100 nucleotides in length. The presence of many regions enriched for A and T interspersed with other regions of higher G+C content has been well established by Bernardi and colleagues (24,25) to occur in grande yeast mtDNA which is of average base composition 17 % G+C (16). It would appear that one particular A+T rich sequence in grande mtDNA has been purified in clone RDIA, although some divergence in the base sequence in RDIA mtDNA from the homologous base sequence in grande mtDNA was reported by Sanders *et al.* (19) (*cf*. P. Borst, this volume) on the basis of the melting temperatures of the homologous (RDIA-RDIA) and heterologous (RDIA- grande) DNA-DNA hybrids. It is interesting that RDIA mtDNA shows no sequence homology (tested by hybridization competition experiments) with two other petite mtDNA's of very low (4 %) G+C content; namely the clones DMI and D310-2A-184 described by Bernardi *et al.* (26) and Mehrotra & Mahler (27), respectively, indicating that the latter two clones have purified different A+T-rich sequences from that in RDIA. In general it may be supposed that the altered buoyant density (28) and base composition (16,29) of some petite mtDNA's results from selective retention of particular regions of grande mtDNA; such changes in the physical properties of petite mtDNA's presumably result from the imbalance in base distribution along the grande mitochondrial genome, as mentioned above (*cf*. also ref. 16).

Slonimski and colleagues have isolated petite clones derived from a ρ^+CM^r ER^r parent. Two clones of genotype ρ^-CM^r ER^o and ρ^-CM^o ER^r are of special interest. These clones respectively contain mtDNA of density 2.5 mg/ml lower, and 4.5 mg/ml higher than the grande (30). Both petite mtDNA's are of very low kinetic complexity compared with grande mtDNA (P. Slonimski, this volume) and the data thus suggests that extensive purification of the base sequences carrying the genetic information for resistance to chloramphenicol and erythromycin respectively has been obtained. The rather low G+C content of the mtDNA of the ρ^-CM^r ER^o clone (14 % G+C) suggests that in fact only a segment of the cistron containing the CM locus has been purified, rather than a complete gene. This point can be made by considering the minimum G+C content (35 %) required of a DNA sequence specifying a protein of a usual type of amino acid composition (29). The whole of a normal gene thus cannot have been purified in this ρ^-CM^r ER^o clone. Alternatively of the CM locus is in fact part of the rRNA cistrons in mtDNA as may be suggested by analysis of mutant mitochondrial ribosomes (L. Grivell, this volume) and the extramitochondrial site of synthesis of mitochondrial ribosomal proteins (31; see ref. 1 for details), the same argument still applies. The rRNA in yeast mitochondria is of 26 % G+C (32,33). It is thus conceivable that in the studies of Slonimski and colleagues the purification of antibiotic resistance genes has proceeded so far that only a segment of the gene has been retained.

PROSPECTS FOR RESEARCH USING MITOCHONDRIAL GENES PURIFIED IN PETITE MUTANTS.

One of the most exciting prospects in this field of research is the use of petite mtDNA's containing defined mitochondrial genes as templates for an *in vitro* transcription and translation system. In this way it may be possible to identify specific mitochondrial gene products being coded by mtDNA, with the criterion being their synthesis *in vitro* from a defined mitochondrial gene. In such studies it will be necessary to have the whole of the relevant gene rather than just a segment containing enough genetic information for an antibiotic resistance marker to be rescued in a cross with a suitable grande strain. It thus seems desirable for petite clones to be isolated carrying a series of adjacent genetic markers which are known to map at close but separate loci. In this way one can be more sure of purifying at least one complete gene. Other criteria of gene purification should be applied to the petite clone, including EthBr target analysis as described herein (*cf.* ref. 3), renaturation kinetics of the petite mtDNA, as well as the extent of sequence homology of the petite mtDNA with grande mtDNA (18,20,21).

Another useful general approach which leads from gene purifications is the study of physical properties of small regions of mtDNA, possibly with a view to base sequence determination. The petite mtDNA's with the most potential in the latter area seem to be those of very low G+C content such as RDIA, on which pyrimidine tract analysis has already been carried out (34).

In conclusion, the unique system for the purification of mitochondrial genes afforded by petite mutants in yeast is potentially a most fruitful area in studies on the biogenesis of mitochondria.

REFERENCES.

1. P. Nagley and A.W. Linnane, in:"Advances in Molecular Genetics", I.R. Falconer & D. Cove, Eds., vol.I. Paul Elek, London, (in press).
2. P. Nagley and A.W. Linnane, J.Mol.Biol. **66**, 181 (1972).
3. P. Nagley, E.B. Gingold, H.B. Lukins and A.W. Linnane, J.Mol.Biol. **78**, 335 (1973).
4. E.B. Gingold, G.W. Saunders, H.B. Lukins and A.W.

Linnane, Genetics 62, 735 (1969).

5. G.H. Rank, Can.J.Genet.Cytol. 12, 129 (1970).
6. G.W. Saunders, E.B. Gingold, M.K. Trembath, H.B. Lukins and A.W. Linnane, in:"Autonomy and Biogenesis of Mitochondria and Chloroplasts", N.K. Boardman, A.W. Linnane and R.M. Smillie, Eds., North Holland Publ.Co., Amsterdam, 185 (1971).
7. M. Bolotin, D. Coen, J. Deutsch, B. Dujon, P. Netter, E. Petrochilo, and P.P. Slonimski, Bull.Inst. Pasteur 69, 215 (1971).
8. B. Ephrussi and S. Grandchamp, Heredity 20, 1 (1965).
9. P. Nagley and A.W. Linnane, Biochem.Biophys.Res. Commun. 39, 989 (1970).
10. G. Michaelis, S. Douglass, M. Tsai and R.S. Criddle, Biochem.Genet. 5, 487 (1971).
11. E.S. Goldring, L.I. Grossman, D. Krupnick, D.R. Cryer and J. Marmur, J.Mol.Biol. 52, 323 (1970).
12. E.S. Goldring, L.I. Grossman, J. Marmur, J.Bacterol. 107, 377 (1971).
13. P.S. Perlman and H.R. Mahler, Nature New Biol. 231, 12 (1971).
14. C.P. Hollenberg, P. Borst and E.F.J. van Bruggen, Biochim.Biophys.Acta 209, 1 (1970).
15. J. Blamire, D.R. Cryer, D.B. Finkelstein and J. Marmur, J.Mol.Biol. 67, 11 (1972).
16. G. Bernardi, M. Faures, G. Piperno and P.P. Slonimski, J.Mol.Biol. 48, 23 (1970).
17. C.P. Hollenberg, P. Borst, R.A. Flavell, C.F. van Kreijl, E.F.J. van Bruggen, A.C. Arnberg, Biochim. Biophys. Acta 277, 44 (1972).
18. M. Fauman, and M. Rabinowitz, FEBS Letters 28, 317 (1972).
19. J.P.M. Sanders, R.A. Flavell, P. Borst and J.N.M. Mol, Biochim.Biophys.Acta (1973) in press.
20. G. Michaelis, S. Douglass, M. Tsai, K. Burchiel and R.S. Criddle, Biochemistry 11, 2026 (1972).
21. P. Gordon and M. Rabinowitz, Biochemistry 12, 116 (1973).
22. M. Cohen, J. Casey, M. Rabinowitz and G.S. Getz, J.Mol.Biol. 63, 441 (1972).

23. M. Cohen and M. Rabinowitz, Biochim.Biophys.Acta 281, 192 (1972).
24. G. Bernardi,G. Piperno and G, Fonty, J.Mol.Biol. 65, 173 (1972).
25. G. Piperno, G. Fonty and G. Bernardi, J.Mol.Biol. 65, 191 (1972).
26. G. Bernardi, F. Carnevali, A. Nicolaieff, G. Piperno and G. Tecce, J.Mol.Biol. 37, 493 (1968).
27. B.D. Mehrotra and H.R. Mahler, Arch.Biochem.Biophys. 128, 685 (1968).
28. J.C. Mounolou, H. Jakob and P.P. Slonimski, Biochem.Biophys.Res.Commun. 24, 218 (1966).
29. L.I. Grossman, D.R. Cryer, E.S. Goldring and J. Marmur, J.Mol.Biol. 62, 565 (1971).
30. G. Michaelis, E. Petrochilo and P.P. Slonimski, Mol.Gen.Genet. 123, 51 (1973).
31. P.J. Davey, R. Yu and A.W. Linnane, Biochem. Biophys.Res.Commun. 36, 30 (1969).
32. M. Fauman, M. Rabinowitz and G.S. Getz, Biochim. Biophys.Acta 182, 355 (1969).
33. I.T. Forrester, P. Nagley and A.W. Linnane, FEBS Letters 11, 59 (1970).
34. C.F. van Kreijl, P. Borst, R.A. Flavell and C.P. Hollenberg, Biochim.Biophys.Acta 272, 61 (1972).

Abbrevations and symbols used:
mtDNA, mitochondrial DNA; EthBr, ethidium bromide; ER, CM, genetic loci in mtDNA specifying resistance of mitochondria *in vivo* towards the antibiotics erythromycin and chloramphenicol, respectively (use of these symbols and of ρ is explained in the text).

INTERACTION OF MITOCHONDRIAL PROTEIN SYNTHESIS ON THE REGULATION OF GENE ACTIVITY IN *SACCHAROMYCES CEREVISIAE*

P.P. Puglisi and A.A. Algeri

Institute of Genetics, University of Parma, Parma, Italy.

Much information is now available about mitochondria with respect to the synthesis of macromolecules, electron transport and energy production, and in spite of several unsolved questions it can be stated that this cellular particle is known in its salient aspects (1). It is therefore possible to approach a problem connected with the mitochondrion, i.e. the basis of the fact that, at least in the yeast *Saccharomyces cerevisiae*, some cellular processes such as the induction of some catabolic enzymes and as sporulation appear impaired in cells that carry the mutation of the cytoplasmic factor ρ (rho).

The data we have obtained suggest that mitochondrial protein synthesis plays a role of significance in the regulation of the activity of nuclear genes connected with the expression of cellular functions such as enzyme induction and sporulation, giving rise to a picture of the mitochondrial role that differs, in some aspects, from its classical one.

ENZYME INDUCTION.

The earlier indications that mitochondria could play a role which is not solely concerned with energy production and respiration were given by Spiegelman, who observed that in spite of the fact that galactose and maltose as well as glucose are fermentable carbon sources, only glucose is utilized by the yeast cells when their mitochondrial functions are impaired by anaerbiosis, 2-4 dinitrophenol or sodium azide (2,3).

The studies on the mechanism of utilization of galactose and maltose, the demonstration that the utilization of these sugars is a process that depends on enzyme induction and the fact that the catabolism of the
sugars mentioned above is impaired also in respiratory deficient mutants, suggest that there could exist
a relationship between some steps of gene regulation
and mitochondrial integrety. The increasing knowledge
on genetics and molecular biology of the mitochondrion
and the possibility to use not only ρ^-mutants, but also highly specific inhibitors of mitochondrial functions and mutants resistant to them prompted us to
analyze whether the block or the failure of the cells
to utilize such carbon sources as galactose or maltose was due to some of the "classical" effects of mitochondrial alteration or that it could be related to
a new cellular role of the particle, linked to mitochondrial functions such as macromolecular synthesis.
For the analysis of the role of mitochondrial integrity on enzyme induction we have taken into considera-
tion the galactose pathway, of which the genetic
structure is well known (4).

As reported in a previous paper (5) regulator constitutive mutants for galactose (i^-) take up galactose at the same rate in ρ^- and ρ^+ condition. This fact,
that holds true also for regulator constitutive mutants for maltose (unpublished results), indicates
that the block in the induction of galactose enzymes
does not rest on the energy impairment or on a block
of the incorporation of the inducer, but suggests that
there exists an interaction of some mitochondrial
function with the genetic regulation of the enzymes of
the galactose pathway.

Our early observations, however, indicated that
among several spontaneous ρ^-mutants, only a fraction
was completely blocked in galactose utilization,
whilst other ρ^-mutants showed only a delay in the time
of the adaptation to the sugar.

Since it is known that different mutations can differ in terms of mitochondrial DNA alteration, and as
a consequence in the pattern of impaired mitochondrial

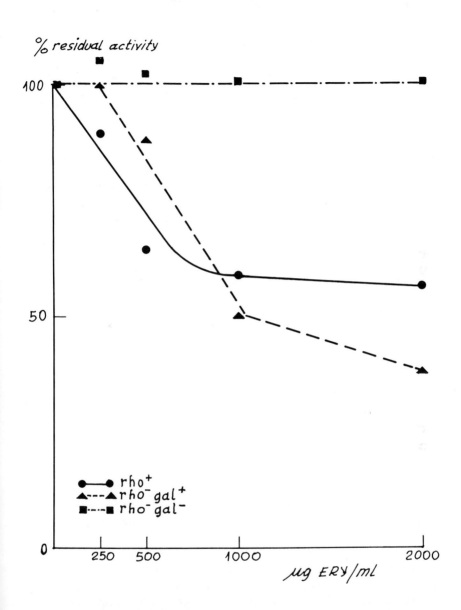

Fig. 1. Dose-effect of erythromycin on cycloheximide-resistant incorporation: ●--● *rho*+; ▲--▲ *rho*⁻ gal⁺; ■--■ *rho*⁻ gal⁻.

functions, we have analyzed several cytoplasmic respiratory deficient mutants both spontaneous and induced by ethidium bromide, 5 fluoro-uracil and UV ra-
diation. Around 50 % of the mutants analyzed grew on
galactose with only a delayed adaptation time, but on
addition of erythromycin to these galactose-positive
respiratory-deficient strains, we observed a complete
block of the growth (6). We have then looked for the
possibility that the galactose-positive respiratory-
deficient mutants possess residual mitochondrial protein synthesis as to explain the fact that they , in
spite of their respiratory deficiency, behave as galactose-positive strains. In fact, as shown in Fig. 1,
galactose positive RD mutants are still active in mitochondrial protein synthesis, as demonstrated by the
presence of erythromycin-sensitive cycloheximide-resistant amino acid incorporation activity which is
absent in galactose-negative respiratory-deficient
mutants.

We have therefore proposed a model (6) in which a
component of the regulatory system, which is endowed
with a positive regulatory role, depends for its function on the erythromycin-sensitive pathway on cellular protein synthesis, $i.e.$ from the mitochondrial
system for protein synthesis.

SPORULATION.

The sporulation process in *Saccharomyces cerevisi-*
ae depends on the heterozygosis of the genes α and a
and is affected by several conditions among which a
relevant role appears to be played by the respiratory
competence. Lindegren and Miller observed tha failure
of sporulation in diploid cells submitted to respiratory inhibitors and Ephrussi and Hottinguer found that
respiratory-deficient yeasts are unable to undergo
sporulation (7). Since the same genetic complement be-
haves in different ways undergoing mitosis or meiosis
and sporulation as effect of environmental conditions,
it must be admitted that the sporulation process is
the result of a regulatory switch-on and -off mechanism of specific genetic determinants.

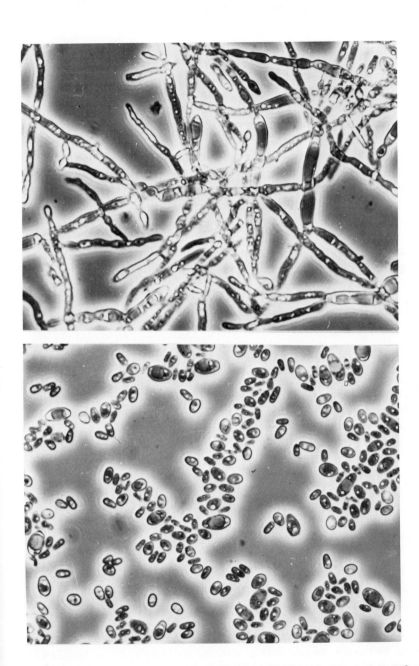

Fig. 2. a) Aerobic cells of *E. capsularis* grown in YM broth glucose 2 %, 28° C; b) Aerobic cells of *E. capsularis* grown in presence of 4,000 µg/ml of erythromycin or chloramphenicol.

For these reasons, in spite of the fact that we do not know the gene products involved in sporulation, although we know that their determinants are probably nuclear in nature, we have studied the effect of erythromycin on the sporulation of erythromycin-sensitive and erythromycin-resistant a/a diploids. The data obtained (8) can be summarized as follows: erythromycin inhibits sporulation of erythromycin-sensitive parent but not of its erythromycin-resistant derivative. Because the only difference between the resistant and the sensitive strain appears to be the response of the mitochondrial protein synthesizing system to the antibiotic, it could be suggested that the inhibition of the sporulation by erythromycin reflects the dependence of the process from mitochondrial protein synthesis.

These results could be explained either by assuming that a gene product endowed with a regulatory role in switch-on and -off of sporulation depends on mitochondrial protein synthesis, or at least one of the sporulation "structural" proteins is synthesized on mitochondrial machinery. Another cellular process that probably rests on the switch-on and -off of genetic determinants as effect of some environmental conditions is the morphogenesis in dimorphic yeasts.

Although the results we have obtained are only preliminary it could be observed that the cells grown in presence of erythromycin or chloramphenicol are almost all of yeast-like type (Fig. 2). This fact, that for the moment is only a suggestive indication, prompted us to study this process and the eventual role in it of the mitochondrial particle.

DISCUSSION.

The data we presented above suggest that there exists a strict dependence of the expression of some nuclear regulatory systems on mitochondrial protein synthesis. It remains to be decided whether the interaction between regulation and mitochondria is a direct and specific one, or that there exists a mitochondrion-dependent product endowed with a generalized regulatotory role upon which the activation of genetic regulatory systems could depend.

In the first case, it must be explained why the mRNA of nuclear origin of a protein endowed with the regulation of a nuclear genetic system is translated on the mitochondrial protein synthesizing system. In fact, although this situation could be envisaged in molecular terms by assuming for instance that this regulatory protein is transcribed from a messenger RNA that interacts only with mitochondrial ribosomes, the evolutionary origin of this situation poses several problems and is at present under investigation. From a general point of view the unusual high (A+T) content of mitochondrial DNA eventually permits the mitochondrial translation of nuclear messengers. In fact, as occurs for the transport of nuclear messengers in eukaryotic cells (9) which are linked to poly-A tails, it could be suggested that the need for the transport of several nuclear messengers across the mitochondrial membranes could have exerted a positive selection pressure on the cells whose mitochondrial DNAs had the higher (A+T) content. The above hypothesis is at present followed as a working hypothesis.

On the other hand the pleiotropic carbohydrate negative phenotype of ρ^- mutants and the increased catabolite repression of such cells could suggest that the effect of the ρ^- condition is primarly concerned with a reduction in the level of cyclic AMP, whose production, according to the effect of erythromycin on enzyme induction and sporulation, seems to be dependent on a protein synthesized on the mitochondrial ribosomes.

REFERENCES.

1. A.W. Linnane *et al.*, Ann.Rev.Microbiol. **26**, 163 (1972).
2. S. Spiegelman, J.Cell.Comp.Physiol. **25**, 121 (1945).
3. J.M. Reiner, Proc.Soc.Exp.Biol.(N.Y.) **63**, 81 (1946).
4. H.C. Douglas and O.C. Hawthorne, Genetics **54**, 911 (1966).
5. P.P. Puglisi and A.A. Algeri, Mol.Gen.Gen. **110**, 110

(1971).
6. P.P. Puglisi *et al.*, in preparation.
7. R.R. Fowell, in:"The yeasts", vol. 1, A.H. Rose and J.S. Harrison, Eds., Academic Press, 303 (1969).
8. P.P. Puglisi and E. Zennaro, Experientia 27, 963 (1971).
9. S.W. Kwan and G. Brawerman, Proc.Nat.Acad.Sci. USA 69, 3247 (1972).

ISOLATION, PURIFICATION, PHYSICO-CHEMICAL CHARACTERISATION AND RECOMBINATION OF YEAST MITOCHONDRIAL DNA SEGMENTS CONFERRING RESISTANCE EITHER TO CHLORAMPHENICOL OR TO ERYTHROMYCIN[*]

H. Fukuhara, J. Lazowska, F. Michel, G. Faye, G. Michaelis, E. Petrochilo and P. Slonimski

Centre de Génétique Moléculaire du C.N.R.S., Gif-sur-Yvette 91190, France.

We have shown that cytoplasmic "petite colonie" mutants (ρ^-) of yeast result from large deletions of the wild type mitochondrial DNA sequence compensated by reiteration of the non-deleted segments leading to gene redundancy; long molecules of mtDNA display numerous periodic repetitions of the informative sequence. Having isolated stable ρ^- clones in which the segment conferring the resistance to chloramphenicol is amplified on the one hand and the segment conferring the resistance to erythromycin on the other, the purified mtDNA sequences have been characterized both by buoyant density and by high resolution analysis of differential denaturation. The C^R segment displays two main peaks at 69.5° and 70.3° while the E^R segment displays peaks at 72.2°, 75.5° and 80.3°. The first has a lower G+C content than the second. There are practically no overlapping sequences between these two segments (1).
The $C^R E^O$ and $C^O E^R$ haploid "petites" have been crossed with each other and the resulting mtDNA in diploid petites was analysed physico-chemically and genetically by a "marker rescue" in triploid crosses. We found that "petites" are able to recombine. The mtDNA of these recombinants is not a mixture of the two paren-

---

[*] Abstract of the paper red by Dr. P. Slonimski. The complete manuscript of the paper was not received by the Editors.

tal types but a novel molecular species of intermediate buoyant density. Recombination of mitochondrial
genes involves physical reunion of the parental molecules (2).

REFERENCES.

1. G. Faye, H. Fukuhara, C. Grandchap, J. Lazowaka,
   F. Michel, J. Casey, G.S. Getz, J. Locker, M. Rabinowitz, M. Bolotin-Fukuhara, D. Coen, J. Deutch
   B. Dujon, P. Netter et P. Slonimski, Biochemie 55,
   in press.
2. G. Michaelis, E. Petrochilo and P. Slonimski,
   Mol.Gen.Gen., in press.

MITOCHONDRIAL GENETIC FACTORS IN THE CELLULAR RESPONSE TO CHLORIMIPRAMINE IN *SACCHAROMYCES CEREVISIAE*

D. Linstead, I. Evans and D. Wilkie

Department of Botany & Microbiology, University College, London, England.

The demonstration of selective inhibition of mitochondria in yeast cells by antibiotics led to the isolation of mitochondrial mutants manifesting resistance to these compounds. The location of resistance factors in mitochondrial DNA was deduced from their concomitant loss with that of the ρ factor in petite mutants. Recombination of resistance factors was seen among the diploid progeny from zygotes in crosses in some of which asymmetrical distribution of reciprocal products and parental types was a feature (1). Attempts were made to relate these distributions to the mechanism of the recombination events themselves, the underlying basic assumption being that the products of such events were randomly distributed to each daughter cell budded off the zygote. More recent work with isolated zygote daugther cells has shown that in some cases a remarkable control can be exerted by the zygote on the distribution of mitochondrial types to daughter cells (2) which, in itself, can explain asymmetry. In further studies (3) a particular distribution pattern leading to asymmetry of parentals and recombinant types with a low frequency of the latter, appeared to be under the control of a nuclear gene. Blockage of this gene in zygotes, either at the level of transcription with thiolutin (4) or translation with cycloheximide, leads to loss of the control resulting in equality of parental types and of recombinants with an increased frequency of the latter. Arrest of protein synthesis in mitochondria with antibiotics in parental cells prior to, during and after zygote formation had no detectable effect on the asymmetrical distribution. As well as providing information on the control of mitochondrial transmission to daughter cells these findings emphasize the necessity to carry out zygote cell lineage analysis before any conclusions may be drawn regarding the mechanism of recombination.

Inhibitors used to obtain mitochondrial mutants are highly specific for mitochondria inhibiting growth of cells in non-fermentable medium, but there is little or no inhibition of cell growth in a glucose medium even at high concentrations of these drugs. Hence genetic changes in mitochondria leading to resistance may be expected to alter organelle characters directly (5). Nuclear changes are well characterised in yeast which also alter mitochondrial constituents, but there is no evidence as far as we are aware, of mitochondrial genetic information specifying or being involved in specifying, cellular characteristics. Our observation that the petite mutation conferred a selective advantage on cells growing in a glucose medium in the presence of chlorimipramine (CI), was of interest in this connection, suggesting that an alteration in mitochondrial DNA resulted in a change in the response of the cell to the inhibitory effects of this drug (7). Since the drug is strongly lipophilic and has been shown to react with erythrocyte membranes (8), the hypothesis was adopted that an alteration in mitochondrial DNA in petite mutants resulted in a change in the properties of cell membranes. Details of experiments designed to test this hypothesis are given here. Reactivity of the drug with mitochondria is also taken into account.

MATERIALS AND METHODS

The effects of CI on yeast cells, isolation of resistant mutants and testing of clones by the drop-out method are described in refs. 9,10. Concentrations of CI routinely used in test series were 10,15,25,30,40, 50 and 100 µg/ml except where otherwise indicated. Maximum tolerance (resistance) is defined as that con-

centration above which in the series there is total
inhibition of cell growth. Medium contained yeast extract (1 %) and glucose (2 %). Standard methods were
used in crossing, sporulating, tetrad analysis and
random ascospore plating (11).

Binding by whole yeast cells of [^{14}C]*imipramine.*
Imipramine has the same effect as CI on yeast cells
but is slightly less toxic. Radioactive imipramine
was obtained from the Radiochemical Centre, Amersham.
Cells were grown to mid-log or stationary phase, harvested and washed twice in cold distilled water then
suspended in ice-cold 5 mM HEPES at pH 6.8. 0.4 ml of
suspension was mixed with 0.1 ml of a [^{14}C]imipramine
stock solution and incubated for 15 min at 30°. Two
0.2 ml samples were then pipetted onto glass fibre filter discs (Whatman GF/C, 2.4 cm diameter), and immediately washed with three 3 ml aliquots of ice cold
5 mM HEPES. The filter discs were air dried and counted in 10 ml aliquots of scintillation fluid. Protein
determinations employed the method of Lowry *et al.*
(12). For dry weight determinations, 1.0 ml aliquots
of cells suspension were filtered onto preweighed
glass fibre filter discs, washed thoroughly with distilled water and dried for 24 h at 50° after which
the discs were reweighed.

In vivo distribution of [^{14}C]*imipramine.* Cells
were grown to stationary phase, harvested, and washed
as before. The yeast pellet was resuspended in 2 mM
HEPES at pH 7.0 and three 5 ml aliquots were incubated for 20 min at 30° in 50 ml conical flasks with
different concentrations of radioactive imipramine.
After incubation, cells were washed twice by centrifugation with 15 ml ice cold isolation medium and
then disrupted in the Braun shaker. Centrifugation of
the homogenate as described below resolved three portions: debris (1000xg pellet), mitochondria (10,000xg
pellet) and supernatant (10,000xg supernatant). The
debris and mitochondrial fractions were resuspended
in small volumes of isolation medium and replicate
samples were removed from all fractions for radioactivity and protein determinations.

Mitochondrial preparations. The procedure was based on the method of Spencer *et al*. (13). Cells after washing were suspended in ice cold isolation medium as mentioned above of composition 0.25 M sucrose, 0.02 M potassium dihydrogen orthophosphate, 0.01 M magnesium chloride, 0.001 M EGTA pH 6.8, 1 ml per g wet weight. Cells were mixed with Ballotini beads (0.45 mm - 0.50 mm diameter), 5 g per g of cells and disrupted by shaking for 15 sec in a CO_2 cooled Braun shaker. The pooled homogenate and bead rinsings (all in isolation medium) were centrifuged for 10 min at 1000xg at 4° to remove unbroken cells and debris. Mitochondria were harvested from the supernatant by centrifugation for 10 min at 10,000xg and then washed by resuspension in ice cold isolation medium and recentrifugation.

Equilibrium dialysis of mitochondria with [14C]*imipramine*. 0.4 ml mitochondrial suspension (in isolation medium, approx. 5 mg mitochondrial protein/ml) and 0.1 ml of a stock solution (used above) of [14C]-imipramine (specific activity approx. 7 mCi/mmol dissolved in 0.25 M sucrose) were sealed into a small bag of Visking tubing (2.55 cm wide, prewashed by boiling in 5 mM EDTA and rinsing in distilled water). The bag was immersed in 50 ml ice cold isolation medium in a plugged 250 ml conical flask which was then shaken gently overnight at 4° on a Mickle rotary shaker. Control dialyses (dialysis bags containing only labelled drug in 0.25 M sucrose) established that equilibrium of the contents of the bag with the external medium occurred in the course of an 18 h incubation at 4°. After dialysis, the contents of bags were sampled for radioactivity (two 0.1 ml samples) and protein (0.1 ml sample) content and the external medium sampled for radioactivity only. Samples were air dried in scintillation vials before radioactivity was determined. In each experiment, duplicate dialyses were done.

Binding of [14C]*imipramine to purified mitochondria* Mitochondria were isolated as described above, incubated *in vitro* with 4 different concentrations of

[^{14}C]imipramine for 5 min at 30° and then washed twice with 20 ml ice cold isolation medium by centrifugation at 25,000x*g* for 1 min. 1 ml samples of the labelled resuspended mitochondria were loaded onto sucrose gradients (12 ml, linear, 20 %-60 % w/v sucrose in 2 mM HEPES pH 6.8) and centrifuged for 60 min at 150,000x*g* at 4°. 0.4 ml fractions were collected and assayed for protein content, radioactivity and succinate-ferricyanide reductase activity using an assay base on that of Estabrook (14).

RESULTS

The effect of the petite mutation on tolerance of various strains to CI was recorded (Table I). It can be seen that most petite mutants of sensitive strains showed an increase in resistance but some showed little or no change in cellular resistance compared with the wild type parent strain. The proportion of

TABLE I

EFFECT OF THE PETITE MUTATION ON CELLULAR RESISTANCE TO CHLORIMIPRAMINE.

Strain	Resistance μg/ml	Origin of petites	Comparative resistance of petites and number in each category			
			No change	2-fold increase	>2-fold increase	decrease
D6	15	acridine*	6	19	0	0
D6-R3	150	acridine	0	0	0	21
D22	15	acridine	0	1	0	0
D22-6R1	25	spontaneous	4	4	1	2
D22-6R1		acridine	3	3	1	10
A18	15	spontaneous	0	0	8	0
A18		acridine	0	9	20	0
45B	15	acridine	1	6	14	0
DV-147	15	spontaneous	2	8	2	0
A7F	25	acridine	0	0	1	0

*20 μg/ml

petites that did show increased resistance, as well as the extent to which resistance was increased, appeared to be strain dependent. When a low-level resistant mutant (D22-6R1) was the parent strain, some petites derived from it were of lower resistance and when a high-level resistant mutant (D6-R3) was used all its petites that were tested were found to have a lower resistance compared with the ρ+ parent. Thus it seems that with increasing cellular resistance of

the parent strain there is an increasing likelihood of
the petite mutation leading to a reduction in the resistance of the cell.

The procedure of plating cells of sensitive strains
on medium containing inhibitory concentrations of the
drug incidentally provides a most efficient technique
for the isolation and scoring of petites in cell samples. Since a proportion of petites do not show a
change in resistance compared with the parent strain,
it seems that loss of the ρ factor is not enough in
itself to account for the tolerance changes seen. It
was further established that respiratory deficiency
per se was not a factor by introducing the nuclear mutation p_7 into some of the sensitive strains by appropriate crosses. Although showing the petite phenotype,
these strains all had the low level of tolerance of
about 15 μg/ml. When the ρ-mutation was induced in p_7-
carrying strains, increased resistance was once more
seen.

TABLE II

EFFECTS ON CELLULAR RESISTANCE TO CHLORIMIPRAMINE (CI) IN PETITE MUTANTS WITH
(+) AND WITHOUT (−) DETECTABLE MITOCHONDRIAL DNA.

Strain	Mitochondrial DNA	CI resistance (μg/ml)
239-1 ρ+	+	10
239-1 ρ−	+	50
Ox2 ρ+	+	10
Ox2 ρ−	+	30
Ox2 ρ−	−	40
Ox2 3ρ−	+	40
Ox2 3ρ−	−	25
D243 ρ+	+	15
D243 ρ−	+	15
D243 ρ−	−	25

A number of ρ-strains which had been classified with
respect to presence or absence of mitochondrial DNA
(kindly supplied by Dr.D.H. Williamson) were tested
for increased drug resistance compared with their respective ρ+ parental strains (Table II). Since DNA-less
petites showed an increase over their sensitive parents, it was concluded that loss or other mutational
change in certain mitochondrial genes which accompani-
ed the change to ρ− were responsible for the altered
response of the cell to CI. Those ρ− mutants not so

changed in their DNA would not show an increase in
drug tolerance.

Genetics of drug resistance in ρ- mutants. It is
well established that mitochondrial genetic factors
can recombine in crosses between ρ+ and ρ- strains and
segregate among the diploid progeny of zygotes as already discussed. ρ- strains showing increased resistance were crossed to sensitive ρ+ strains.

TABLE III

INHERITANCE OF CHLORIMIPRAMINE RESISTANCE IN DIPLOID ZYGOTE CLONES FROM CROSSES
BETWEEN ρ+ AND ρ- STRAINS.

Cross	Number of clones analysed	Resistance (μg/ml) of clones and number [**]			
		15	25	50	100
D22 x D6 [*] (15) (15)[*]	25	25	0	0	0
D22ρ- x D6 [*] (25)[*]	37	17	11 (2 ρ-)	9 (all ρ-)	0
45Bρ- x D22 [*] (150)[*]	27		24	3	0

[*] Resistance of strain in μg/ml. [**] Some clones from ρ+ x ρ- crosses contained a
proportion of cells (10^{-2}-10^{-3}) capable of growing into colonies at the next
higher concentration to that listed for the clone as a whole.

Various resistance levels were obtained in diploid segregants indicating control by mitochondrial factors
as exemplified in the results shown in Table III.
Diploid cells of a zygote clone from the cross D22ρ-
x D6 were sporulated and scored for resistance. In 4
tetrads out of 6 analysed, all 4 ascospores gave haploid cultures resistant to 25 μg/ml; 1 tetrad segregated 3 resistant to 25 to 1 resistant to 50 μg/ml
and the sixth tetrad segregated 2 resistant to 25 to
2 resistant to 50 μg/ml CI. In the corresponding cross
between the parental (ρ+) strains, both of which showed the low tolerance of 15 μg/ml, diploids all showed
this tolerance level as did ascospore cultures derived from them. These striking differences in resistance inherited by both vegetative and sexual progeny
resulted from the introduction of the ρ- mutation in
D22. The marked increase in resistance in 45B following the ρ-mutation (from 20 to 150) was not transmitted to the diploid progeny analysed in the cross 45Bρ-
x D22, although comparatively high resistance levels

were seen in diploid segregants. It was concluded that the particular arrangement of mutant mitochondrial factors in 45Bρ- leading to high level resistance was not re-established in diploid segregants. It may be reiterated that segregation in diploids is a feature of cytoplasmic inheritance.

Analysis of the spontaneous resistant mutant D6-R3. When D6-R3 was crossed to the sensitive strain D22, all zygote diploid clones when tested were resistant to 50 μg/ml with a proportion of cells capable of growing at higher concentrations of CI up to 150 μg/ml, the latter being the resistance level of all cells of D6-R3. Random ascospores obtained from a random sample of diploid cells in the cross, gave a 1 : 1 segregation for resistance to 50 μg/ml to sensitivity to 50 μg/ml. (Table IV). This suggests that resistance to 50 μg/ml is controlled by a nuclear gene but that the resistance to 150 μg/ml seen in D6-R3 is achieved by the modifying influence of other factors.

TABLE IV

INHERITANCE OF CHLORIMIPRAMINE RESISTANCE OF D6-R3 AND D6-R3ρ- IN CROSSES WITH STRAIN D22: ANALYSIS OF RANDOM ASCOSPORE CLONES

Cross	Total clones tested		Chlorimipramine concentration (μg/ml)					
			0	25	50	75	100	150
D6-R3								
x	80	clones	80	65	38	15	10	4
D22		growing						
D6-R3ρ-								
x								
D22								
Diploid 1*	145	clones	145	116	53	19	8	4
Diploid 2	62	growing	62	53	40	30	20	12

*Diploid 1, zygote clone with constituent cells mainly resistant to 50 μg/ml
Diploid 2, cells mainly resistant to 150 μg/ml.

That these are probably mitochondrial is indicated by the fact that the petite mutants of strain D6-R3 mainly show resistance to 50 μg/ml. When D6-R3ρ- was crossed to D22 modifying factors were re-introduced, presumably by mitochondrial recombination. It was possible by selecting diploid segregants with high levels of resistance, to increase the inheritance of high level resistance among the ascospores (Table IV).

186

Genetic analysis using DNA-less petites. In cross-es between ρ- strains in which the ρ- strain is DNA-less, recombination in zygotes is presumably not possible (except perhaps between ρ+ genomes) and segregation of various resistance levels would not be seen among the diploid progeny. This was seen to be the case when DNA-less mutants of DV147 and 0x2 were used (see Table II). In crosses to sensitive ρ+ strains, the diploids obtained tended to be of uniform resistance, did not inherit the higher resistance levels of the ρ- mutants but were similar in resistance to diploids obtained from control crosses between ρ+ parental strains.

Effect of anaerobic growth and severe glucose repression. ρ+ and ρ- strains, pre-grown anaerobically were exposed to CI under anaerobic conditions. All ρ+ strains were about 3 times more sensitive than under aerobic conditions while ρ- mutants were similarly affected but to a lesser extent. In other words, ρ- strains still showed a selective advantage under anaerobiosis. ρ+ strains in these experiments were of the sensitive type. Thus anaerobic growth which in many ways resembles the petite condition, does not confer increased resistance to CI.

Culture medium normally contained 2 % glucose which can exert some measure of mitochondrial repression. It was found that strains grown in 10 % glucose were considerably more resistant to CI than with 2 % glucose (Table V). The increase seen in resistance was more pronounced the more sensitive the strain tested. 10 % glucose, which is known to cause a mark-

TABLE V

EFFECT OF GLUCOSE REPRESSION ON CHLORIMIPRAMINE TOLERANCE IN 22 STRAINS.

Glucose %	Chlorimipramine (µg/ml)			
	0	25	50	75
	number of strains growing			
2	22	16	6	1
10	22	21	19	15

ed repression of mitochondrial activities with much reduced levels of ATPase, thus produces increases in CI tolerance of similar magnitude to those produced

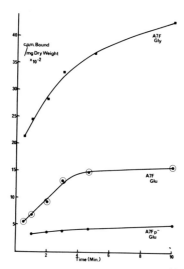

Fig. 1. Kinetics of uptake of [^{14}C]imipramine (50 μM) by whole cells of strain A7F and A7Fρ⁻ (see Table I). Gly: 4 % glycerol medium, mid log phase cells; glu: glucose medium, late log phase cells.

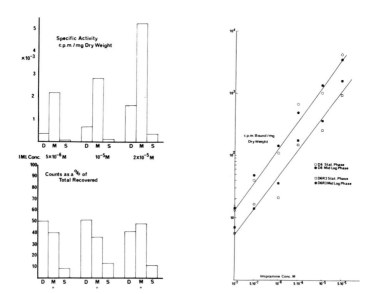

Fig. 2. *In vivo* distribution of [^{14}C]imipramine. Upper: specific activity of three subcellular fractions, debris (D), mitochondria (M) and supernatant (S); lower: distribution of the total radioactivity recovered between the three fractions. IMI = imipramine.

Fig. 3. Titration of whole cells of strain D6 and resistant mutant D6-R3 with [^{14}C]imipramine. Each point is the average of two duplicate experiments.

by petites grown on 2 % glucose medium.

Drug binding to yeast cells and cell fraction. The
effects described above relate to inhibition of cell
growth by CI in a glucose medium, the assumption being
that membrane associated systems were the targets of
drug activity. Our earlier finding that cells utilizing a glycerol carbon source were arrested at lower
concentrations of the drug than in glucose, suggested
that mitochondria were more susceptible to drug reactivity than other cell systems (10). Attempts were
made to correlate the differences in drug resistance
between A7F and various mutants derived from it, with
differences in cellular binding of the drug. Results
shown in Table VI and Fig. 1 clearly show a correlation in this respect. Using the sensitive strain 188,

TABLE VI

IN VIVO BINDING OF [^{14}C]IMIPRAMINE IN STRAIN A7F AND SPONTANEOUS RESISTANT MUTANTS
DERIVED FROM IT.

Strain	Cellular resistance	cpmbound/mg dry weight*
A7F	25 µg/ml	693 ± 103
A7F-214	40 µg/ml	543 ± 39 (p= 0.01)
A7F-285	50 µg/ml	417 ± 46 (p= 0.01)

*Labelling carried out over the period 0-3 min. Means are from 10 samples.

in vivo distribution of labelled drug was studied in
cell fractions. The results (Fig. 2) show that the
mitochondrial portion has a high affinity for imipramine but it cannot be excluded that components with
a high drug affinity might exist in the debris fraction, for example plasma membrane associated with
cell wall. Indeed, nearly half of the total imipramine recovered is bound to the debris. Isopyknic gradient centrifugation of isolated mitochondria showed
that radioactivity sedimented with fractions having
succinate-ferricyanide reductase activity.

In a further series of experiments, binding of
[^{14}C]imipramine to whole cells and isolated, purified
mitochondria of D6-R3 was determined. The results
(Figs. 3 and 4) show clearly that the cells of the
resistant mutant bind 1/3 - 1/4 of the amount of drug
bound by the sensitive parent strain. In the experi-

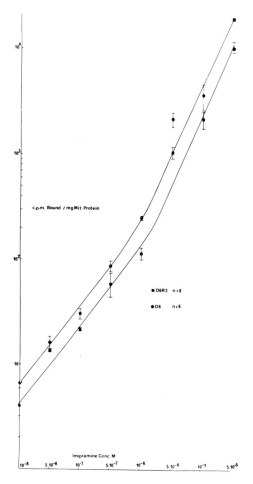

Fig. 4. Titration of mitochondria isolated from strains D6 and D6-R3 with $[^{14}C]$-imipramine. Each point represents the mean $\pm$ standard deviation of the indicated number of duplicates. Differences in binding are significant at the 1 % level by students T-test at concentrations other than 10^{-8}, 5×10^{-8} and 10^{-5} M.

ments with isolated mitochondria, again drug binding
to the mitochondrial fraction showing marker enzyme
activity was again demonstrated. The technique of e-
quilibrium dialysis was also used together with the
standard method to determine drug binding. The results
obtained indicated a consistent and significant difference in binding affinity between mitochondria of
D6 and D6-R3. This difference is not as great as that
between whole cells.

DISCUSSION

Genetic analysis indicates that mitochondrial factors have a controlling influence on the response of
yeast cells to CI. In the case of the mutant D6-R3,
resistance is apparently achieved by the combined activity of nuclear and mitochondrial factors. The
change to resistance is seen to affect both the cellular and mitochondrial binding capacity of the drug
and we suggest that the binding shown by whole cells
is a combination of the binding affinities of both
mitochondria and non-mitochondrial components. Attempts are being made to isolate plasma membranes as
the likely non-mitochondrial element and data on the
comparative binding of $\rho-$ and respective $\rho+$ mitochondria will soon be available.

In considering the case of increased resistance
following the petite mutation, certain points may be
made. Petite mitochondria do not synthesize proteins
and since some petites do not have altered resistance,
it would appear that mitochondrial translation products are not involved in the changes seen. Transcription products, however, may be involved since $\rho-$
mitochondria that have DNA can synthesize RNA. These
products may have a role in regulating nuclear factors that maintain the normal features of cell membranes. Deletion of the corresponding DNA regions in
the mitochondrion would be expected to lead to an alteration in these features. Glucose repression may
also affect transcription of these mitochondrial factors and lead to similar changes in response to CI.
Anaerobic growth on the other hand, affects in a

gross way the overall composition of cell membranes
which may explain the general increase in cell sensitivity to CI.

In summary, our findings indicate the involvement
of mitochondrial factors in specifying cell characteristics related to drug reactivity. The precise nature of these characters has yet to be determined.

REFERENCES

1. D.Y. Thomas and D. Wilkie, Biochem.Biophys.Res.
   Commun. 30, 368 (1968).
2. D. Wilkie, in: Mitochondria/Biomembranes, p. 85
   North Holland, Amsterdam (1972).
3. M.F. Waxman, N. Eaton and D. Wilkie, Mol.Gen.Genet.
   submitted.
4. A. Jiminez, D.D. Tipper and D. Davies, in the
   press.
5. L.A. Grivell, L. Reijnders and H. de Vries, FEBS
   Letters 16, 159 (1971).
6. J.H. Parker and F. Sherman, Genetics 62, 9 (1969).
7. D. Wilkie and A.R. Hughes, Heredity 24, 290 (1969).
8. W.O. Kwant and P. Seeman, Biochim.Biophys.Acta
   193, 388 (1969).
9. A.R. Hughes and D. Wilkie, Biochem.Pharmacol. 19,
   2555 (1970).
10. O. Linstead and D. Wilkie, Biochem.Pharmacol. 20,
    839 (1971).
11. R.K. Mortimer and D.C. Hawthorne, in: The Yeasts
    (A.H. Rose and J.S. Harrison, Eds.). p. 386
    Acad.Press London (1969).
12. O. Lowry, N. Rosebrough, P. Farr and R. Randall,
    J.Biol.Chem. 193, 265 (1951).
13. C. Spencer, S. Symons and R. Brunt, Arch.Mikro-
    biol. 15, 246 (1971).
14. R. Estabrook, J.Biol.Chem. 236, 3051 (1961).
15. A. Tzagoloff, J.Biol.Chem. 246, 3050 (1971).

MITOCHONDRIAL GENETICS

Anthony W. Linnane, Neil Howell and H.B. Lukins

Biochemistry Department, Monash University, Clayton, Victoria, Australia.

Six years ago, we reported the first isolation and characterization of a cytoplasmically-inherited antibiotic resistance mutation in yeast (1,2). Since then, a number of other mutations of this type have been isolated in this laboratory as well as others and the discovery of these has allowed the study of yeast mitochondrial DNA (mtDNA) as a genetic system. Initially, mutants were isolated which showed increased resistance to inhibitors of mitochondrial protein synthesis such as erythromycin, paromomycin, spiramycin, chloramphenicol and mikamycin (3-6). More recently, mutants resistant to inhibitors of oxidative phosphorylation such as oligomycin, triethyltin and venturicidin have been reported (7-9).

Our approach to the problem of mitochondrial genetics has been to progressively study mitochondrial mono-, bi- and trifactorial crosses, analysing the transmission and recombination of mitochondrial genes in populations of zygotes, single zygotes and zygotic bud cells. In this paper we take the opportunity to retrospectively and prospectively consider some earlier and some recent work from our laboratory.

Monofactorial crosses. The characteristics of the transmission and segregation of single mitochondrial markers can be summarized:

1) When a strain carrying a single mitochondrial marker is crossed to a number of different strains, the frequency of transmission of the single determinant varies according to the particular strain used in the cross (5,10).

2) When a series of strains, isogenic with the ex-

ception of genetically separable mitochondrial markers, are crossed to the same sensitive strain, the transmission frequencies of the different markers may vary. For example, there is 50 % and 95 % transmission of the (ery 1-r) and (spi 4-r) markers, respectively, in crosses with the sensitive strain D253-9C (5).

3. When progeny cell samples from single yeast zygotes, rather than a large pool of zygotes, are analysed for marker transmission, marked variation among different zygotic clones is observed. Thus, the transmission of a mitochondrial marker from the zygote to its progeny cells is not an event with a fixed probability but, rather, may vary from 0-100 % in an approximately Gaussian fashion (10). Similar observations have been made in other laboratories (4,11).

This last conclusion is a particularly important one as it follows that mitochondrial genetics is a problem of population genetics and analysis of mitochondrial genetic behaviour in progeny derived from a large number of zygotes has led to some interesting results. However, it is important also to carry out analysis of zygotic clones since population analysis yield average values and do not necessarily describe the genetic events occurring in a single zygote. The importance of this point will become more apparent in later sections of this paper.

Bifactorial crosses. The use of two mitochondrial antibiotic resistance markers in crosses provides information not only on the frequency of transmission of the two individual markers but also enables mitochondrial gene recombination to be followed. In addition to the frequencies of transmission of individual markers, which are observed in monofactorial crosses, another two parameters can be defined for bifactorial crosses; one is the frequency of recombinant types among the cross progeny and the other is the polarity of recombination which is defined as the ratio of the two recombinant types (R1/R2). Recently, the results of a series of yeast crosses involving four bifactorial combinations of mitochondrial resistance markers with over a dozen antibiotic sensitive strains of di-

TABLE I

MITOCHONDRIAL BIFACTORIAL CROSSES

	RESISTANT PARENT [OLI 1-R SPI 4-R]				[OLI 1-R ERY 1-R]			
Sensitive Parent	Polarity $\frac{R1}{R2}$	Recomb. Freq. (%)	Transmission [OLI 1-R] (%)	Transmission [SPI 4-R] (%)	Polarity $\frac{R1}{R2}$	Recomb. Freq. (%)	Transmission [OLI 1-R] (%)	Transmission [ERY 1-R] (%)
Non and Low Polar					**Non Polar**			
a L2200	1.7	19.4 ± 1.4	63	58	1.10	22.6 ± 1.8	51	50
α L2300	1.7	22.2 ± 2.8	43	38	0.97	22.9 ± 2.5	54	54
a 101 - 5D	1.4	14.9 ± 3.5	63	60	0.84	17.3 ± 3.8	43	45
a 56 -28D	2.1	16.3 ± 3.2	62	56	0.83	18.2 ± 3.6	54	56
α D515- 1B	—	—	—	—	0.66	13.3 ± 3.6	55	58
					Low Polarity			
a L 300	—	—	—	—	0.55	20.4 ± 3.6	33	39
a D 609- 28C	1.3	14.5 ± 3.5	51	49	0.54	18.8 ± 3.4	42	48
a D 605- 2A	1.3	14.1 ± 4.0	72	70	0.53	17.4 ± 4.1	52	58
a D 545- 6A	—	—	—	—	0.49	15.2 ± 2.2	66	71
a D 565- 1D	—	—	—	—	0.46	16.0 ± 3.3	40	46
α L 410	—	—	—	—	0.46	13.1 ± 2.0	75	80
a 73	—	—	—	—	0.39	18.5 ± 1.7	54	62
α D 587- 48	—	—	—	—	0.35	12.2 ± 2.6	50	56
High Polarity					**High Polarity**			
a 652	0.0025	40.3 ± 1.8	53	94	0.26	26.8 ± 2.0	51	67
a D 253 - 4B	0	43.1 ± 2.6	42	85	0.042	24.8 ± 2.2	38	61
a D 253 - 4D	0	72.6 ± 3.9	23	96	0.061	17.2 ± 2.5	47	62
a D 253 - 9C	0.0028	55.6 ± 2.8	34	89	0.035	23.6 ± 1.8	26	48
α D 253 - 9D	0	72.0 ± 3.3	23	95	0.052	13.3 ± 2.5	52	64

Two parental strains, one carrying the mitochondrial gene combination [oli 1-r spi 4-r] and the other the combination [oli 1-r ery 1-r] were crossed to each of the sensitive strains listed. The progeny of the crosses were analysed for polarity of recombination (the ratio of the two recombinant classes R1 & R2), the frequency of recombination (the proportion of the total recombinant genotypes among the progeny) and the transmission of the individual resistance markers.

Crosses classed as non-polar do not significantly differ from 1.0 in terms of polarity of recombination.

verse origin were reported (12). Table I is a summary of our results for two of the marker combinations examined. Each of the four parameters defined above varied significantly depending both upon the mitochondrial resistance markers followed and the parental sensitive strain used. It is particularly important to note that polarity of recombination, although having a discrete and reproducible value in any single cross, can assume many different values and for these crosses ranges from 2.1 to less than 0.003. The evidence for multiple polarity values is most compelling for those crosses involving the (oli 1-r ery 1-r) markers in which an almost continuous range of values from 1.10 to 0.035 is observed. For a detailed discussion of correlates between the parameters observed in these bifactorial crosses, see ref. 12.

Clearly, the properties of mitochondrial genetics are complex but recent experiments have clarified the situation, at least in part, by resolving the nature of the genetic determinants of polarity and frequency of recombination. Thus, the polarity of recombination is determined by mitochondrial genes while the frequency of recombination is genetically determined primarily by the mitochondrion but is also subject to nuclear genetic influence. Table II outlines the experimental procedures used and data obtained supporting this conclusion. In a cross of a tester strain carrying the (oli 1-r ery 1-r) markers and antibiotic sensitive strain L2300, the polarity of recombination among the cross progeny is 1.0 while the frequency of recombination is 23 %. However, when a tester strain is crossed to strain D545-6A, the polarity of 0.49 and recombination frequency is 15 %. To analyse the genetic factors determining these differences, the sensitive strains L2300 and D545-6A are crossed and a number of diploid clones isolated and sporulated. Spore cultures from the resulting tetrads were then crossed to the (oli 1-r ery 1-r) tester strains and these crosses analysed for their mitochondrial genetic behaviour. The results (Table II) show that tetrads from some diploids had all four spores yielding low polari-

TABLE II

CYTOPLASMIC GENETIC DETERMINATION OF POLARITY AND FREQUENCY OF RECOMBINATION

Cross		Polarity of Recombination	Recombination Frequency	Meiotic Segregation
Tester Strain [OLI 1-R ERY 1-R]	x L2300	1.0	23%	
Tester Strain [OLI 1-R ERY 1-R]	x D545-6A	0.49	15%	

Tetrad Spore Cultures From L2300 x D545-6A

Cross		Polarity of Recombination	Recombination Frequency	Meiotic Segregation
Tester Strain [OLI 1-R ERY 1-R]	x A1	0.61	15%	0 L2300 :4 D545-6A
"	x B1	0.46	11%	
"	x C1	0.58	11%	
"	x D1	0.63	15%	
"	x A2	1.2	23%	4 L2300 : 0 D545-6A
"	x B2	1.4	22%	
"	x C2	1.6	20%	
"	x D2	1.0	20%	

The antibiotic sensitive strains L2300 x D545-6A were crossed and the resultant diploids sporulated to yield tetrads in which the four spores are designated A, B, C & D. These sensitive spore cultures were then crossed to the tester strains and the crosses analysed for the polarity and recombination values characteristic of strains L2300 or D545-6A as shown in the upper half of the Table. Two types of tetrads were found to result from the cross L2300 x D545-6A; in one type (A2, B2..)only the genetic characteristics of strain L2300 were recovered while in the other type (A1, B1..) only the genetic characteristics of strain D545-6A were found.

ty (0.5–0.6) and 11–15 % recombination in crosses to
the tester strains. Other diploids yielded tetrads in
which all spores had no polarity of recombination (1.0–
1.6) and 20–23 % total recombination. The non–Mendelian segregation in tetrads and vegetative segregation among diploids indicate that cytoplasmic genes
determine the polarity and frequency of mitochondrial
gene recombination for strains L2300 and D545-6A. While there is some variation in the mitochondrial genetic behaviour among spore cultures of a tetrad, the
behaviour is not significantly different from the parental phenotype. Hence, the mitochondrial genetic
characteristics of these strains behave as discrete
hereditary characteristics; apparently, a single cytoplasmic gene in each strain determining both polarity and frequency of recombination. However, when
a similar type of analysis involving sensitive strains
L2200 and D253-9D is performed, evidence for a nuclear
genetic influence on the frequency of mitochondrial
gene recombination is obtained. In crosses to the tester strains, L2200 has a polarity of 1.10 and recombination frequency of 22 % while the values for D253-
9D are 0.52 and 13 %. When these two sensitive strains
are crossed and isolated diploids sporulated, tetrads
are recovered in which all four spores show no polarity of recombination in backcrosses to the tester
strains. However, only two spores in each tetrad have
22–23 % recombination while the other two have 13–15 %
recombination; these results show that strain D253-9D
carries a nuclear gene influencing mitochondrial gene
recombination.

Genetic analysis of the type described above have
been applied to a series of strains (L2200, L2300,
L410, D545-6A, D609-28C, D587-4B, 652 and D253-9D with a
number of different recombination frequencies and polarities and a summary of these analyses presented in
Table III. For all these strains, it is concluded that
polarity of recombination (*viz.* no low or high) is under cytoplasmic genetic control while frequency of recombination is also determined by cytoplasmic genes
but can be modified by nuclear genes. Furthermore,

TABLE III.

SUMMARY OF STUDIES ON GENETIC CONTROL OF MITOCHONDRIAL GENE RECOMBINATION.
The data summarized in this table are derived from experiments of the type outlined in Table II. The values in parentheses following each strain are the polarity of recombination and frequency of recombination characteristic of that strain
as determined by crosses with antibacterial resistant tester strains. The sensitive strains as listed were crossed and diploids sporulated to yield tetrad spore
cultures. The segregation in these tetrads of the characteristics of polarity and
frequency of recombination was determined by crosses to the antibiotic resistant
tester strains.

Cross		Meiotic Segregation	
		Polarity	Recombination
L2200 (1.10 − 23 %) x L410 (0.46 − 13 %)		Cytoplasmic	Cytoplasmic
L2300 (1.0 − 23 %) x D545 − 6A (0.49 − 15 %)		Cytoplasmic	Cytoplasmic
L2300 x D609 −28C (0.54 − 19 %)		Cytoplasmic	−
L2200 x D587 − 4B (0.35 − 12 %)		Cytoplasmic	Cytoplasmic
L2300 x 652 (0.26 − 27 %)		Cytoplasmic	−
L2200 x D253 (0.035 − 17 %)		Cytoplasmic	Nuclear

these analyses show that there are a number of distinct, heritable polarity of recombination values and we conclude that several mitochondrial genes (or alleles of the same gene) control this phenomenon. This conclusion differs with that of Slonimski and coworkers (13) who concluded that mitochondrial gene polarity was determined by a ω^+/ω^- sex factor system.

The cytoplasmic genes which determine polarity and frequency of recombination are located within the mt-DNA itself; this conclusion is based on the observation that the cytoplasmic gene determining no polarity in strain L2200 is eliminated by ethidium bromide treatment, a mutagen used to selectively delete mitochondrial genes (2).

One of the present dilemmas in the understanding of this genetic system comes from a zygotic clonal analysis of some of the crosses shown in Table I. Clearly, there are mitochondrial genes which determine both polarity and frequency of recombination and the values obtained are characteristic of a particular gene and are highly reproducable from one cross to the next. However, it will now be shown that these values for polarity and frequency of recombination are *population* values and do not predict the values for zygotic clones. Rather, these mitochondrial genes determine the Poisson mean of any genetic parameter about which the values of the zygotic clones are distributed but they do not determine a unique value for the individual zygotes. Recently, we reported on the mitochondrial genetic analysis of zygotic clones isolated from a non-polar bifactorial cross and as with the monofactorial crosses, there was a marked interclonal variability for any of the parameters analysed (14). Thus, although population analysis gave a polarity ratio of R1/R2 of 1.3 (that is, non polar), analysis of clones from single zygotes yielded polarity values ranging from 8.2 to 0.005 (Table IV). About 50 % of these clones had R1 > R2 while the other 50 % had R1 < R2. Similarly, even though the population value in a high polarity cross was only 0.035, there was a range of polarity values for zygotic clones from 6.0 to zero, but only

TABLE IV

DISTRIBUTION OF MITOCHONDRIAL POLARITY VALUES AMONG SINGLE YEAST ZYGOTES

(a)

NON-POLAR CROSS	POLAR CROSS
R_1/R_2	R_1/R_2
$R1/_0$ (3)	-
8.2	6.0
3.3	0.71
2.6	0.49
1.9	0.27
1.8	0.26
1.7	0.13
——	0.11
0.89	0.07
0.71	0.04
0.38	0.03
0.21	0.02
0.10	$0/R_2$ (31)
0.04	
0.02	
0.005	
$0/R_2$ (3)	
Average 0.89 (20)	0.037 (42)

TRANSMISSION[b]

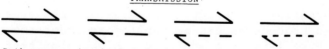

(a)   Both crosses involved analysis of recombination between the [oli 1-r] and [ery 1-r] markers.   The numbers in parenthesis are the number of zygotic clones showing that particular polarity.

(b)   A diagramatic representation of polarity in zygotic clones of bifactorial crosses.   The solid arrows at the far left represent the approximately equal occurrence of both directions of polarity among clones.   The middle pairs of arrows represent low polarity crosses in which clones with one direction of polarity significantly exceed clones with the other direction.   The diagram at far right represents a high polarity cross in which very few clones show the other direction.

201

one clone had R1 > R2 with 49 having R1 < R2 (31 had
R1 = 0) as shown in Table IV. These findings will be
reported in greater detail elsewhere (Howell, Linnane
and Lukins, manuscript in preparation).

The two crosses showing polarity values in zygotic
clones (Table IV) represent the two extremes of polarity of recombination. For non-polar crosses, zygotic
clones showing either direction of polarity occur with
equal frequency, this is diagrammatically illustrated
by the arrows at the bottom far left of Table IV. In
contrast, zygotic clones from a high polarity cross
show only one direction of polarity although clones
showing the other direction are occasionally observed
(arrows at bottom far right). However, between these
two extremes are other crosses, such as the low polarity crosses of Table I, in which the preference in
zygotic clones for one direction of polarity is significant but not as strong as for the high polarity
crosses so that clones with both directions of polarity are found but one direction predominates (middle
sets of arrows). Thus, we envisage that there are not
two simple classes of mitochondrial genetic crosses
but many classes, or perhaps there are not distinct
classes but a continuous range of values, depending on
the degree of preference for one direction of polarity.

Trifactorial crosses. The principal advantage of
mitochondrial trifactorial crosses is that classical
genetic procedures can be applied to order the markers. Fig. 1 is an example of a high polar trifactorial cross where all three pairs of reciprocal recombinants showed marked polarity values. Moreover, by
classical three-point mapping procedures (15) it is
evident that the (ery 1-r) marker is the central gene
in the order (oli 1-r ery 1-r cap 1-r). This order is
inferred since the R5/R6 class of recombinants occurs
with a much lower frequency than either the R1/R2 or
R3/R4 classes and is presumed to represent a double
recombination event while the latter classes were presumably generated by single events (if the mtDNA molecules recombine as circles, these frequencies would

H 513   [OLI 1-R   ERY 1-R   CAP 1-R]   X   D253-9C [OLI-S   ERY-S   CAP-S]

TRANSMISSION		RECOMBINATION

% CAP-R = 68 ± 3 $\dfrac{R1}{R2}$ %
$\begin{array}{ccc} \text{OLI-S} & \text{ERY-R} & \text{CAP-R} \\ \text{OLI-R} & \text{ERY-S} & \text{CAP-S} \end{array}$ = $\begin{array}{c} 12\cdot8 \\ 0\cdot3 \end{array}$ (13·1)

% ERY-R = 42 ± 3 $\dfrac{R3}{R4}$ %
$\begin{array}{ccc} \text{OLI-S} & \text{ERY-S} & \text{CAP-R} \\ \text{OLI-R} & \text{ERY-R} & \text{CAP-S} \end{array}$ = $\begin{array}{c} 25\cdot2 \\ 0\cdot1 \end{array}$ (25·3)

% OLI-R = 30 ± 2 $\dfrac{R5}{R6}$ %
$\begin{array}{ccc} \text{OLI-R} & \text{ERY-S} & \text{CAP-R} \\ \text{OLI-S} & \text{ERY-R} & \text{CAP-S} \end{array}$ = $\begin{array}{c} 0\cdot1 \\ 0\cdot7 \end{array}$ (0·8)

POLARITY OF TRANSMISSION   POLARITY OF RECOMBINATION

UNAMBIGUOUS   MAP ORDER

OLI-R   ERY-R   CAP-R →

Fig. 1. Example of a high polarity bifactorial cross. The transmission frequencies are expressed as the mean *per cent* (± 95 % confidence interval) of progeny diploids carrying the particular resistance marker. The frequencies of reciprocal recombinant types are shown at the right and all classes show marked polarities of recombination. The relative frequency of occurrence of recombinant class R5/R6 is taken as arising from a double recombination event while the classes R1/R2 and R3/R4 are generated by single events. Thus, a unique order of the three markers is obtained by three-point analysis and is shown at the bottom. The same order was obtained from the relative frequencies of transmission.

represent double and quadruple events but the interpretation is the same).

An order for these markers can also be obtained from the transmission values, again the (ery 1-r) marker is central while the (cap 1-r) marker shows the highest transmission. Thus, in high polarity crosses, mitochondrial markers are easily and unambiguously mapped. This type of analysis has been extended to a number of other antibiotic resistance markers and a map of mitochondrial genes constructed (Fig. 2).

In contrast to the high polarity crosses, non-polar trifactorial crosses do not yield an unambiguous order of mitochondrial markers by either three-point analysis of transmission frequencies. Fig. 3 is an example of a non-polar cross; all three pairs of recombinant types show no significant polarity and there are also no significant differences in transmission frequencies among the three markers. No pair of recombinants occurs with a low frequency expected of a double recombination event in a unique linear order of the

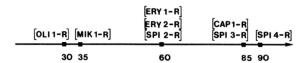

Fig. 2. A map of mitochondrial antibiotic resistance markers based on the frequencies of transmission in high polarity crosses. The map was derived from analysis of marker transmission frequencies in a series of crosses involving strains isogenic with the exception of two or three mitochondrial resistance mutations which were all crossed with the antibiotic sensitive strain D253-9C. The markers are defined by resistance to the antibiotics oligomycin (OLI), mikamycin (MIK), erythromycin (ERY), spiramycin (SPI) or chloramphenicol (CAP). The numbers beneath the line are the absolute frequencies of marker transmission expressed as the *per cent* of progeny diploids carrying the particular resistance determinant.

H 513   [OLI 1-R   ERY 1-R   CAP 1-R]   X   H5136   [OLI-S   ERY-S   CAP-S]

TRANSMISSION   RECOMBINATION

% CAP-R = 45 ± 3	$\frac{R1}{R2}$	% OLI-S OLI-R	ERY-R ERY-S	CAP-R = CAP-S	$\frac{6 \cdot 9}{5 \cdot 4}$ (12·3)
% ERY-R = 43 ± 3	$\frac{R3}{R4}$	% OLI-S OLI-R	ERY-S ERY-R	CAP-R = CAP-S	$\frac{3 \cdot 4}{2 \cdot 4}$ (5·8)
% OLI-R = 42 ± 3	$\frac{R5}{R6}$	% OLI-R OLI-S	ERY-S ERY-R	CAP-R = CAP-S	$\frac{1 \cdot 8}{2 \cdot 6}$ (4·4)

NO POLARITY OF TRANSMISSION   NO POLARITY OF RECOMBINATION

NO UNIQUE MAP ORDER

OLI-R   ERY-R   CAP-R

Fig. 3. Example of a non-polar trifactorial cross. The transmission frequencies are expressed as the mean *per cent* (+ 95 % confidence interval) of progeny diploids carrying the particular resistance marker. The frequencies of reciprocal recombinant types are shown at the right; no class of recombinant shows a significant polarity value. The recombination frequencies for classes R3/R4 (5.8 %) and R5/R6 (4.4 %) are statistically indistinguishable and, thus, do not allow an ambiguous order of markers to be derived by three-point analysis. Moreover relative marker transmission frequencies also do not permit a unique order to be established.

three markers; rather, of the reciprocal recombinant classes, R3/R4 and R5/R6 have the lower values which are not statistically distinguishable so that an unambiguous order in this cross is not obtained. An ambiguity of mitochondrial markers in low polarity trifactorial crosses has also been reported by Kleese *et al.* (16).

An interpretation linking ambiguous map orders in non-polar trifactorial crosses with the observation of different directions of polarity among zygotic clones of non-polar bifactorial crosses is presented in Fig. 4. We envisage that in single zygotes there is an ordered transmission of markers to the progeny cells and that for non-polar crosses there is more than a single possible origin of transmission on each parental mitochondrial genome and that single zygotes may have any one of several possible orders of marker transmission. Therefore, in the progeny of many zygotes, there is no class of recombinants which would have arisen solely as the result of double recombination events. Rather, any recombinant type can have arisen from either a double or single event depending upon the order of transmission in the zygote. For high polarity trifactorial crosses, one can derive a unique map order (by both marker transmission frequencies and three-point mapping procedures) since there is a preferred origin of transmission in zygotes and, hence, a class of recombinants generated solely by two recombination events occurs. It is assumed in this model that the mitochondrial genome is circular but that markers are transmitted to zygote progeny in such a manner that a linear map is obtained.

The hypothesis of multiple origins of transmission has been supported by recent experiments done in our laboratory. Zygotic clones of a non-polar trifactorial cross were examined for the relative order of the markers based on transmission frequencies. The results are shown in Fig. 5. Of the 24 clones analysed, seven had equivalent transmission of all three markers (within the limits of statistical resolution) while another 10 had equivalent transmission of two markers.

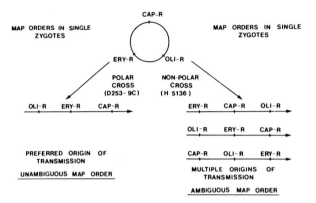

Fig. 4. A model to explain the results of highly polar and non-polar trifactorial crosses. It is envisaged that in single zygotes there is a polarized transmission of mitochondrial markers to the progeny cells. For non-polar crosses, there are several possible origins of marker transmission such that zygotic clones from this cross will show several different orders of mitochondrial markers. In contrast, for high polarity crosses there is a preferred origin of transmission such that zygotic clones have a single order of markers.

ZYGOTE	%[OLI 1-r]	%[ERY 1-r]	%[CAP 1-r]	
5310-2	8±3	2±2	0·4±0·8	CAP ERY — OLI
-19	58±5	29±5	25±5	
-32	45±6	29±5	30±6	
-8	15±4	29±3	42±6	OLI ERY CAP
-3	41±6	78±5	88±4	OLI ERY CAP
-12	62±7	44±7	69±7	ERY OLI CAP
-28	26±5	40±6	49±6	OLI ERY CAP
-21	50±6	68±5	55±6	OLI CAP ERY
-7	0	96±2	100	ERY CAP

Fig. 5. Transmission frequencies (± 95 % confidence intervals) of mitochondrial markers in zygotic clones derived from the non-polar trifactorial cross shown in Fig. 3. The order of marker transmission for these clones, based on the relative transmission frequencies, are diagrammatically represented by the arrows at the right. Note that no one order of mitochondrial markers is found among all the zygotic clones.

Another seven clones yielded a unique marker order which in each case was (oli 1-r ery 1-r cap 1-r). Of the clones with equivalent transmission of two markers, four showed equivalence between the (cap 1-r) and (ery 1-r) markers, five between the (cap 1-r) and (oli 1-r) markers and one showed equivalence of the (oli 1-r) and (ery 1-r) markers. Such results are not compatible with all clones having the same order of mitochondrial markers and, thus, support our model of multiple origins of transmission in non-polar crosses.

Gene retention in multiple marked strains. Another approach directed at establishing linkage relationships between mitochondrial markers has involved the analysis of mitochondrial marker retention following petite mutation in multiply marked strains. Since petite mutations involve deletions of the mtDNA (for review, see ref's. 17,18), this experimental approach is analogous to deletion analysis. One interesting result from these studies (Molloy, Linnane and Lukins, manuscript in preparation) is that the deletion process is not random but, rather, appears to involve loss of specific regions of the mtDNA. As an example, of the 226 petite strains isolated from a strain carrying the (oli 1-r ery 1-r cap 1-r) markers, 108 had lost all three, 80 had retained all three, 18 were of the type (ery 1-r cap 1-r) and 20 were of the type (oli 1-r). Thus, the (cap 1-r) and (ery 1-r) markers are extremely difficult to separate by deletion although either is easily separated from the (oli 1-r) marker. However, all three markers are readily separated from each other by recombination (Fig. 1 and 3). The petite mutations is phenotypically characterized by a loss in mitochondrial protein synthesis so that it is possible to envisage that the (cap 1-r) and (ery 1-r) mutation occurs within the same cistron, in this case possibly the gene coding for the 21S RNA of the large mitoribosomal subunit, and that the petite deletions generally cover the entire cistron. The correlation between physical map distances obtained by recombination analysis warrants further study.

DISCUSSION.

The work of our laboratory, briefly reviewed here, as well as that of other laboratories has defined many of the basic features of yeast mitochondrial genetics. As yet, however, there has been only limited success in integrating these features into a satisfactory theoretical framework. In discussing mitochondrial gene recombination, Slonimski and colleagues (13) have introduced the concept of mitochondrial sex factors and have pointed out the similarities between their model and the transmission and recombination of bacterial genes in Hfr x F⁻ conjugation. Our results, on the other hand, indicate a number of similarities between mitochondrial genetics and the genetics of coliphages such as the T-even and *lambda* viruses (for reviews, see ref's. 15,19,20). In the following paragraphs, we discuss the appropriateness of phage genetic systems as models for yeast mitochondrial genetics.

Population genetics. The most striking result in our studies was the finding that zygotic clones showed a marked variation for any mitochondrial genetic parameter analysed and, thus, mitochondrial genetics is a problem in population genetics. The population nature of the phage systems has long been recognized and is observed when a bacterium is multiply infected with genetically different phages which are replicated to establish a genetic pool and then subsequently undergo recombination in this pool and may be withdrawn from the pool to form mature phage particles. Phage genomes in this pool can either undergo no rounds of recombination, only one round or several rounds, but once withdrawn from the pool, the possibility of further recombination ceases. Yeast zygotes contain many mitochondria from each parent and each mitochondrion may contain several mtDNA molecules (for reviews, see ref's. 17,18). These mitochondrial genomes must become physically associated for recombination to occur and, hence, a pool of genomes must be formed. The nature of this pool and its possible role in mitochondrial gene transmission and recombination is discussed below.

Switch areas. Phage genomes in the replicative pool undergo, on average, from one (T-odd and *lambda*) to five (T-even) rounds of recombination during the normal lytic cycle. Moreover, these recombination events are not distributed randomly throughout the genome but are clustered in specific genetic regions (switch areas); these regions are the chromosomal termini in the case of T4 infection, With the mitochondrial markers (oli 1-r spi 4-r) in high polarity crosses, it is observed that the recombination frequency greatly exceeds the frequency of the minority parent (*viz.* the (oli-s spi-s) parent) among the progeny and this result is that expected if there are mitochondrial switch areas. However, it should be pointed out that similar genetic behaviour could occur of there is a single, *obligatory* recombination event near these markers. It is significant to note that the (spi 4-r) marker is very close to the origin of transmission (Fig. 2) while the (oli 1-r) and (ery 1-r) markers are more distal, thus there is evidence that multiple rounds of recombination occur near the origin of transmission in high polarity crosses. As might be predicted, high polarity crosses involving the (oli 1-r ery 1-r) markers do not result in the frequency of recombination exceeding the frequency of occurrence of the minority parent among the progeny; this type of behaviour is expected for regions undergoing at most a single round of recombination. Another characteristic of multiple rounds of recombination among phage genomes is high negative interference between closely linked markers; consequently this phenomenon should be looked for in mitochondrial genetic crosses.

Polarity. Phage genetic crosses can be non-polar or polar depending on the experimental conditions. If the multiplicities of infection of the two parental phages are equal then the cross appears to be non-polar when a mass lysate (analogous to a population of yeast zygotes) is analysed. In contrast, if the multiplicities of infection are different (*e.g.* a ratio of 10:1), then there will be a polarity of recombinant types among the progeny (about 10:1 in this example). However, the polarities of recombination for the phage crosses are population values and single burst experiments (analogous to yeast zygotic clonal analysis) show marked variation about the mean. The occurrence in phage of population polarity values about which there is clonal variation is strikingly similar to the results reported here for yeast mitochondrial crosses. It should be remembered, however, that for yeast mitochondrial crosses, the polarity value may be less than 0.002; it is difficult to envisage a model for mitochondrial genetics in which the ratio of the two parental genomes in the zygote is 0.002; is is much more likely that there is a selection process which operates to produce such non-reciprocity of recombinant types.

Thus far we have simply compared the basic features of yeast mitochondrial genetics to the coliphage genetic systems. Since, however, the molecular mechanisms governing the processes in the mitochondrial system are not yet known, it seems of some heuristic value to extrapolate further from the phage genetic systems to the mitochondrial systems in an attempt to provide some possibilities for such mechanisms. We envisage that the ratio of the two mitochondrial parental types in the zygote pool determines the polarity of mitochondrial gene recombination. The dilemma is in explaining how a mitochondrial gene determines this population value for polarity and why there is such a broad distribution about it upon clonal analysis. As pointed out above, the number of mitochondrial genomes in a zygote must be quite large, if all these genomes had an equal probability for replication and entry into the pool, then a large variability among zygotic clones would not seem likely. It is more plausible that there is a selection process whereby only a few genomes are chosen for replication or entry into the pool. If one parental type has a greater selective advantage, the cross will show significant polarity of recombinant types among the progeny. The greater the difference in selective advantage of one parental type, the more polar the cross; if the overall selective advantages of the two parental types are equal, then the population polarity would be one (*i.e.*, non-polar). One interesting possibility to consider may be that only membrane-bound mitochondrial genomes are replicated and that there are a limited number of such attachment sites; this phenomenon does occur in ΦX-174 infection (21). Thus, it has been shown that in *E. coli* infected by ΦX-174, the phage replicate to form a pool of parental replicative forms and that some of these (from one to four depending on the *E. coli* strain used) become membrane-associated and it is only these attached genomes which produce progeny phage genomes.

Mitochondrial genes determine the polarity of recombination but the frequency of recombination we have shown to be determined by both mitochondrial and nuclear genes. How these processes occur is unknown at present but continuing the ΦX-174 analogy, it may be envisaged that these genes determine the number of attachment sites or the association between the membranes and these sites.

The term "origin" of mitochondrial marker transmission has been used throughout this paper in a purely operational sense to indicate that in high polarity crosses and zygotic clones generally there is a linear order of mitochondrial markers with specific and characteristic transmission values. This origin could be an origin of replication, a switch area, an Hfr-like sex factor, the site of a chromosome terminus or several of these. Until the mechanism of mtDNA replication is elucidated, the resolution of this question seems unlikely but continuing the phage genetic analogy, it might be envisaged that the selection process described above also establishes a switch area of multiple recombination events. Recombination between a selected and an unselected genome would be apparently non-reciprocal because only the selected genome would be replicated. For non-polar crosses, the switch area could be established at any one of several genetic regions (explaining the presence of multiple origins) whereas high polarity crosses would show a single region for the switch area. In this discussion, we have proposed the suitability of phage genetic systems as a model for the understanding of yeast mitochondrial genetics.

REFERENCES.

1. A.W. Linnane, J.M. Haslam, Curr.Top.Cell Reg. 2, 101 (1970).
2. A.W. Linnane, J.M. Haslam, H.B. Lukins, P. Nagley, Ann.Rev.Microbiol. 26, 163 (1972).
3. D.Y. Thomas, D. Wilkie, Biochem.Biophys.Res.Commun. 30, 368 (1968).
4. D. Coen, J. Deutsch, P. Netter, E. Petrochilo, P.P. Slonimski, Soc.Exp.Biol.Symp. 24, 449 (1970).
5. M.K. Trembath, C.L. Bunn, H.B. Lukins, A.W. Linnane, Mol.Gen.Genet. 121, 35 (1973).
6. N. Howell, P.L. Molloy, A.W. Linnane, H.B. Lukins, Mol.Gen.Genet., in press (1973).
7. P.R. Avner, D.E. Griffiths, Eur.J.Biochem. 32, 301 (1973).
8. P.R. Avner, D.E. Griffiths, Eur.J.Biochem. 32, 312 (1973).
9. C.L. Bunn, C.M. Mitchell, H.B. Lukins, A.W. Linnane, Proc.Nat.Acad.Sci. USA 67, 1233 (1971).
10. G.W. Saunders, E.B. Gingold, M.K. Trembath, H.B. Lukins, A.W. Linnane, in:"Autonomy and Biogenesis of Chloroplasts and Mitochondria", N.K. Boardman, A.W. Linnane, R.M. Smillie, Eds., North Holland Publ., Amsterdam, 185 (1971).
11. G.H. Rank, N.T. Bech-Hansen, Genetics 72, 1 (1972).
12. N. Howell, M.K. Trembath, A.W. Linnane, H.B. Lukins, Mol.Gen.Genet. 122, 37 (1973).
13. M. Bolotin, D. Coen, J. Deutsch, B. Dujon, P. Netter, E. Petrochilo, P.P. Slonimski, Bull.Ins.Pasteur 69, 215 (1971).
14. H.B. Lukins, J.R. Tate, G.W. Saunders, A.W. Linnane, Mol.Gen.Genet. 120, 17 (1973).
15. W. Hayes, in:"The Genetics of Bacteria and their Viruses", Blackwell Scientific Publications, Oxford (1968).
16. R.A. Kleese, R.C. Grotbeck, J.R. Snyder, J. Bac-

teriol. <u>112</u>, 1023 (1972).

17. P. Borst, Ann.Rev.Biochem. <u>41</u>, 333 (1972).
18. P. Nagley, A.W. Linnane, in:"Advances in Molecular Genetics", I.R. Falconer and D. Cove, Eds., Paul Elek Ltd., London, in press (1973).
19. G. Mosig, Adv.Genet. <u>15</u>, 1 (1970).
20. E. Signer, in:"The Bacteriophage Lambda", A.D. Hershey, Ed., Cold Spring Harbor Laboratory, New York, 139 (1971).
21. R.L. Sinsheimer, R. Knipper, T. Komano, Cold Spring Harbor Symp.Quant.Biol. <u>33</u>, 443 (1968).

MITOCHONDRIAL GENES AND ATP-SYNTHETASE

D.E. Griffiths, R.L. Houghton and W.E. Lancashire

Department of Molecular Sciences, University of Warwick, Coventry, England.

In our laboratory we have been carrying out a systematic biochemical genetic study of the oxidative phosphorylation complex in yeast utilising mitochondrial drug resistant mutants (1-5). The rationale of these investigations has been discussed previously and is based on the following premises:
1. Specific inhibitors of the oxidative phosphorylation complex have specific inhibitor sites associated with specific protein subunits of the oxidative phosphorylation complex;
2. Inhibitor resistant mutants should exhibit a modified inhibitor binding site and/or modified enzyme activity;
3. Demonstration that the resistance allele is cytoplasmically inherited and located on mtDNA is good *a priori* evidence that a mitochondrial gene product is a component of the oxidative phosphorylation complex.
The investigation is analogous to investigations of ribosome structure and function utilizing mutants resistant to antibiotics which inhibit protein synthesis. The mitochondrial oxidative phosphorylation complex is particularly suitable for these studies as we are dealing with a system which interacts with several specific inhibitors which interact with different specific components in the complex. The multi-enzyme nature of the complex is illustrated in Fig. 1 which emphasises the integration of phosphate transport, adenine nucleotide translocation and proton and K^+ transport and the specific inhibitors of these systems.
The study of mitochondrial drug resistance may pro-

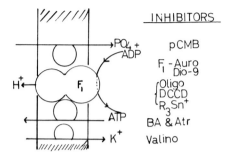

Fig. 1. Schematic representation of the oxidative phosphorylation complex and the specific site of action of various inhibitors.

vide a means of establishing which components are synthesized in the mitochondrion and also provide information on the interaction of binding sites and their organization in the mitochondrial membrane. In particular it may provide new information on components of enzyme complexes which have not been adequately characterized as yet. A list of mitochondrial inhibitors and the mitochondrial mutants that have been isolated is summarized in Table I together with the available information as to the nuclear or mitochondrial origin

TABLE I

MITOCHONDRIAL INHIBITORS AND MITOCHONDRIAL DRUG RESISTANT MUTANTS.

Inhibitor	Site	Resistant mutant	Resistant allele
Mercurials	Phosphate carrier	No	?
Aurovertin	F_1 ATPase	Yes	Nuclear
Dio-9	F_1 ATPase	Yes	Nuclear
Oligomycin)	Membrane	Yes	Mitochondrial
Rutamycin)	Subunits of	Yes	Mitochondrial
Peliomycin)	OS ATPase	Yes	Mitochondrial
Ossamycin)		Yes	Mitochondrial
Venturicidin	"	Yes	Mitochondrial
DCCD	"	Yes	?
Triethyl tin	"	Yes	Mitochondrial
Triethyl tin	ADP translocase	Yes	Mitochondrial
Bongkrekic acid	ADP translocase	Yes	(Mitochondrial)?
uncouplers	?, proton transport	Yes	?
"1799"	?, proton transport	Yes	Mitochondrial
Valinomycin	?, K^+ transport	Yes	(Mitochondrial)

of the resistance allele. This paper will deal with studies of oligomycin, venturicidin and triethyl tin resistance and their relationship to the hydrophobic components of ATP-synthetase shown by Tzagoloff (6) to be synthesized on mitoribosomes and known to be the site of inhibitor sensitivity.

The isolation of specific oligomycin and triethyl tin-resistant mutants and their genetic characterization as cytoplasmic mutants in which the resistance determinant is located on mtDNA has been described previously (1-4). Venturicidin resistant mutants were isolated by similar techniques and characterized as described previously (2,4). Allelism tests on specific oligomycin mutants indicated the presence of at least two cytoplasmic loci conferring the oligomycin resistance phenotype (4). Detailed genetic analysis (7) of mitochondrially coded OL^R mutants isolated at Warwick and at Gif show that they belong to two mitochondrial loci designated OL_I and OL_{II}. It is not yet possible to decide whether the two OL loci represent independent cistrons coding for distinct gene products or for sites within a single cistron. However, it is inferred that the OL_I and OL_{II} loci represent two distinct genes (7).

BIOCHEMICAL STUDIES.

We have previously presented evidence that the oligomycin resistance phenomenon is expressed *in vitro* at the inner mitochondrial membrane level (1). Studies of mitochondrial ATPase and P_i-ATP exchange reactions in mitochondria, submitochondrial particles and solubilized ATPase preparations show that preparations from mutant strains have a *decreased* sensitivity to oligomycin (*i.e.* increased resistance). These findings suggest that we are not dealing with a permeability barrier at the plasma membrane or mitochondrial membrane level but that a modification of the binding sites for oligomycin is involved (1,8).

Table II shows that OL^R mutants exhibit a decreased sensitivity to oligomycin at the mitochondrial ATPase level. This effect is specific for oligomycin and ana-

TABLE II

EFFECT OF INHIBITORS OF ATP-SYNTHETASE [I_{50} VALUES (µg/mg PROTEIN)].
Values quoted are I_{50} values for the effect of inhibitors on mitochondrial ATPase
(1). pH 9.5, 30°. (R) and (S) signify the *in vivo* resistance or sensitivity to the
inhibitor.

	oligo-mycin	dio-9	auro-vertin	triethyl-tin	venturi-cidin
D22 OL^S	2.1(S)	49(S)	60-70(S)	0.13(S)	0.13(S)
CLASS I					
D22 B9 OL^R	5.4(R)	65(R)	-	0.10(R)	0.14(R)
D22 A7 OL^R	3.3(R)	50(R)	-	0.35(R)	0.19(R)
CLASS II					
D22 A16 OL_I^R	17.0(R)	36(S)	-	0.26(S)	0.25(S)
D22 A21 OL_I^R	19.0(R)	26(S)	40-50(S)	0.17(S)	0.17(S)
D22 A13 OL_{II}^R	5.5(R)	38(S)	-	0.17(S)	0.19(S)
D22 A15 OL_{II}^R	9.7(R)	19(S)	-	0.15(S)	0.18(S)

logues and is not observed with triethyl tin or venturicidin. Of particular interest are the different
levels of sensitivity exhibited by OL_I and OL_{II} mutants. OL_I mutants exhibit a 9-15 fold increase in
resistance to oligomycin whereas OL_{II} mutants exhibit
only a 3-5 fold increase in resistance. This parallels
the situation *in vivo* where OL_I mutants are 3-8 fold
more resistant than OL_{II} mutants.

A further means of discriminating between OL_I and
OL_{II} mutants comes from examination of the pH profiles
of ATPase activity of OL_I and OL_{II} mutants. The pH
profile for mitochondrial ATPase activity exhibits a
main peak at pH 9-9.5 and a minor peak at pH 6.2-6.5
the pH 9.5/pH 6.5 activity ratios being in the range
5-6. Profiles of OL_{II} mutants are, in general, similar to the wild type with pH 9.5/6.5 ratios of 4-6.
pH profiles of OL_I mutants show marked differences
with an increase in pH 6.5 activity such that the pH
9.5/6.5 ratios are decreased to 1.8-2.2. The mutation
in OL_I mutants is of such a nature to cause a conformational change in the ATPase complex or it has changed a charged group at the oligomycin binding site
(possibly a histidine residue ?).

SPECIFICITY OF RESISTANCE *IN VIVO* AND *IN VITRO*.

The specificity of the OL^R mutants with respect to inhibitors such as triethyl tin, venturicidin, aurovertin and Dio-9, *in vivo* and *in vitro*, have been discussed previously (Table II and ref. 1). Three other antibiotics which show similar structural and inhibition properties to oligomycin have been tested for cross-resistance. Rutamycin, Peliomycin and Ossamycin show marked cross-resistance both *in vivo* and *in vitro* and show marked similarities in their mode of action to oligomycin. Venturidicin shows no cross resistance to oligomycin in all OL^R mutants examined except for one particular mutant, D22 A19. However, it should be pointed out that apparent differences in the phenotype of resistance *in vivo* and *in vitro* are not indicative of the genetic localization or allelism relationship of the mutations concerned. In addition, heteroalleles at the same locus can exhibit different phenotypes and this is well described in studies of yeast

TABLE III

CROSS RESISTANCE PATTERNS FOR TRIETHYL TIN RESISTANCE (T^R) AND VENTURICIDIN RESISTANCE (V^R).
All strains are sensitive to inhibitors when tested *in vivo*. (ATPase assay, pH 9.5, 30°; *cf.* ref.1).

Strain	Antibiotics				
	triethyl-tin	venturi-cidin	oligo-mycin	"1799"	bong-krekic
D22	S	S	S	S	S
D22/EC1 [T^R11]	R	S	S	S	S
D22/EC6 [T^R16]	R	R	S	R	R
D22/69 [V^R69]	R	R	S	R	S
D22/70 [V^R70]	R	R	S	R	S

TABLE IV

RECOMBINATION GROUPS OF CYTOPLASMIC OLIGOMYCIN AND VENTURICIDIN MUTANTS.

GENETIC LOCUS			
OL I		OL II	
D22/A16	[OL^R]	D22/A13	[OL^R]
D22/A21	[OL^R]	D22/A14	[OL^R]
55R5-3C/1	[OL^R]	D22/A15	[OL^R]
D22/A19	[V^R, OL^R]	D22/A18	[OL^R]
D22/61	[V^R, OL^R]	D22/C4	[OL^R]
D22/60	[V^R]	D22/B21	[OL^R]

ribosomal mutants by Linnane and coworkers. These points become self-evident on examination of results obtained with venturicidin resistant mutants (Tables III and IV).

In summary, oligomycin resistance can be correlated with two loci on mt-DNA, OL_I and OL_{II}. The resistance phenomenon can also be correlated with changes in the oligomycin binding site at the mitochondrial, submito chondrial and solubilized ATPase level.

TRIETHYL TIN RESISTANT MUTANTS.

Triethyl tin salts under defined conditions have a mode of action similar to oligomycin and are potent inhibitors of oxidative phosphorylation. Specific triethyl tin resistant mutants have been isolated and genetically characterized as cytoplasmic mutants in which the resistance allele is located on mtDNA (3,5, 9). Detailed genetic analysis of these mutants by Mr. Lancashire in collaboration woth Dr. Slonimski have shown that the tin resistance locus (T) is different from the oligomycin resistance loci OL_I and OL_{II}. It is thus reasonable to conclude that the T locus repre- sents an independent cistron coding for a gene pro- duct distinct from the product(s) of the OL_I and OL_{II} loci. This finding is in agreement with the total lack of cross-resistance shown by OL^R and T^R mutants. Also, in contrast to the situation with oligomycin resis- tant mutants it has not been possible to demonstrate increased resistance to triethyl tin at the mitochon- drial level as the I_{50} values for triethyl tin inhibi- tion of mitochondrial ATPase are not significantly different in T^R mutants and wild type. The large majo- rity of T^R mutants examined are cross-resistant to venturicidin but again no resistance to venturicidin is shown *in vivo* (Table III).

VENTURICIDIN RESISTANT MUTANTS.

It has been shown above that all T^R mutants are cross-resistant to venturicidin. Also the majority of venturicidin mutants are cross-resistant to triethyl tin . Recombination analysis shows that the V^R mutants cannot be separated from T^R mutants and V^R mutants map at the T^R locus. This class of cytoplasmic (mitochondrial) mutants of phenotype $V^R T^R$ will be designated V_{II}.

A further class of venturicidin resistant mutants has been isolated which are cross-resistant to oligomycin, ($V^R O^R$ mutants). Recombination analysis of these mutants show that this class of cytoplasmic (mitochondrial) V^R mutants map at the OL_I locus (Table IV); a wide variety of *in vitro* and *in vivo* phenotypes have been observed. Of particular interest is the mutant $V^R 60$ which is venturicidin resistant and oligomycin sensitive and maps at the OL_I locus. Mutants of this type will be of particular value in further genetic studies of the OL_I and OL_{II} loci. A diagram of the current "map" of the mitochondrial genome is given in Fig. 2.

Studies to date have established the presence of at least two independent cistrons and possibly a third cistron coding for distinct gene products. The nature of these gene products is under investigation in our laboratory and subunit 9 of the ATP-synthetase preparation of Tzagoloff (6) is being purified from several mutants with a view to peptide mapping and amino acid analysis to establish an amino acid change in mutant subunits.

While the correlation of a mitochondrial gene product with a mtDNA sequence is a desirable objective,

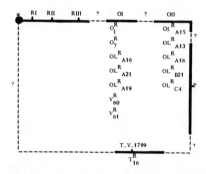

Fig. 2. Current form of the mitochondrial genome. (*cf*. P.R.Avner, D.Coen, B.Dujon and P.P.Slonimski (1973), Mol.Gen.Genetics, in press and ref.9).

possibly the more important uses of genetic studies of mitochondrial drug-resistant mutants lie in the possibility of defining unknown mitochondrial gene products associated with the oxidative phosphorylation complex. For example we have described the isolation of a valinomycin resistant mutant which exhibits a cytoplasmic mode of inheritance (5) and the demonstration of valinomycin binding sites in mitochondrial and other cell membranes is an intriguing possibility. Studies with bongkrekic acid have shown that the majority of triethyl tin resistant mutants (T^R mutants) are cross-resistant to bongkrekic acid (10). This raises the possibility that the mitochondrial ADP translocase enzyme contains a mitochondrially determined component. A further possibility is that there is a common protein subunit component of ATP-synthetase and ADP translocase sites in view of the interaction between the inhibition sites of triethyl tin and bongkrekic acid.

ACKNOWLEDGEMENTS.

This work was supported by research grants from the Science Research Council and the Medical Research Council.

REFERENCES.

1. D.E. Griffiths, P.R. Avner, W.E. Lancashire and J.R. Turner, in:"Biochemistry and Biophysics of Mitochondrial Membranes", G.F. Azzone, E. Carafoli, A.L. Lehninger, E. Quagliariello and N. Siliprandi, Eds., Academic Press, New York, 505 (1972).
2. P.R. Avner and D.E. Griffiths, FEBS Letters 10, 202 (1970).
3. W.E. Lancashire and D.E. Griffiths, FEBS Letters 17, 209 (1971).
4. P.R. Avner and D.E. Griffiths, Eur.J.Biochem. 32, 312 (1973).
5. D.E. Griffiths, in:"Mitochondria/Biomembranes" FEBS Symposia n. 28, North Holland, Amsterdam, 95 (1972).
6. A. Tzagoloff and P. Meagher, J.Biol.Chem. 247, 594 (1972).

7. P.R. Avner, D. Coen, B. Dujon and P.P. Slonimski, (1973) in press.
8. D.E. Griffiths, R.L. Houghton and J.R. Turner, (1973) in press.
9. W.E. Lancashire and D.E. Griffiths, (1973) in preparation.
10. D.E. Griffiths, W.E. Lancashire and K. Cain (1973) unpublished observations.

CHARACTERIZATION OF A RESPIRATION COMPETENT, OLIGOMYCIN RESISTANT MUTANT OF *SCHIZOSACCHAROMYCES POMBE* LACKING CYTOCHROME *b*-566 BUT STILL BINDING ANTIMYCIN.

Wolfhard Bandlow,

Institute for Genetics, University of Munich, Maria-Ward-Str. 1a, Munich, Germany.

The usefulness of mutants of the respiratory chain in the examination of the function and biogenesis of the respiratory chain complexes has been pointed out by Schatz (1). However, all such mutants except of the antimycin-resistant mutants of *Candida utilis* describes by Butow *et al.* (2) and Grimmelikhuijzen *et al.*(3) show a variety of pleiotropic effects which complicate the localization of the primary effect (4-6). This communication described an oligomycin-resistant mutant, 0 77, of the fission yeast *Schizosaccharomyces pombe* specifically lacking cytochrome *b*-566 (but retaining cytochrome *b*-558) and yet capable of growing on non-fermentable sources.

This yeast, for which no mitochondrial inheritance has been reported, is petite negative (7). Random spore and tetrad analysis shows that oligomycin resistance in mutant 0 77 is chromosomally inherited. Its growth rate in liquid medium with glycerol as the carbon source is about one-half of the wild type. Intact mitochondria were prepared from spheroplasts after enzymatic digestion of the cell walls by a crude lytic enzyme mixture of *Arthrobacter luteus* (8) for experiments on the P/O ratio with NADH as substrate. These studies indicate the absence of one of the coupling sites in the mutant. The mutant mitochondria exhibit respiratory control ratios of about 1.6 with pyruvate, α-oxoglutarate, and malate as substrates, as compared to values up to 2 for the wild type. The magnesium-dependent ATPase of the mutant is not sensitive to

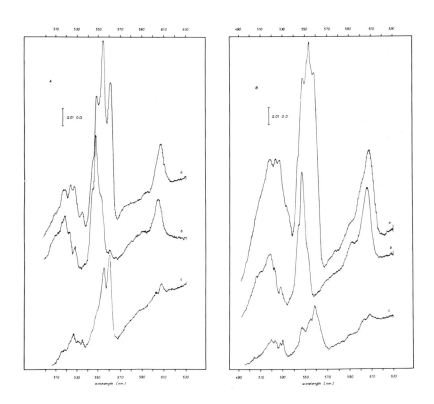

Fig. 1. Difference spectra measured at the temperature of liquid nitrogen of wild type (A) and mutant (B). 8.4 and 8.0 mg of mitochondrial protein (suspended in 0.25 M mannitol, 20 mM Tris, 1 mM EDTA, pH 7.4) was used per cuvette, respectively. The path length was 3 mm. a: dithionite reduced *minus* oxidized particles; b: ascorbate plus TMPD plus FCCP *minus* oxidized; c: succinate plus ascorbate plus TMPD plus FCCP *minus* ascorbate plus TMPD plus FCCP.

oligomycin up to 20 µg of oligomycin per mg protein
(Bandlow and Metzke, unpublished). Another interesting aspect of this mutant, besides oligomycin resistance, is the alteration of the cytochromes of the
respiratory chain. Low temperature spectra of substrate and dithionite reduced mitochondrial particles
clearly show that the cytochrome b absorption band at
560 nm, which is present in the wild type, is absent
in the mutant (Fig. 1.). This band is a composite of
the α bands of b-566 (absorbing at 561.5 nm in the
wild type at liquid nitrogen temperature) and b-562
(exhibiting a maximum absorption at 559 nm). The mutant shows bands for cytochrome b only at 559 and
555.5 nm (Fig. 1 B). Sato *et. al.* (9,10) hypothesize
that b-566 and b-558 comprise a single species with a
dual absorption band, while Wikström (11) and Slater
and Lee (12) who found a different reduction behaviour of the two absorption bands under certain conditions, suggest that they are two different species.
The results described here (*cf.* ref. 6) suggest a
structural or environmental difference between these
two b-cytochromes. Since the mutant grows on glycerol,
it can be concluded that cytochrome b-566 is not obligatory for electron transport from substrate to oxygen and can be bypassed at least in the mutant; however, the possibility of b-558 of the mutant being
different from the cytochrome of the wild type cannot
be excluded. In other mutants lacking b-566, electron
transport capacity through Complex III has been lost.
The residual b-cytochromes are CO reactive, not reduced by QH_2-3, NADH, or succinate, and do not bind antimycin (*cf.* ref. 6). Therefore, it was of interest to
see whether (i) NADH or succinate dependent respiration in this mutant is still antimycin sensitive, and
(ii) antimycin still binds to mutant mitochondrial
particles.

Fig. 2 shows the titration curves of succinate oxidation with antimycin. It can be seen that, when expressed as percentage of the wild type, the succinate
oxidation rate and inhibition in the mutant are 28 %
and 68 %, respectively. Nevertheless, sigmoidal inhi-

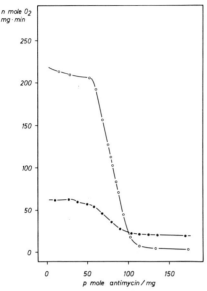

Fig. 2. Titration curve of specific succinate oxidation rate by antimycin, measured at 25° C in a closed vessel equipped with a Clark-type electrode. The final concentrations used were 7.4 mM succinate, 30 μM FCCP in a buffer containing 0.25 M sucrose, 100 mM potassium phosphate, 1 mM EDTA, final pH 7.2. Antimycin was added as a methanolic solution. The methanol concentration was kept constant in all incubations. The preincubation time was 8 min. O—O: wild type; ●—●: mutant O 77.

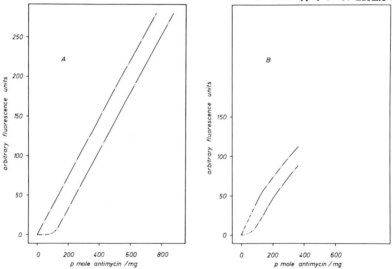

Fig. 3. Binding of antimycin to oxidized wild type and mutant mitochondrial particles. The binding was determined by the quenching of the fluorescence of the antimycin-serum albumin complex by mitochondrial particles as described (6,13). A: wild type (2.0 mg/ml); B: mutant O 77 (3.1 mg/ml).

bition curves are found in both cases. Fig. 3 shows that oxidized mutant particles still bind antimycin as determined by the quenching of the fluorescence of antimycin-BSA complex by stoichiometric binding of antimycin to mitochondrial particles (13). The dissociation constant, calculated from a Scatchard plot, is 2.6×10^{11} M in the mutant, somewhat higher than 1.5×10^{12} M in the wild type. Thus, despite the lack of b-566, mitochondrial particles of mutant O 77 still bind antimycin. This finding seems to settle the question whether the antimycin binding site is directly associated with b-566 as was earlier suggested by Slater *et al.* (14) because the addition of antimycin to anaerobic beef heart particles in the presence of NADH induced the reduction of a b species with a peak at 566 nm. However, recently it has been shown by Berden *et al.* (15) that all b-cytochromes are affected by the antimycin binding. Therefore, it now appears that none of the cytochromes is directly engaged in the binding of antimycin because (i) the drug has multiple effects despite the 1:1 stoichiometry for antimycin binding sites per Complex III, and (ii) antimycin binds to O 77 mitochondrial particles inspite of the absence of b-566. This is further substantiated by the report that the b-cytochromes are structurally identical (16).

Thus, it is concluded that in mutant O 77, the mutation primarily affects a chromosomally coded protein of CF_0 by which the ATPase gains oligomycin insensitivity. The non-integration of functional b-566 seems to be a consequence of the structural alteration of this membrane factor which is possibly a part of the environment of this special b-cytochrome but not of b-558 and b-562 retain their electron transport capacity while Complex III retains the ability of binding antimycin.

ACKNOWLEDGEMENTS.

Part of this work was done at the B.C.P. Jansen Institute, Amsterdam, The Netherlands. The author wishes to express his gratitude to Professor E.C. Slater for his hospitality, constructive critisism and many valuable suggestions. Lytic enzyme of *Arthrobacter luteus* was a generous gift of Y. Yamamoto. This work was supported by the Deutsche Forschungsgemeinschaft.

REFERENCES.

1. E. Ross, E. Ebner, R.O. Poyton, T.L. Mason, B. Ono and G. Schatz, this volume.
2. R.A.Butow and M.Zeydel, J.Biol.Chem. 243, 2545 (1968).
3. C.J.P.Grimmelikhuijzen and E.C.Slater, Biochim. Biophys.Acta 305, 67 (1973).
4. F.Sherman and P.P.Slonimski, Biochim.Biophys.Acta 90, 1 (1969).
5. A.Goffeau, A.M.Colson, Y.Landry and F.Foury, Biochem.Biophys.Res.Commun. 48, 1448 (1972).
6. W.Bandlow, K.Wolf, F.Kaudewitz and E.C.Slater, Biochim.Biophys.Acta, submitted for publication.
7. C.J.E.A.Bulder, J.Microbiol.Serol.Antonie van Leeuwenhoek, 30, 442 (1969).
8. K.Kitamura, T.Kaneko and Y.Yamamoto, Arch.Biochem. Biophys. 154, 402 (1971).
9. N.Sato, D.F.Wilson and B.Chance, Biochim.Biophys. Acta 253, 88 (1971).
10. N.Sato, T.Ohnishi and B.Chance, Biochim.Biophys. Acta 275, 288 (1972).
11. M.F.K.Wikström, Biochim.Biophys.Acta 253, 332 (1971).
12. E.C.Slater and I.Y.Lee, 2nd Int.Symp.on Oxidases and Related Oxidation-Reduction Systems, Memphis, Tennessee, in press.
13. J.A.Berden and E.C.Slater, Biochim.Biophys.Acta, 256, 199 (1972).
14. E.C.Slater, C.P.Lee, J.A.Berden and H.J.Wegdam, Nature 226, 1241 (1970).
15. J.A.Berden and F.R.Opperdoes, Biochim.Biophys.Acta 267, 7 (1972).
16. J.A.Berden, Site II of the respiratory chain. Studies on the cytochromes b and c_1 in mitochondria and sub-mitochondrial fragments. Ph.D. thesis, University of Amsterdam, Gerja, Waarland (1972).

MITOCHONDRIAL ENZYMES IN MAN-MOUSE HYBRID CELLS

V. van Heyningen, I.W. Craig, W.F. Bodmer

Genetics Laboratory, Department of Biochemistry, University of Oxford.

Man-mouse somatic cell hybrids (1,2) usually lose human chromosomes, while the mouse chromosomes are retained. Most of the loss appears to take place during early divisions following fusion, so that by the time the hybrid clone is a few hundred cells in size the chromosome complement is relatively stable, though so-some segregation may still occur.

The fate of the cytoplasmic components of the two species, such as the mitochondria, can also be followed conveniently if species specific differences permit identification of the two parents' contribution.

Thus in human-mouse cell hybrids the presence of human enzymes or non-enzyme proteins which are distinguishable from their mouse homologues can be correlated with the presence of other enzymes or proteins and specific chromosomes remaining after the initial loss of human chromosomes. Theoretically the segregation of the mitochondria of the two parent species could be observed as well as interaction of the organelle of one species with at least some nuclear coded products of the other. Somatic cell hybrids therefore provide a suitable system for the study of the genetic control of mitochondrial biogenesis in higher organisms.

It has already been shown independently by two groups (3,4) that all the human-mouse hybrid lines examined by them (ten in all) had apparently lost the human mitochondrial DNA (mtDNA), whilst retaining the mouse component. However Dawid and Coon (5) have apparently obtained rat-human hybrids in which chromosomes are lost from either parent. In these, 40-60 genera-

tions after fusion when the chromosome content of the
hybrids is fairly stable, mtDNA from either or both
parents has been reported. There appears to be a
strong positive correlation between the proportion of
rat mtDNA and the proportion of rat nuclear DNA.

We have examined nine independently isolated man-
mouse hybrid lines and a number of subclones from them
for the presence of three mitochondrially located human enzymes that are usually assumed to be nuclear coded (6): NAD-malate dehydrogenase (E.C. 1.1.1.37), citrate synthase (E.C. 4.1.3.7.) and aspartate aminotransferase (E.C. 2.6.1.1.). Our studies confirm the
nuclear control of all three enzymes and give information on the linkage of mitochondrial malate dehydrogenase and citrate synthase with other markers since
the hybrids were also examined for the presence of a
large number of cytoplasmic enzymes (7). Three of the
hybrid lines we analysed were amongst those shown by
Clayton *et al.* (3) to have no human mtDNA detectable by
CsCl centrifugation.

Citrate synthase (CS) and mitochondrial NAD-malate
dehydrogenase (mMOR) are consecutively acting enzymes
of the citric acid cycle. Mitochondrial aspartate aminotransferase (mAAT) is one of the enzymes linking the
cycle to amino-acid metabolism. CS, mMOR and mAAT are
all believed to be matrix enzymes (6). Citrate synthase is believed to have no cytoplasmic counterpart, but
malate dehydrogenase and aspartate aminotransferase
both have cytoplasmic analogues, sMOR and sAAT respectively. Human and mouse homologues of sMOR and sAAT
are easily distinguishable electrophoretically on
starch or cellulose acetate gels (7,8,9). Citrate synthase from the two species can now also be resolved
on the latter medium (8). However, we have not been
able to separate mouse from human mMOR or mAAT by
electrophoresis in any of several buffer systems between pH 5.2 and 9.0 on polyacrylamide, starch or cellulose acetate. We were thus forced to consider other
approaches to distinguishing these mouse and human enzymes. One approach would be the analysis of hybrids in
which one parent is at least heterozygous for an elec-

trophoretically mutant form of the enzyme in question. However such variants are not readily available for mMOR or mAAT in mouse inbred lines (10,11) and hence, so far, in cell lines; and in human populations the frequency of heterozygotes for such variants is very low (12,13). It is for these reasons that we were led to an immunological analysis.

Cells. The parentage of the hybrid lines analyzed is shown in Table I. The cells were grown in Biocult RPMI 1640 medium supplemented with 10 % foetal-calf serum and the components of Littlefield's (14) HAT selective agent: hypoxanthine, aminopterin and thymidine. Parental cells were grown in the same medium without the HAT components. Two of the hybrid lines: 3W4 and HORP, were subcloned in both selective and non-selective medium and some revertants back-selected for growth in 6-thioguanine were also studied. Whole cell extracts were made by sonicating cells at a concentration of approximately 10^8 cells/ml in a protective lysis buffer (8). Clarified sonicates were stored in liquid nitrogen. Some extracts used for mAAT analysis were prepared by freezing and thawing in a dry ice bath cells suspended in 2 % aqueous sodium deoxycholate containing 10^{-4} M phenyl methyl sulphonyl fluoride.

Mitochondria. Mitochondria were prepared from two mMOR positive hybrids: the line 2W1, using 1.5 x 10^8 cells and the subclone HORP 25 using 6 x 10^8 cells. Freshly harvested cells were washed in phosphate buffered saline and suspended in 10 ml STE (0.25 M sucrose, 0.01 M Tris-HCl, 0.001 M EDTA pH 7.0 with acetic acid). The suspension was homogenized with nine strokes of a Broeck ground glass homogenizer. Nuclei and debris were removed at 3000 rpm and mitochondria were pelleted and washed using a 15,000 rpm spin. They were resuspended in 0.8 ml STE and layered onto a 19 ml linear sucrose gradient (0.88 - 1.88 M) and sedimented at 24,000 rpm for 90 min in a swinging bucket rotor. Pellets were recovered at 33,000 rpm from the eight, appropiately diluted fractions of the gradient. The pellets were sonicated in 0.2 ml lysis buffer for 1 min. The sonicated fractions were assayed spectropho-

TABLE I.

PARENTAGE OF HYBRID LINES.

Hybrid Lines	Human Parent	Mouse Parent
1W1, 2W1, 3W4, 4W10 (34)	Normal lymphocytes (MAR)	1R, 8-azaguanine resistant L-cell
HORL, HORP made as in (34)	Normal lymphocytes (HOW)	1R, 8-azaguanine resistant L-cell
4.12Z, 4.31Z, 4.42.7Z (17)	Normal lymphocytes (BRE)	1T, bromodeoxyuridine resistant subclone of 3T3

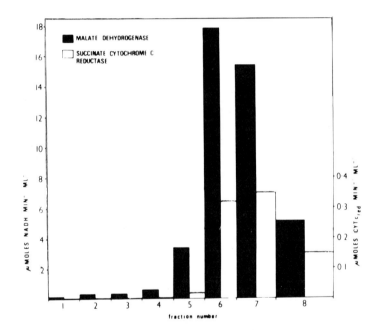

Fig. 1. HORP25 (subclone of HORP see Table I) mitochondrial preparation: analysis of sucrose gradient fractions. Fraction numbers are from bottom (1) to top (8) of the gradient. The varying widths of the columns of the histogram accurately denote the relative fraction sizes.

tometrically for succinate-cytochrome c reductase (15) and malate dehydrogenase. Fig. 1 shows the cosedimentation of these two activities on the sucrose gradient. Immunologically, the human mMOR is also to be found in the top of four fractions (5,6,7 & 8).

Electrophoresis. Electrophoresis for citrate synthase and malate dehydrogenase was carried out on Cellogel (Reeve Angel cellulose acetate as described previously (8,16). Fig. 2 shows the results of electrophoretic CS analysis on some of the hybrids. It demonstrates that most hybrids which have human CS show only the slow mouse band and a band intermediate in mobility between mouse and human. This is probably a man-mouse heteropolymeric band (A^HA^M) produced by random combination in proportion to their concentrations of mouse (A^M) and human (A^H) subunits (8,17). The observed unequal amounts of A^M and A^H is most simply explained by invoking a gene dosage effect: all of the heteropoloid mouse cell's CS loci (perhaps as many as six) are probably retained in the hybrid whilst human chromosome loss most likely leaves only one human CS gene.

Enzymes. Human mMOR was purified from placental mitochondria as described elsewhere (8). A sample containing only human cytoplasmic enzymes (sMOR and sAAT) was also prepared by the same ion exchange procedure which is designed to separate the electrophoretically similar mMOR and mAAT from their more negatively charged cytoplasmic counterparts.

Human (placenta) and mouse (livers, kidneys,spleens and hearts of Swiss mice) mitochondrial AAT were purified by a scaled down version of the method of Scandurra and Cannella (18). The inclusion of a one minute sonication step after mincing and the omission of the Sephadex G-100 step were the only deviations from their procedure. mMOR and mAAT preparations that were free of their cytoplasmic analogue when tested by electrophoresis on Cellogel were used in immunization schedules.

Immunization. In order to produce antisera specific for the human enzymes and not cross-reacting with the

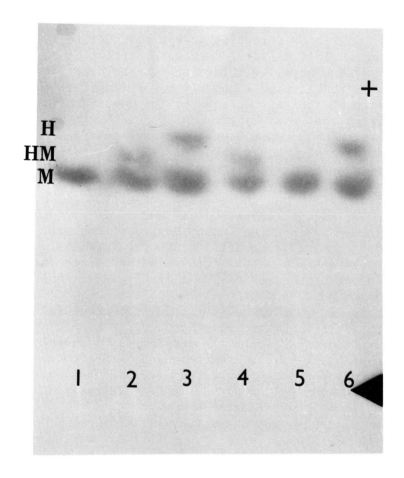

Fig. 2. Citrate synthase electrophoresis on Cellogel (photograph of the gel). Samples: 1, 4W10; 2, 3W4; 3, 1R + BREL; 4, HORP; 5, HORL; 6, 1R + BREL. The arrow indicates the origin. H, HM and M refer respectively, to human, human–mouse heteropolymer and mouse bands.

mouse counterparts, mice were immunized with human m-MOR and mAAT according to the schedule described elsewhere (8). C3H mice were used for the mMOR and Swiss mice for mAAT. In both cases ascites producing sarcoma cells (19) were used to increase the total antibody yield. Immunization with human mAAT produced auto-antibodies to mouse mAAT. To investigate this further four Swiss mice were also immunized with mouse mAAT to see if auto-antibodies would be produced. No sarcoma was used in this case. After four mouse mAAT injections which elicited no response, two of the mice were boosted with human mAAT.

Ouchterlony double diffusion. This method, using 1.5 % agar in 0.9 % saline on glass slides, was employed to screen hybrid cell extracts for human mMOR and mAAT. For AAT diffusions the addition to the agar of polyethylene glycol 6000 (BDH) at 4 % w/v was found advantageous for good resolution by increasing the efficiency of precipitation (20). Diffusion was carried out at 4° C for 24-36 h followed by washing for 24 h in 0.9 % saline. Precipitin lines were stained for the appropriate enzyme activity. The malate dehydrogenase stain is tetrazolium based (8). The aspartate aminotransferase stain is 0.1 M Tris-HCl pH 7.4, 0.034 M aspartate, 0.017 M 2-oxoglutarate and 0.5 % Fast Blue 2B salt (Sigma).

Fig. 3 shows the analysis of hybrid lines for human mMOR. It will be observed that there is no cross-reaction with the mouse parents 1R and 3T3, nor with the human cytoplasmic enzyme, and therefore the analysis is relatively simple.

However the antiserum to human mitochondrial aspartate aminotransferase was found to cross-react with the mouse homologue, though not with the human or mouse cytoplasmic analogue. This cross-reaction, manifested by a spur of partial identity between human and mouse precipitin lines can be absorbed out by the addition to the serum of mouse mAAT. On analysis of the hybrid lines with the absorbed antiserum to human mAAT only one of the nine lines appeared to possess this human mitochondrial enzyme. This paucity of mAAT-posi-

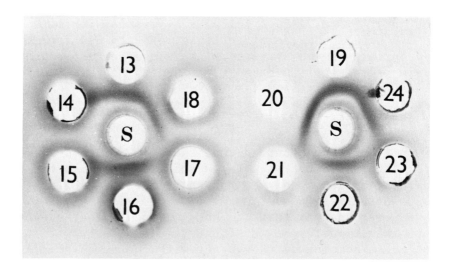

Fig. 3. Ouchterlony double diffusion in agar, using specific mouse anti-human mMOR serum; stained for malate dehydrogenase activity. S in the centre well refers to the antibody containing sarcoma induced ascites fluid. Identification of other wells: 1, BREL; 2, sMOR; 3, mMOR (immunizing antigen); 4, 2W1; 5, 1R; 6, 1W1; 7, 1R; 8, HORL; 9, mMOR; 10, 4W1O; 11, BREL; 12, 3W4; 13, BREL; 14, 4.42Z; 15, 3T3; 16, 4.31Z; 17, 3T3; 18, 4.12Z; 19, BREL; 20, mMOR; 21, sMOR; 22, HORP; 23, 1R; 24, HORP.

tive lines in not particularly surprising even assuming random loss of chromosomes during hybrid evolution (one other hybrid, not on the panel, Daudi x A9 (35), shown by Clayton *et al.* (3) to have no human mtDNA was also found to have human mAAT).

The production of antisera by two mice, that cross-react with mouse enzyme is of interest as it implies that an autoantibody has been induced. Enzyme prepared from Swiss mice does not in Swiss mice elicit an immune response demonstrable by the Ouchterlony technique. A single human enzyme boost does not affect this lack of response. This suggests that the autoantibody produced is specific for a region of the enzyme which is the same in mouse and man, but that the auto-response depends on human-mouse species differences in other parts of the molecule acting as carrier. Similar phenomena have been observed with other systems (21). Autoantibodies to mitochondrial components are observed in certain disease states, such as primary biliary cirrhosis (22).

DISCUSSION.

Table II summarizes our findings on repeated testing of nine hybrid lines, together with the human mitochondrial DNA data, where available.

Three major implications for the control of CS and mMOR and mAAT are suggested by the data on Table II.
1. Of the three hybrid lines known to have no human mtDNA as shown up by isopycnic centrifugation, one has mMOR, CS and mAAT one has only the first two mitochondrial enzymes and one has none of them. Thus the presence of human mtDNA is not required for the expression of at least some nuclear coded human mitochondrial functions. Whether this is because there is no strong control by mitochondrial DNA on the expression of all nuclear-coded products which eventually become incorporated into the organelle, or because such control (2,3) exists but is not species specific, we cannot say on the basis of the present data. It would appear from the data on Dawid and Coon's (5) hybrid lines and was also suggested by Clayton *et al.* (5) that

239

TABLE II.

DISTRIBUTION OF HUMAN ENZYMES IN HYBRID LINES.
Mouse activity is always seen to be present, except in the case of the immunological technique for mMOR, which only scores for the human activity.

Hybrid Line	LDHB/Pep B	CS	mMOR	mAAT	mtDNA (3)	sMOR	sAAT
1W1	−	−	−	−	NT	−	−
2W1	−	−	+	−	NT	−	−
3W4	+	+	+	−	−	−	−
4W10	−	−	−	−	−	−	−
HORL	−	−	−	−	NT	−	−
HORP	+	+	+	−	NT	+	+
4.12Z	+	±*	+	+	−	+	−
4.31Z	+	+	+	−	NT	−	−
4.42.7Z	−	−	−	−	NT	+	−

+= presence of human activity; −= absence; NT= not tested.
* weak activity possibly because of extreme gene dosage imbalance, 4.12Z
has a doubled chromosome complement from the mouse parent, 3T3.

TABLE III.

DISTRIBUTION OF HUMAN ENZYMES IN SUBCLONES.

Line	No. of Clones	LDHB	CS	mMOR	sMOR	sAAT
HORP	1	+	+	+	+	+
	1	+	+	+	+	−
	2	+	+	+	−	−
	1	+	+	+	−	+
	2	−	−	+	+	+
	1	−	−	+	−	+
	1	+	+	−	−	−
3W4	10	+	+	+	−	+
	1	+	+	+	−	−
	2	−	−	+	−	−
	1	−	−	+	−	+

+= presence of human enzyme; −= absence.

some species specific control is exercised on mtDNA replication, as there is a strong correlation between the species whose nuclear DNA is predominantly retain- ed and the one whose mtDNA is found in interspecific hybrids. This is in sharp contrast with the species independent control of nuclear DNA and RNA synthesis which was implied by the ability of much reduced hu- man—mouse hybrids to express human functions (2,4) and by observations in human-chick and mouse-chick hetero- karyons (25). Despite the possibility of species spe- cific control of mitochondrial DNA synthesis we have found that in two human mMOR containing hybrids (2W1 and HORP 25) at least some of the human mMOR is loca- ted in the mouse-mtDNA controlled mitochondria (Fig. 1).

2. There is no relationship between the expression of cytoplasmic and mitochondrial MOR nor the two AATs in the hybrid cells examined. This absence of linkage is in agreement with genetic data on mutant forms of m- MOR, sMOR, mAAT and sAAT in both man and mouse (10,11, 12,26). The absence of close *structural* similarities between these mitochondrial enzymes and their cyto- plasmic analogues is emphasized by the repeatedly ob- served lack of immunological cross-reaction between the two forms both for MOR and AAT (27,28).

3. A strong positive correlation between the presen- ce of human citrate synthase and mitochondrial malate dehydrogenase is seen in the hybrid lines examined. The unrelated enzyme functions lactate dehydrogenase B (LDHB) and peptidase B, the loci for which have been shown to be linked to each other (17,29) and are pro- bably on chromosome 12 (30) also show a marked posi- tive correlation with the expression of CS and mMOR. The only exception to a complete correlation between the presence and absence of the four enzymes, which is expected if they are all coded for by genes on the sa- me chromosome, is the line 2W1. None of the other 28 enzymes tested for (7) show any obvious association with these four enzymes.

The analysis of 23 subclones (Table III) from two hybrid lines (3W4 and HORP) which originally possessed

all four enzymes, supports the linkage of CS and LDHB/
PepB without exeption. On the other hand mMOR does not
segregate with the other three enzymes in these clones, and pending chromosome analysis of appropriate
lines and subclones, is assumed not to be coded for on
chromosome 12.

Possibility of Charge Conservation. In the course
of screening the electrophoretic properties of mitochondrial enzymes the difficulty of separating mouse
and human forms of several enzymes was striking.Others
(7,9,12) have encountered similar difficulties. The
mitochondrial form of enzymes which possess a cytoplasmic counterpart, migrates at around physiological
pH only a short distance to one or other side of the
origin, and is always more cathodal than the cytoplasmic form. It is possible that there is a strong selective conservation of charge properties amongst these
enzymes (MOR, AAT, NADP isocitrate dehydrogenase).
Conservation of charge amongst these mitochondrial enzymes is perhaps also suggested by the low frequency
of electrophoretic variants in the populations so far
examined (31). It is not impossible that the similar
and apparently invariable charge of these enzymes at
physiological pH values is a requirement for their localization into mitochondria. There are exceptions to
this suggested charge conservation. Mitochondrial NADP
malate dehydrogenase migrates to the anode in the mouse and to the cathode in man at pH 7.0 (32,33). However, this enzyme shows a high level of polymorphism
(30 % heterozygotes for the variant in the populations
examined (33). Moreover the theory does not appear to
hold for enzymes like citrate synthase which have no
cytoplasmic analogue.

The existence of similar difficulties in resolution
of the different species contribution in interspecific
hybrids, makes the immunological approach a powerful
tool in somatic cell genetics. Somatic cell hybrids
for their part offer an interesting system for the study of mitochondrial biogenesis.

ACKNOWLEDGEMENTS.

This work was supported in part by a grant from the Medical Research Council. Veronica van Heyningen is a recipient of an MRC Scholarship for Training in Research Methods. We thank Dr. David Finnegan for help with the subcloning and Art Reingold for help with some of the mitochondrial preparations. Fig. 2 and 3 and Tables I and II are reproduced by kind permission of Nature.

REFERENCES.

1. H. Harris, in:"Cell Fusion", Oxford University Press (1970).
2. B. Ephrussi, in:"Hybridization of Somatic Cells", Princeton University Press (1972).
3. D.A. Clayton, R.L. Teplitz, M. Nabholz, H. Dovey and W. Bodmer, Nature 234, 560 (1971).
4. B. Attardi and G. Attardi, Proc.Nat.Acad.Sci. USA 69, 129 (1972).
5. I. Coon and H. Coon, Communicated at a Symposium on mitochondrial Biogenesis and Bioenergetics at Meet.Fed.Eur.Biochem.Soc. 8 (1972).
6. M. Ashwell and T.S. Work, Ann.Rev.Biochem. 39, 251 (1971).
7. Unpublished results obtained in collaboration with H. Harris, D. Hopkinson, S. Povey and their colleagues.
8. V. v. Heyningen, I.W. Craig and W.F. Bodmer, Nature 242, 509 (1973).
9. P. Meera Khan, Arch.Biochem.Biophys. 145, 470 (1971).
10. J.B. Shows, N.M. Chapman and F.H. Ruddle, Biochem. Genet. 4, 707 (1970).
11. R.J.de Lorenzo and F.H. Ruddle, Biochem.Genet. 4, 259 (1970).
12. R. Davidson and J. Cortner, Science 157, 1569 (1967).
13. E. Hackel, D.A. Hopkinson and H. Harris, Ann.Hum. Genet. 35, 491 (1972).
14. J.W. Littlefield, Cold Spring Harbor Symp.Quant.

Biol. 29, 161 (1964).

15. M. Rabinowitz and B. de Bernard, Biochim.Biophys. Acta 26, 22 (1957).
16. I.W. Craig, Biochem.Genet., in press.
17. A.S. Santachiara, M. Nabholz, V. Miggiano, A.J. Darlington and W. Bodmer, Nature 227, 248 (1970).
18. R. Scandurra and C. Cannella, Eur.J.Biochem. 26, 196 (1972).
19. E.C. Herrmann and C. E gle, Proc.Soc.Exp.Biol. Med. 98, 257 (1958).
20. J. Harrington, J. Fenton and J. Pert, Immunochem. 8, 413 (1971).
21. P. Bretscher, Transplant.Rev. 11, 217 (1971).
22. D. Doniach, Brit.Med.Bull. 28, 145 (1972).
23. Z. Barath and K. Küntzel, Proc.Nat.Acad.Sci.USA 69, 1371 (1972).
24. G. Elicieri and H. Green, J.Mol.Biol.41, 253 (1969).
25. H. Harris, Proc.Roy.Soc.B. 166, 358 (1966).
26. R. Davidson, J. Cortner, M. Rattazzi, F. Ruddle and H. Lubs, Science 169, 391 (1970).
27. G.B. Kitto and N.O. Kaplan, Biochem. 5, 3966 (1966).
28. J.S. Nisselbaum and O. Bodansky, J.Biol.Chem. 239, 4232 (1964).
29. F. Ruddle, V. Chapman, T.R. Chen and R. Klebe, Nature 227, 251 (1970).
30. R.H. Ruddle, in:"Advances in the Biosciences", 8, 299 (1972), Workshop on Mechanisms and Prospects Exchange (Pergamon).
31. H. Harris and D.A. Hopkinson, Ann.Hum.Genet. 36, 9 (1972).
32. T. Shows and F. Ruddle, Science 160, 1356 (1968).
33. P.T. Cohen and G.S. Omenn, Biochem.Genet. 7, 303 (1972).
34. M. Nabholz, V. Miggiano and W. Bodmer, Nature 223, 258 (1969).
35. T. Caspersen *et al.*, Exp.Cell Res. 65, 475 (1971).

TEMPERATURE-DEPENDENT CHANGES IN LEVELS OF MULTIPLE-LENGTH MITOCHONDRIAL DNA IN CELLS TRANSFORMED BY A THERMOSENSITIVE MUTANT OF ROUS SARCOMA VIRUS

Margit M.K.Nass

Department of Therapeutic Research, University of Pennsylvania School of Medicine, Philadelphia, Pa. 19104, U.S.A.

The occurrence of unicircular and catenated dimeric and oligomeric circular DNA is common in microbial systems, such as bacterial plasmids, bacteriophages and certain animal viruses, and in some cases the oligomer levels can be experimentally influenced (1-4). Circular mitochondrial DNA (mtDNA) of animal cells has also been shown to contain a small percentage of double- and multiplelength DNA forms whose origin and biological significance has not yet been resolved (see refs. 5-8 for reviews). The catenated dimer (trimer and higher oligomers) consisting of two (or more) interlocked monomers has been identified as a minor component in all cell types examined. The other form of dimer, a unicircular molecule, is less common. It has been identified in mtDNA of human leukemic leucyocytes, where its level correlated with the severity of chronic granulocytic leukemia (9), in mouse L-cells, where its level could be regulated by growth conditions (10,11), and in thyroid cells of presumably non-malignant origin (12).

We have shown previously a significant increase in the content of dimeric and oligomeric mtDNA if mouse L-cells were allowed to grow from logarithmic into stationary phase and if protein synthesis was inhibited by amino acid starvation or treatment with cycloheximide and puromycin (4-8 fold accumulation), (10, 11). Return to normal conditions of growth restored the formation of DNA monomers. A marked increase in oligomer content occurred in baby hamster kidney cells transformed with polyoma virus (11). Relatively little or no effects on dimer or oligomer content were found in cells treated with several mitotic spindle inhibitors, chloramphenicol, rifampicin, or mengovirus. We observed marked variations in the extent to which oligomer accumulation could be induced, depending on cell type (11).

The studies to date have shown that (1) the unicircular dimer is neither detectable in all types of malignant cells nor is it unique to malignant cells; (2) catenated dimers may exist in high frequencies in many malignant cells but at low levels in others. These investigations have suffered primarily from lack of a well-controlled system where normal cells and their direct malignant derivatives can be compared. The present study utilizes an ideally controlled experimental system of chick embryo fibroblasts made malignant by infection with temperature-sensitive mutants of oncogenic viruses. Upon infection with the thermosensitive mutant T5 of Rous sarcoma virus (13), cell transformation becomes manifested morphologically and functionally at 36° C, the permissive temperature. The cells round up, grow in soft agar, and have an increased rate of deoxyglucose uptake (13) and an altered glycoprotein pattern (14). At 41° C, the non-permissive temperature, the cells have the phenotypic appearance and behavior of normal cells. The same batch of cells can thus be induced to express malignant properties, then reversed to normal and back to malignant again, simply by temperature switches. Furthermore, the transformed cells can be compared to the uninfected cells from which they have been derived, and to cells infected with the wild-type virus. This paper demonstrates a direct correlation of multiple-length mtDNA accumulation with the phenotypic manifestation of oncogenic virus transformation. The results are summarized from a more detailed presentation elsewhere (15).

METHODS, RESULTS AND DISCUSSION

246

Chick embryo fibroblasts were plated in roller bottles and infected with either the Schmidt-Ruppin strain of Rous sarcoma virus, subgroup A, which is the wild-type virus, or the temperature-sensitive mutant of this virus, T5 (13,16). This mutant can transform the cells at 36° but not at 41°C; however, it can infect and replicate at both temperatures. After infection the cells were grown at 36° untill morphological transformation of the cells was apparent after 5 to 6 days. The cells were then subcultured in roller bottles and incubated at 36 or 41° C in the presence of isotope for 2 to 3 days. Fig. 1 shows the morphological appearance of control and virus-infected cells grown at the two temperatures.

A comparison of single-length and multiple-length forms of mtDNA was made in a double-label experiment with cells transformed by wild-type and by T5 virus, using band sedimentation (10) to identify the monomeric and dimeric DNA. Mitochondria and mtDNA were isolated as described (15,17). Fig. 2 shows that in wild-type virus infected cells the $[^{14}C]$-labelled dimer peak is of equal size at both temperatures whereas in T5-infected cells the size of the $[^{3}H]$-labelled dimer peak is temperature-dependent. At 36° C the dimer peak is equal to that of wild-type virus infected cells, but at 41° C it is reduced to about one-half (about 5 % of the monomer peak).

A more quantitative measure of oligomer content was obtained by electron microscopy where both covalently closed and nicked circular DNA forms can be scored. $[^{3}H]$thymidine-labelled mtDNA was isolated from control and transformed cells and the combined peaks I and Im from CsCl-ethidium bromide gradients (15,17) were analysed by the DNA spreading technique in 50 % formamide (18). The catenated dimeric form shown in Fig. 3 is the major type in the multiple-length DNA population. Oligomers (catenated trimers, tetramers) constitute less than 5 % of the dimer population. The unicircular dimer as opposed to the catenated dimer has not been identified with certainty in these cells.

Table I shows a comparison of the levels of multi-

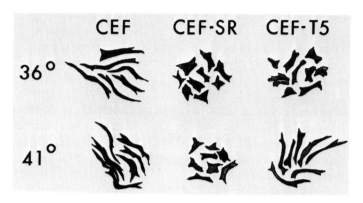

Fig. 1. Morphological appearance of control chick embryo fibroblasts (CEF) and of CEF infected with wild-type Schmidt-Ruppin Rous sarcoma virus (SR) and with the temperature-sensitive mutant of this virus (T5), grown at permissive (36°C) and non-permissive (41°C) temperatures.

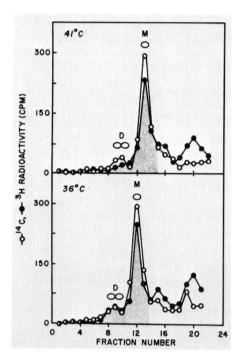

Fig. 2. Co-sedimentation in sucrose gradients of [^{14}C] labeled mtDNA derived from CEF-SR (O——O) and [^{3}H] labeled mtDNA derived from CEF-T5. The cells were grown at 41 and 36°C for 48 h. The isotope concentrations were 0.1 µC [^{14}C] thymidine and 1.0 µC [^{3}H] thymidine per ml of growth medium. D: dimeric DNA; M: Monomeric DNA.

248

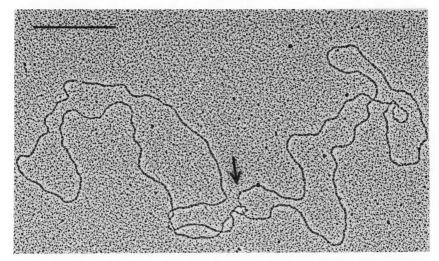

Fig. 3. Electron micrograph of catenated dimeric mtDNA from CEF-T5 grown at 36°C. The contour length of the monomeric form spread in 50 % formamide is 5.51 μ (15). Arrow points to region where the 2 submolecules interlock. Bar represents 0.5 μ.

TABLE I.

LEVELS OF MULTIPLE-LENGTH MITOCHONDRIAL DNA IN CHICK EMBRYO FIBROBLASTS INFECTED WITH WILD-TYPE (SR) AND TEMPERATURE-SENSITIVE MUTANT (T5) OF RSV.

Cells	Temperature of culture °C	Fraction of cells synthe-sizing DNA %	2-Deoxyglucose uptake cpm/mg protein	Total dimers + oligomers %*
CEF	36	10.6	8200	6.5
CEF	41	11.2	7400	6.4
CEF-SR	36	10.1	57100	13.2
CEF-SR	41	10.2	51500	12.1
CEF-T5	36	11.3	71800	11.5
CEF-T5	41	10.1	12200	6.4

* Total number of DNA molecules scored: CEF, 1013-1075; CEF-SR, 371-439; CEF-T5, 1364-1477.

ple-length mtDNA in the 3 cell types grown in roller bottles at 36° and 41° C. The content of dimeric and oligomeric mtDNA was found to be 2 to 2.5 times higher in SR-RSV-infected cells than in uninfected cells. In T5-infected cells, the temperature dependence of this stimulation is again apparent. As a control for the phenotypic manifestation of cell transformation the rate of 2-deoxyglucose uptake, shown to be elevated in RSV transformed cells (13), is increased 7- to 9-

fold when transformation is expressed (Table I). The
cells used for deoxyglucose transport were subcultured in 25 cm² Falcon flasks in parallel with the roller
bottles. The 2- to 3-fold accumulation of oligomeric
mtDNA does not appear to be due to significantly altered rates of cell growth since the fractions of
cells synthesizing DNA per 1 h, as determined by auto-
radiography (13,15) are similar in all cases (Table I).
The rates of nuclear DNA synthesis have also been
shown to be similar under these conditions (15).

TABLE II.

EFFECT OF TEMPERATURE SHIFTS ON CONTENT OF MULTIPLE-LENGTH MITOCHONDRIAL DNA
IN T5-INFECTED CHICK EMBRYO FIBROBLASTS.

Temperature of culture °C	2-Deoxyglucose uptake %	Total DNA molecules Nr.	Total dimers + oligomers %
36 (96 h)*	71800	247	14.1
41 (96 h)*	12200	254	7.1
36 to 41 (24 h)	27100	218	8.0
41 to 36 (24 h)	53500	251	15.1

*Cells analyzed for mtDNA were grown for 96 h, cells analyzed for deoxyglu-
cose transport for 48 h (see Table I).

Table II shows that temperature shifts can revert
the levels of dimeric and oligomeric mtDNA. The apparent final levels are reached within 24 to 48 h at the
new temperature. Control experiments with deoxyglucose transport are also included in the table.

In order to determine if the content of multiple-
length mtDNA in uninfected cells can be increased experimentally to or beyond the 11.5 to 15 % level observed in the transformed cells, chick embryo fibroblasts were treated with cycloheximide (25 µg/ml) for
24 h. This inhibitor of cytoplasmic protein synthesis
was shown previously to cause accumulation of multiple
length mtDNA in various cell types (10,11). Table III
shows that the level of dimeric and oligomeric mtDNA
was increased 5-fold. A relatively insignificant 1.6-
fold change was observed with 9-β-D-arabinofuranosyl
adenine (ara-a), which had not previously been tested
in this respect. With both inhibitors, nuclear DNA but
not mtDNA synthesis was inhibited (15). A small size
increase was observed in mtDNA of T5-infected cells

TABLE III.

EFFECT OF CYCLOHEXIMIDE AND 9-ß-D-ARABINOFURANOSYL ADENINE ON CONTENT OF MULTIPLE-LENGTH MITOCHONDRIAL DNA IN CHICK EMBRYO FIBROBLASTS.

Conditions	Total DNA molecules Nr.	Total dimers + oligomers %
Control	356	5.8
Cycloheximide (25 μg/ml, 24 h)	307	30.9
Ara-a (5 x 10⁻⁴M, 24 h)	319	9.2

as compared with control cells (15). Whether this is due to experimental factors or reflects a significant alteration of the mitochondrial genome is under investigation. The main conclusion drawn from the present experiments is that the formation of multiple-length mtDNA reversibly parallels the phenotypic manifestation of oncogenic virus-induced transformation.

The mechanism that may link mtDNA replication and virus transformation can merely be speculated upon. It is well established that there are many differences in the structure and function of cellular membranes in normal and malignant cells (*e.g.*, see ref. 19). Such changes include sugar transport (13) as well as alterations of glycoprotein (14) content. The differences in glycoprotein patterns are not confined to surface membranes but also include mitochondrial and other intracellular membranes (20). Since the evidence suggests the existence of a mitochondrial DNA-inner membrane complex (5,7,21), a conformational and/or functional alteration of the membrane may affect the attachment site of mtDNA. This may cause a reduction in the activity or content of enzymes that are involved in DNA replication, such as an endonuclease or "nickase" which may cleave newly replicated DNA structures into monomeric molecules (*e.g.* 22,23). Alternatively, the postulated distortion of the DNA-membrane complex may render the replicating DNA less accessible to, or susceptible to, the action of such enzymes. It is also possible that changes in the membrane affect intramitochondrial transport of specific "nickases" or of factors involved in the control of mtDNA replication.

Since cycloheximide, an inhibitor of cytoplasmic but not mitochondrial protein synthesis, causes oligomer accumulation (refs. 10,11 and Table III), whereas chloramphenicol, a selective inhibitor of mitochondrial protein synthesis, does not induce significant oligomer formation (10,11), the responsible factor may involve a product of the extramitochondrial protein synthetic system. The majority of mitochondrial enzymes are probably specified by nuclear genes, synthesized completely or in part on cytoplasmic ribosomes and are then transported into the mitochondrion (see ref. 6). It is possible that a protein factor or enzyme(s) involved in the replication of mtDNA, such as the above-mentioned "nickase", is manufactured in the cytoplasm and then transferred into the organelles. Inhibition of cytoplasmic protein synthesis may lead to a deficiency of such enzyme (s), either by interference with its synthesis or by perturbation of its transfer into or integration with the mitochondrial structure. Whether any recombinational events play a role in the formation of oligomeric mtDNA is not known. Comparative studies of mtDNA structure and replication and the intra- and extramitochondrial protein synthetic apparatus in control and wild-type virus- or T5-infected cells are in progress.

ACKNOWLEDGEMENTS

This work was supported by grant NP-93 from the American Cancer Society and grant RO1-CA13814 from the U.S. Public Health Service. I thank Drs. I. Macpherson, G.S.Martin, M.Weber, L.Warren for samples of RSV virus, and Mrs. D.Reinhart, A.Trischitta and H.Williams for excellent technical assistance.

REFERENCES

1. W. Goebel and D.R. Helinski, Proc.Natl.Acad.Sci. U.S.A. 61, 1406 (1968).
2. R. Jaenisch and A.J. Levine, J.Mol.Biol. 73, 199 (1973).
3. F. Cuzin, M. Vogt, M. Dieckmann and P. Berg, J. Mol.Biol. 47, 317 (1970).

4. W. Goebel, Eur.J.Biochem. 15, 311 (1970).

5. M.M.K. Nass, in: Hormones and Cancer (K.W. McKerns Ed.) Academic Press, New York, in press.

6. P. Borst, Ann.Rev.Biochem. 41, 333 (1972).

7. M.M.K. Nass, Science 165, 25 (1969).

8. C. Paoletti and G. Riou, Bull.du Cancer 57, 301 (1970).

9. D.A. Clayton and J. Vinograd, Proc.Natl.Acad.Sci. U.S.A. 62, 1077 (1969).

10. M.M.K. Nass, Nature 223, 1124 (1969).

11. M.M.K. Nass, Proc.Natl.Acad.Sci.U.S.A. 67, 1926 (1970).

12. C. Paoletti, G. Riou and J. Pairault, Proc.Natl. Acad.Sci.U.S.A. 69, 847 (1972).

13. G.S. Martin, S. Venuta, M. Weber and H. Rubin, Proc.Natl.Acad.Sci.U.S.A. 68, 2739 (1971).

14. L. Warren, D. Critchley and I. Macpherson, Nature 235, 275 (1972).

15. M.M.K. Nass, Proc.Natl.Acad.Sci.U.S.A., in press (1973).

16. G.S. Martin, Nature 227, 1021 (1970).

17. M.M.K. Nass, Exp.Cell Res. 72, 211 (1972).

18. M.M.K. Nass and Y.Ben-Shaul, Biochim.Biophys.Acta 272, 130 (1972).

19. M.M. Burger, Fed.Proc. 32, 91 (1973).

20. G. Soslau, J.P. Fuhrer, M.M.K. Nass and L. Warren, submitted.

21. G.C. Van Tuyle and G.F. Kalf, Arch.Biochem.Biophys. 149, 425 (1972).

22. D. Dressler, Proc.Natl.Acad.Sci.U.S.A. 67, 1934 (1970).

23. J.A. Kiger,Jr. and R.L. Sinsheimer, J.Mol.Biol. 40, 467 (1969).

PROPAGATION AND RECOMBINATION OF PARENTAL mtDNAs IN HYBRID CELLS.

Igor B. Dawid, Ivan Horak and Hayden G. Coon,

Department of Embryology, Carnegie Institution of Washington, Baltimore, Maryland 21210,
Laboratory of Cell Biology, National Cancer Institute, Bethesda, Maryland 20014.

The lack of genetic tools has been a serious restriction in the study of mitochondrial biogenesis and function in animal cells. One approach to the introduction of genetics into this field involves the use of hybrid somatic cells. Before the usefulness of such cells can be evaluated several facts must be established experimentally. First among these it must be demonstrated that cells can be generated which contain various combinations of the mitochondrial and nuclear genomes from different cells and different animal species. We report that such hybrid cells were obtained, and that in many of these cells the parental mtDNAs or parts thereof are physically linked in a way which is best described as recombination.

CELL FUSION AND THE PROPERTIES OF HYBRID CELLS.

Hybrid cells were produced with the aid of Sendai virus from cells of freshly dissociated tissues of the embryonic rat or mouse, and the human cell line VA-2. Hybrid colonies were grown in HAT medium under conditions described (ref. 1; Coon *et al.*, to be published). After about 25 population doublings the chromosomes of each cell strain were inspected and the hybrid nature of the cells was established. After additional 15 to 35 doublings samples were taken for karyological and biochemical analysis. Some of the cell strains were cultured further in the ways described

below. Hybrid cell strains derived from freshly liberated, primary rodent cell (either embryonic or adult)
and a human cell line, VA-2, show the heretofore un-
common property of segregating either human or rodent
chromosomes. This conclusion is supported by both ka-
rotype and isoenzyme studies that will be presented
elsewhere (Coon and Minna; Minna and Coon, Horak, and
Dawid; to be published), and by the analyses below.

DISTINGUISHING mtDNAs BY MOLECULAR HYBRIDIZATION.

Our method for distinguishing the mtDNAs from different animals is based on the extensive differences
in nucleotide sequence between these DNAs. We prepared complementary RNA (cRNA) transcribed by *E.coli*
RNA polymerase from pure mtDNA and labeled one cRNA
with [^{3}H] (e.g., rat [^{3}H]mt-cRNA), the other with
[^{32}P] (e.g., human [^{32}P]mt-cRNA). These cRNA's were
mixed and hybridized to a series of filters containing mixtures of rat and human mtDNA in known proportions. After the filters were processed the [^{3}H] [^{32}P]
ratio of RNA bound to each filter was plotted against
the percentage of rat mtDNA in the mixture of DNA on
the filter. In this way a standard curve was obtained
(Fig. 1). When filters containing DNA of unknown composition were hybridized in the same vial the isotope
ratio of RNA bound to these filters gave a quantitative measure of the proportion of rat mtDNA in that
DNA sample. The proportion of nuclear DNA (nDNA) of
two species in a hybrid cell sample could be estimated by the same method, using cRNA transcribed from
nuclear DNA. While the latter assay is less accurate,
we found that the results obtained from it correlated
quite well with the ratio of chromosomes of the two
species. Additional details and technical considerations of these assays are presented elsewhere (Coon,
Horak, and Dawid, to be published).

RAT - HUMAN HYBRID CELLS.

Among 33 rat-human strains 19 contained both rat
and human nDNAs as well as both mtDNAs when first tested, some 40 to 60 population doublings after fusion.

Another 14 strains contained only human DNA at this point; these were segregants rather than parentals, since the cell lines had been found to be hybrid with respect to morphology of chromosomes at an earlier point in their history. No pure rat segregants were found at this time. After additional periods of growth some of the cell strains contained pure rat mtDNA, but no strains were observed which had lost all human chromosomes. A positive but loose correlation prevailed between the percentage of mtDNA and nDNA of the same species in each cell strain.

To ascertain that both rat and human mtDNAs were present in the same cell in the hybrid cell strains we derived sets of subclones from several strains. After these subclones had grown up they were tested for their mtDNA and nDNA composition. Both rat and human mtDNAs, as well as both nuclear DNAs, were found in most members of five sets of subclones. Only in 5 out of 40 individual subclones all human mtDNA had been lost, but human chromosomes (nuclear DNA) were still present. These experiments show that individual cells in hybrid strains at a time 40 to 60 doublings after fusion, contained both rat and human mt-DNA sequences.

MOUSE-HUMAN HYBRID CELLS.

Our results with these cells are summarized in Fig. 2. In this graph the percentage of mouse mtDNA in a particular cell strain is plotted against the percentage mouse nDNA in the same strain. Arrows connect measurements made at one time to measurements on the same cell strain about 25 doublings later. Open circles denote subclones. Several facts emerge. First, all cell strains which contain more than 50 % mouse nDNA (and chromosomes) contain pure mouse mtDNA. This may be a general property of mouse-human hybrid cell and may explain the results of Clayton *et al.* (2) and of Attardi and Attardi (3); these authors found exclusively mouse mtDNA in hybrid cells containing a predominance of mouse chromosomes. Second, in spite of this apparent "strength" of the mouse mtDNA we have

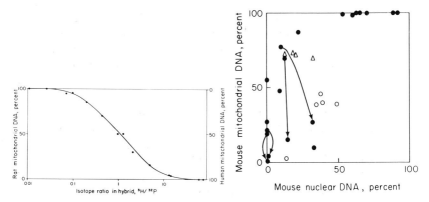

Fig. 1. Calibration curve for the determination of the proportion of rat and human mtDNAs. A set of filters were loaded with known mixtures of the two mtDNAs. They were hybridized with a mixture of [^{3}H]human-mt-cRNA and [^{32}P]rat-mt-cRNA. The [^{3}H]/[^{32}P]ratio of RNA bound to each filter was plotted against the percentage of rat mtDNA on the filter.

Fig. 2. DNA composition of mouse-human hybrid cell strains. The percentage of mouse mtDNA is plotted against the percentage of mouse nDNA for each hybrid cell strain. The percentage of human DNA is 100 minus percent mouse DNA. Each point represents an independently derived hybrid cell strain, except for points connected by arrows; these show the change in DNA composition during continued growth of a strain. Open circles denote subclones, i.e., clones derived from hybrid cells. Triangles: data based on chromosome counts rather than nDNA determinations.

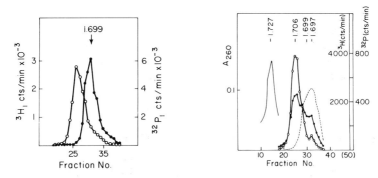

Fig. 3. Separation of mouse and human mtDNAs by banding in CsCl. A mixture of the two whole-cell DNAs was centrifuged in CsCl, the gradient fractionated, and each fraction adsorbed to a filter. The filters were hybridized with a mixture of [^{3}H]-human-mt-cRNA and [^{32}P]mouse-mt-cRNA. Open circles: [^{3}H]; closed circles: [^{32}P].

Fig. 4. Buoyant density analysis of mtDNA from hybrid cell strain VME 15. This strain contained 27 % mouse mtDNA. The analysis was carried out, and the symbols have the same meaning, as described in the legend to Fig. 3.

found that cells which have lost all the mouse chromosomes lose all mouse mtDNA as well. This loss may be slow, however: we observed several strains which contained no measurable mouse nDNA (which means less than 5 %) but did contain large amounts of mouse mtDNA (Fig. 2). These strains lost their mouse mtDNA during the next 25 generations of growth (Fig. 2). Third, cell strains containing a low moderate proportion of mouse chromosomes are relatively stable in retaining both mouse and human mtDNAs.

APPARENT RECOMBINATION OF RODENT AND HUMAN mtDNA IN HYBRID CELLS.

In a cell containing sequences of mtDNA from two different species several possible arrangements may be considered. One arrangement would have separate mitochondria from each species coexisting independently in the same cell. Since mitochondria appear to fuse *in vivo* (*cf*. ref. 4) this possibility is not very likely. Second, intact molecules of human and of rodent mtDNA could coexist in the same organelles. Third, the different mtDNAs could interact and be linked to each other in various ways. To distinguish between the last two possibilities we carried out experiments based on the technique illustrated in Fig. 3. Human and mouse mtDNAs differ in density by about 8 mg/cm^3. A mixture of the two mtDNAs (actually whole-cell DNA of which the mtDNA forms a small fraction) was banded in a CsCl gradient, the gradient was fractionated, and each fraction adsorbed to a membrane filter. The filter set was hybridized with a mixture of differentially labeled cRNA, as described above for the quantitative assay of human and rodent mtDNAs. Each mtDNA binds its own cRNA and there is very little cross-hybridization (Fig. 3). There is some overlap between the two bands, due to the small density difference between the DNAs.

The DNA from several hybrid cell strains was analyzed in such gradients. Most strains showed a pattern substantially different from the control. Figure 4 gives an example, using a cell strain which contained

27 % mouse mtDNA. The human cRNA ($[^3H]$cRNA) hybridized to a band with a peak about at the density of human mtDNA and a shoulder towards the light side. The mouse $[^{32}P]$cRNA hybridized with sequences banding at about the same density as the human sequences, and also with a lighter band. The latter band had a density equal to or slightly higher than the density of pure mouse mtDNA. We interpret this experiment to indicate the presence of molecules which consist predominantly of human mtDNA sequences and of a smaller amount (10 to 20 %) mouse mtDNA sequences, physically linked together. Such molecules would band at the density of human mtDNA (or perhaps 1 mg/cm^3 lower, which cannot be reliably determined in preparative gradients), but they would bind significant amounts of $[^{32}P]$mouse-cRNA. Such molecules should account for the main peak which binds both types of cRNA. In addition, this peak might contain "clean" human mtDNA molecules. Our present technology does not allow the distinction between a peak containing only molecules which are 90 % human and 10 % mouse, from a peak in which half the molecules are pure human, the other half are 80 % human and 20 % mouse. In addition, the hybrid DNA sample contains molecules which band at a density close to that of mouse mtDNA. This material could be pure mouse mtDNA, or it could contain recombinants of the opposite composition, i.e., molecules composed of mostly mouse sequences and some human sequences. Again any combination of the above is also possible.

Alternate explanations must be considered for the experiment in Fig. 4. Instead of true recombination, i.e., the covalent linkage of parts of mouse mtDNA to parts of human mtDNA, we might be dealing with very special catenanes (*cf.* ref. 5). These catenanes would have to involve differently-sized human and mouse partners, since the linked ("recombinant") molecules contain more human sequences than mouse sequences (Fig. 4). Thus, a full-size human circle would have to be linked to a small circle derived by a large deletion from mouse mtDNA. This, and other interpretations appear less likely than standard recombination

TABLE I.

RECOMBINANT mtDNA IN RODENT-HUMAN HYBRID CELLS.
Strains carrying a letter at the end of their identification are subclones derived from hybrid cell populations.The basis of identification of recombinant
mtDNA is discussed in the text.

| Cell Strain | Percent of | | Recombinant mtDNA |
	Mouse mtDNA	Mouse nDNA	
VMSC 10F	7	N.D.	+
VMSG 16	20	1	+
VME 15	27	32	+
VMSC 10G	28	N.D.	+
VMPOC 15F	39	49	+
VMPSC 10D	39	35	+
VMPE 18F	40	39	+
VMPOC 18E	46	38	+
VMPth 18	48	9	+
VMPOC 15L	69	32	−
VMPOC 15B	73	20	−
VMSC 10E	71	13	−
VMSC 10B	73	19	−
	Percent of		
	Rat mtDNA	Rat nDNA	
VRE 36A	15	58	+
VRSC 21-5S	84	89	+
VRSC 21-5N	89	90	+
VRSC 21-8	95	81	+
VRSC 33	24	8	−

but will have to be investigated in future experiments.

A number of mouse-human and rat-human hybrid cell
strains were analyzed by density gradient centrifugation (Table I). Cases with and without apparent recom-
bined mtDNA molecules were found. The limited resolution of the method makes it impossible to exclude a
low level of recombination. Therefore, those hybrid
strains classified as "nonrecombinant" may actually
contain recombinant molecules, but these molecules
would have to be a minority in the population. With
this reservation in mind we classified the results of
these experiments in Table I. The following generali-
zations can be made. First, apparent recombination of
parental mtDNAs in these hybrid cells is not a rare

event. Second, recombination was observed both in "high-human"and in "high-rodent" hybrids. Third, there is heterogeneity in the mtDNA population of cells containing recombinant molecules. Gradients like the one shown in Fig. 4 do not have sufficient resolution to assess this heterogeneity in detail. The minimum heterogeneity needed to explain Fig. 4 would invoke only two types of molecules: "high-human recombinants" (with about 90 % human, 10 % mouse sequences), and "high-mouse recombinants". Alternately, the gradient can be explained in terms of only one type of recombinant molecules in addition to pure human and pure mouse mtDNA molecules. It is also possible that this - and other similar gradients - actually contain a very heterogenous population of molecules ranging from pure human via many types of recombinants to pure mouse mtDNA molecules. Whatever its extent, this heterogeneity is remarkable in relation to the established homogeneity of mtDNA in the cells of any one animal species (*cf.* ref. 6).

The studies reported here demonstrate that mtDNAs, or sequences of mtDNAs, from different animal species can coexist in the same cell and replicate for extended periods. Furthermore, such mtDNAs frequently become linked to each other, most likely in a covalent way best described as recombination. These experiments may provide one basis for the application of genetic techniques to the study of mitochondrial biogenesis.

REFERENCES.

1. H.G. Coon & M.C. Weiss, Proc.Nat.Acad.Sci.U.S.A. 62, 852 (1969).
2. D.A. Clayton, R.L. Teplitz, M. Nabholz, H. Dovey & W. Bodmer, Nature 234, 560 (1971).
3. B. Attardi & G. Attardi, Proc.Nat.Acad.Sci U.S.A. 69, 129 (1972).
4. A.L. Lehninger, in The Mitonhondrion, Ed. W.A. Benjamin, New York, p. 30 (1965).
5. B. Hudson & J. Vinograd, Nature 216, 647 (1967).
6. P. Borst, Ann.Rev.Biochem. 42, 333 (1972).

MITOCHONDRIAL MUTATIONS IN PARAMECIUM: PHENOTYPICAL CHARACTERIZATION AND RECOMBINATION

André Adoutte[*]

Laboratoire de Génétique - Université Paris-Sud, 91405 Orsay (France).

As previously done in yeast (1,2,3), mutants resistant to various antibiotics inhibiting mitochondrial protein synthesis have been isolated in Paramecium and shown by genetical analysis (4,5) and by microinjection techniques (6) to be located in mitochondrial DNA. The antibiotics used as selective agents (erythromycin, chloramphenicol, spiramycin) are known to act on the 50S subunit of the bacterial ribosome (7). In this paper we show that the resistant mutants belong to several distinct phenotypical classes and present preliminary evidence suggesting the occurence of genetic recombination between two mutants belonging to very distinct classes.

CHARACTERIZATION OF THE MITOCHONDRIAL MUTANTS

Several phenotypic classes are found among antibiotic - resistant mutants. From the same wild-type stock of *Paramecium aurelia* (species 4), some 100 mutants have been isolated (either spontaneously or after U.V. mutagenesis) using either erythromycin, spiramycin or chloramphenicol as a selective agent. About 50 of these mutants have been classified according to several criteria: a) level of resistance (*i.e.* growth rate in various concentrations of each of the three antibiotics); b) thermosensitivity (*i.e.* growth rate and survival at 36° C; at this temperature the wild-type strain grows quite well and undergoes 5 fissions

[*]Present address: Centre de Génétique Moléculaire - C.N.R.S. - 91190 Gif-sur-Yvette (France).

per day whereas several mutant strains are affected); c) transmission efficiency of the mutated mitochondria during vegetative growth, when confronted to wild-type mitochondria in "mixed" cells (see below).

TABLE I

PHENOTYPIC CLASSIFICATION OF THE MITOCHONDRIAL MUTANTS

The level of resistance of a strain to the three antibiotics tested is measured as its daily growth rate in the selective medium, recorded from single cell isolations. All tests are performed at 27° C. +++: 4 to 5 fissions/day (equal to growth in the absence of antibiotic); ++: 3 fissions/day; +: 1 to 2 fissions/day; ε: 1 fission/day or less.

In the column growth at 36° C: + is the wild-type growth (5 fissions/day); <+ indicates that the cells grow slowly and die after 7 to 8 days; <<+ indicates that the cells grow slowly and die after 5 to 6 days; <<<+ indicates that the cells grow very slowly and die after 3 to 4 days.

"Transmission efficiency" is measured by confronting mutant and wild-type mitochondria, usually in "doublet" cells (14). <+ indicates that mutant mitochondria are eliminated after 30-40 fissions; <<+ indicates that mutant mitochondria are eliminated after 20-30 fissions; <<<+ indicates that mutant mitochondria are eliminated after 15-20 fissions; ? indicates that the test has not been performed.

Several mutants have been isolated and characterized by Dr. Annie Sainsard. ERY=erythromycin; SPI=spiramycin; CAP=chloramphenicol; MIK=mikamycin.

Class	Number of mutants	Selective agent	Name of one representative strain	Resistance to ERY 400 μg/ml	SPI 100 μg/ml	CAP 800 μg/ml	MIK 200 μg/ml	250 μg/ml	Growth at 36° C	Transmission efficiency
E^R										
A-Highly E^R	15	ERY or SPI	E^R_{102}, s^R_1	+++	+++	++	−	−	+	< +
B-Moderately E^R										
B_1	7	SPI	E^R_{111}, E^R_{37}	+	++	+	ε	?	<<+	<<+
B_2	6	SPI	E^R_{104}	+	++	+	−	?	< +	?
B_3	1	SPI	E^R_{110}	+	++	+	−	+	+	?
C-Weakly E^R	6	SPI	E^R_1	−	++	−	ε	−	<<<+	<<<+
D-Very weakly E^R	1	ERY	E^R_{50}	−	+	−	+	++	+	?
s^R	1	SPI	s^R_7	−	ε		?	?	+	?
c^R	3	CAP	c^R_4	−	ε	−		−	+	+
wild-type, stock d4-2				−	−	−	−	−	+	+

The results are reported in Table I. Several observations can be made:

1) At least 8 phenotypical classes can be distinguished on the basis of the criteria listed. The phenotypes range from "highly erythromycin-resistant – chloramphenicol-sensitive" to "very weakly erythromycin-resistant-highly chloramphenicol resistant" with several intermediate types. Highly erythromycin-resistant strains are always chloramphenicol sensitive. As the level of resistance to erythromycin decreases, the level of chloramphenicol-resistance increases. There appears to exist a reverse correlation between the levels of resistance to the two antibiotics.

2) The great majority of mutants isolated for resistance towards erythromycin are cross-resistant to spiramycin and *vice versa* and the level of resistance is also usually homologous (*i.e.* high E^R are also high S^R, etc.). Some rare exceptions are found (S^R class).
3) As found for E^R mutants in species 1 of *Paramecium aurelia* (18), one very weakly erythromycin-resistant mutant (E^R_{50}) has been found to be highly resistant to mikamycin.
4) The 3 C^R mutants isolated are thermoresistant. Among the other types no clear correlation is found between level of resistance and thermosensitivity.

These results show several striking similarities with those reported for yeast (8,9,10 and this volume) which suggests a common molecular basis for these mutations in the two organisms. In yeast, evidence that one erythromycin-resistant mutant possesses altered mitochondrial ribosomes has been provided by Grivell *et al.* (11) and the demonstration has now been extended to a range of E^R, C^R and S^R mutants (8). Direct evidence for an alteration of the mitochondrial ribosomes in an E^R mutant of *Paramecium* has also been obtained by Tait (12). Furthermore evidence is accumulating in favor of the location of the alterations in the ribosomal RNA. (8,13 and this volume). If this is indeed the case, our results would indicate that a relatively high number of sites can be modified in the ribosomal RNA to yield antibiotic-resistant ribosomes displaying a range of pleiotropic phenotypes, and would provide a genetic approach to the study of RNA- protein interactions in the mitochondrial ribosome.

The mutations are slightly deleterious in Paramecium. Two lines of evidence indicate that most antibiotic-resistant mutations have some negative effects on mitochondria in *Paramecium*:
a) Mutated mitochondria are overcome by wild-type mitochondria in "mixed" cells.

Two different types of mitochondria can be brought together within the same cell using the cytoplasmic exchanges that occur at conjugation between mates.

Numerous combinations between different mutated mitochondria and wild-type mitochondria have been studied.
It was found that, during vegetative growth of the
"mixed" cells in normal medium, mutant mitochondria
are always progressively lost, the loss being more or
less rapid depending on the mutant considered (14).
Mutant mitochondria thus appear to replicate more slowly than wild-type mitochondria. It must be noted that
this test is very sensitive and allows the distinction
between mitochondria differing only very slightly in
replication rate.
b) This counter-selection may be due to unpaired mitochondrial protein synthesis in the mutants.
   Several double-resistant ($C^R E^R$) strains have been
isolated by selecting for E^R mutations in C^R strains.
Most of the double-resistant strains show several
clear-cut phenotypic defects. They are slow growing
(*i.e.* generation time of 10 h *vs* 6 h for the wild-
type strain), highly thermosensitive, their mitochondria are very quickly eliminated in mixed cells (14).
Furthermore one of these mutants (CE^R_2) shows, in normal medium, a decrease in the cytochrome *c* oxidase absorption peak as well as a decrease in the amount of
mitochondrial cristae as observed at the EM (15). These are two characteristic effects of erythromycin or
chloramphenicol on sensitive mitochondria (15,16).
Thus the double-resistance mutations mimick, in some
ways, the effects of inhibition of mitochondrial protein synthesis. We assume therefore that the accumulation of mutations in the mitochondrial ribosome leads
to a less efficient mitochondrial protein synthesis
which, in turn, accounts for the various phenotypical
characteristics of the double-resistant strains. E^R
strains also show some slight defects (15,17).

PRELIMINARY EVIDENCE FOR RECOMBINATION BETWEEN MITO-
CHONDRIAL MARKERS.

   The search for recombination between mitochondrial
markers has been carried out mainly by using the C^R
and E^R mutations since these are the most clearly different phenotypic categories . Attempts to isolate

other types of mitochondrial mutations have been un-
successful so far.

We have tried for a long time to isolate double-
resistant (C^RE^R) strains from various C^RE^S x C^SE^R combinations but no clear-cut conclusions could be drawn
from these experiments due to the relative high frequency of spontaneous E^R mutations in C^R strains. We
therefore tried to isolate double-sensitive clones out
of the same type of combinations. The isolation of
double-sensitive cells is a more reliable criterion
for the occurence of some genetic exchange between C^R
and E^R mitochondria since several pure C^R and E^R
strains kept in the laboratory for more than three
years, some of them in continuous growth, have never
yielded sensitive revertants.

One C^RE^S x C^SE^R combination has been extensively
studied: 29-7E^R_6 x d4-2 8C^R_4 (29 and d4-2 designate 2
geographically different stocks of the same species
(4) of *Paramecium aurelia*. Strain 29-E^R_6, isolated by
Dr. Annie Sainsard, is a highly erythromycin-resistant
strain, showing no cross-resistance to chloramphenicol
(class A type, Table I), whereas strain 8C^R_4 shows a
weak cross-resistance to erythromycin). Starting with
a pair that had undergone cytoplasmic exchange, clones
were obtained from each ex-conjugant. These clones were subcloned in normal medium every 10 fissions by
transfer of one cell and were tested at regular intervals. C^R mitochondria show a selective advantage over
E^R ones; originally E^R cells become progressively enriched with C^R mitochondria during vegetative growth;
by about the 60th fission after conjugation subclones
derived from the E^R ex-conjugant appear to be almost
"pure C^R". Such clones have been further subcloned in
normal medium, and in a first experiment, all the lines (30/30) that were maintained, starting from one
E^R ex-conjugant, yielded double-sensitive (C^SE^S) cells.
It must be pointed out that these double sensitive
cells do not result from a sudden transformation of
the C^RE^S cells into C^SE^S ones but appear as the term
of a progressive dilution of the C^R mitochondria by a
new (C^SE^S) mitochondrial type (By the 80th, 90th, ...

fission, the cells display an increasing lag before
starting to grow in chloramphenicol-containing medium,
indicative of a decrease in the amount of C^R mitochondria per cell. However, once the cells have started
growing in chloramphenicol they recover their full resistance even if transfered back and maintained for
numerous generations in normal medium, again typical
of mixed $C^R + C^S$ CELLS).

The double-sensitive cells have been back-crossed
to wild-type and $8-C_4^R$ strains and the C^SE^S character
behaved in all respects as a purely cytoplasmic one.
The cross has been repeated and again in one pair extensively studied several subclones yielded pure sensitive cells by about the 120th fission after going
through the characteristic phases just described. 15
independent clones of the E^R parent and 30 of the C^R
parent where grown in parallel as controls and none
yielded sensitive cells (Table II). The double-sensitive mitochondria, therefore, do not appear to be due
either to reversion of one of the two parental types
or to the effect of some nuclear "modifier". It can
be concluded that double-sensitive mitochondria most
probably arose from recombination between C^RE^S and
C^SE^R mitochondria.

The difficulties in isolating recombinants in *Paramecium* suggest the following comments:
a) The very high number of generations needed to isolate pure sensitive cells and the fact that in one
case, only some subclones from the same E^R ex-conjugant clone finally yielded C^SE^S cells (Table II) may
be attributed to three factors. 1. The very high number of mitochondria per cell in *Paramecium* (several
thousands) renders the sorting out of cells pure for
one mitochondrial type a lengthy process. This problem would be overcome if the mitochondrial type that
is looked for had a clear-cut selective advantage over
the other types. In the situation studied here it is
known, on the contrary, that C^SE^S mitochondria have
only a very slight selective advantage over C^RE^S ones
(14). 2. An important drift is introduced by the subcloning operated every 10 fissions. This fact, added

TABLE II

THE PROCESS OF APPEARANCE OF $C^S E^S$ CLONES IN CROSS 29-E_6^R x d_{4-2} C_4^R.
This table indicates the results of the tests performed during the vegetative growth of one E^R ex-conjugant (30a) that has received a small amount of cytoplasm from its C^R partner. Starting from the ex-conjugant clone, 3 subclones were first maintained by regular subcloning of one cell every 10 fissions. After 65 fissions the 3 subclones were further subcloned each into 5 clones which were again maintained by subcloning every 10 fissions. At successive stages, tests are carried out in 100 μg/ml erythromycin (E) and 200 μg/ml chloramphenicol (C) and the results are given under the respective columns E and C. + indicates that the cells grow (i.e. are resistant); - indicates that the cells are blocked (i.e. are sensitive); t indicates that the cells start growing after a lag of one to a few days (indicating that the cells are mitochondrially mixed).

Full results have been given for the ex-conjugant whereas for the two control parental strains (29 E_6^R and d_{4-2} $8C_4^R$) they have been abridged, all subclones yielding identical results. Cells of the C^R partner behaved like the C_4^R control.

Number of fissions in normal medium after conjugation

	25 E	25 C	45 E	45 C	65	125 E	125 C	145 E	145 C	180 E	180 C
30a (ex E_1^R conjugant)	+	-	+	-		-	-	-	-	-	-
						-	+	-	+	-	+
						-	+	-	+	-	+
						-	-	-	-	-	-
	+	t	-	+		-	-	-	-	-	-
						-	+	-	+	-	+
						-	+	-	+	-	+
						-	+	-	+	-	+
	+	t	-	+		-	+	-	+	-	+
						-	+	-	+	-	+
						-	+	-	+	-	+
						-	-	-	-	-	-
						-	+	-	+	-	+
29 E_6^R	+	-	+	-		+	-	+	-	+	-
	+	-	+	-		+	-	+	-	+	-
	+	-	+	-		+	-	+	-	+	-
d_{4-2} C_4^R	-	+	-	+		-	+	-	+	-	+
	-	+	-	+		-	+	-	+	-	+
	-	+	-	+		-	+	-	+	-	+

to the preceeding one, may explain that certain subclones remain indefenitely C^R. 3. Finally, it is possible that in the precise combination studied the frequency of recombination is low. Quantitative estimates are difficult to make yet.

b) The failure to isolate double-resistant recombinants may be due to the fact that double-resistant mitochondria arising by recombination are non-functional since, as described above most $C^{R}E^{R}$ mitochondria isolated by mutation are defective.

Mitochondrial combinations in which both parental types have a clear-cut selective disadvantage with respect to wild-type mitochondria are now under study.

ACKNOWLEDGEMENTS

I thank Mme Michèle Rossignol for excellent assistance and Dr. Janine Beisson for support throughout this work and many helpful comments on the manuscript.

REFERENCES

1. A.W. Linnane, G.W. Saunders, E.B. Gingold and H. B. Lukins, Proc.Natl.Acad.Sci.U.S.A. 59, 903 (1968).
2. D.Y. Thomas, and A. Wilkie, Genet.Res. 11, 33 (1968).
3. D. Coen, J. Deutsch, P. Netter, E. Petrochilo and P.P. Slonimski, In: Twenty-fourth Symp. Soc.Exp. Biol. (Cambridge University Press, London, 1969).
4. G.H. Beale, Genet.Res. 14, 341 (1969).
5. A. Adoutte and J. Beisson, Mol.Gen.Gen. 108, 70 (1970).
6. G.H. Beale, J. Knowles and A. Tail, Nature 235, 396 (1972).
7. S. Pestka, Ann.Rev.Microbiol. 25, 487 (1971).
8. L.A. Grivell, P. Netter, P. Borst and P.P. Slonimski, Biochim.Biophys.Acta, 312, 358 (1973).
9. P.P. Slonimski, G. Perrodin, M. Bolotin, D. Coen, J. Deutsch, B. Dujon, P. Netter and E. Petrochilo, Biochimie (in the press).
10. C. Bunn, G.H. Mitchell, H.B. Lukins and A.W. Linnane, Proc. Natl.Acad.Sci.U.S.A. 67, 1233 (1970).
11. L.A. Grivell, L. Reijnders and H. de Vries, FEBS Letters 16, 159 (1971).
12. A. Tait, FEBS Letters 24, 159 (1971).
13. G. Faye, H. Fukuhara, C. Grandchamp, J. Lazowska, F. Michel, J. Casey, G.S. Getz, J. Locker, M. Rabinowitz, M. Bolotin-Fukuhara, D. Coen, J. Deutsch,

B. Dujon, P. Netter and P.P. Slonimski, Biochimie 55 (in the press).

14. A. Adoutte and J. Beisson, Nature 235, 393 (1972).
15. A. Adoutte, M. Claisse and M. Balmefrezol, Unpublished data.
16. A. Adoutte, M. Balmefrezol, J. Beisson and J. Andre, J.Cell Biol. 54, 8 (1972).
17. A. Adoutte, A. Sainsard, M. Rossignol and J. Beisson, Biochimie 55 (in the press).
18. G.H. Beale, and G. Spurlock, Mol.Gen.Gen. (In the press).

Part II

Characteristics of the Mitochondrial Protein Synthetic Machinery

MUTATIONS AFFECTING MITOCHONDRIAL RIBOSOMES IN YEAST

L.A. Grivell

Section for Medical Enzymology, Laboratory of Biochemistry, University of Amsterdam, Eerste Constantijn Huygensstraat 20, Amsterdam, The Netherlands.

In the last few years, considerable interest has developed in the isolation and characterisation of mutants of mtDNA. Interest in such mutants stems, naturally enough, from our curiosity to know more about the function of this DNA, since apart from the two ribosomal RNAs (rRNAs) and some tRNAs, no other gene products have yet been positively identified. In this article the properties of several yeast mutants, resistant to antibiotics as a result of mutation in mtDNA, are described. Identification of the component(s) altered in such mutants provides one of the most promising approaches yet available for the discovery of new gene products of mtDNA.

MUTANTS RESISTANT TO CHLORAMPHENICOL, ERYTHROMYCIN AND/OR SPIRAMYCIN

Isolation and general properties. Mutants resistant to chloramphenicol, erythromycin and/or spiramycin are isolated on the basis of their ability to grow on non-fermentable media in the presence of antibiotic. Selection of both fast and slow growing colonies has given a spectrum of mutants with differing properties (1). Plates are unsuitable for quantitation of these differences since small effects on growth rate lead to large variations in final colony size. Further, local changes in pH, antibiotic concentration, etc. lead to changes in the response of cells which are difficult to interpret. Characterisation is best performed by measurement of logarithmic growth in liquid glycerol-antibiotic media. Results summa-

TABLE I

PROPERTIES OF ANTIBIOTIC-RESISTANT MUTANTS: GROWTH ON ANTIBIOTIC-CONTANING MEDIA
Data taken from ref. 1.

	Concentration (mg/ml) of antibiotic giving 50 % inhibition		
	CAP	ERY	SPI
$C^S E^S$	2.4	0.046	0.39
E^R_{514}	2.4	>5	>8
E^R_{354}	1.2	>5	6.7
E^R_{353}	>4.0	2.2	1.0
E^R_{553}	2.0	4.2	0.31
C^R_{321}	>4	0.078	1.4
C^R_{323}	>4	0.14	1.5
S^R_{352}	>4	0.17	>8

rised in Table I show that in this way four classes of erythromycin-resistant mutant can be distinguished. E^R_{514} and E^R_{354} are cross-resistant to spiramycin, but differ in their level of resistance; E^R_{353} is moderately resistant to chloramphenicol and marginally resistant to spiramycin; E^R_{553} displays a low resistance to erythromycin only. Two types of chloramphenicol-resistant mutants have been isolated. Both display low cross-resistance to erythromycin and spiramycin and are distinguishable by their quantitative response to these antibiotics. Spiramycin-resistant mutant 352 also carries resistance to chloramphenicol and to erythromycin. None of the mutations appear to be deleterious to mitochondrial function, since in the absence of antibiotics all mutants exhibit growth rates similar to that of the wild type.

Genetic characterisation. Although nearly all of the mutants are pleiotropic, none of the observed cross-resistances segregate in genetic crosses. It is likely, therefore, that they result from a single primary mutation. Netter *et al.* (2) have amassed sufficient recombination data to permit assignment of several of the mutants to three loci on a linkage map of mitochondrial genes (Fig. 1). Mutants resistant to chloramphenicol map close to ω, the gene influencing mitochondrial recombination. Mutants resistant to erythromycin and/or spiramycin map at the other two loci. An unusual feature of this map is that mutants of widely differing phenotype are to be found at each locus. It is possible, however, that present genetic techniques are incapable of the necessary resolution.

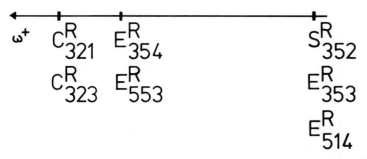

Fig. 1. Preliminary linkage map of mitochondrial antibiotic resistance genes. From ref. 2.

Biochemical properties. Resistance in these mutants may originate from possession of altered mitochondrial ribosomes, or from an altered permeability of the mitochondrial membrane, such that the antibiotic is unable to interact with otherwise sensitive ribosomes. Distinction between these two alternatives is easily possible by a comparison of the antibiotic response of intact mitochondria with that of purified ribosomes (3). For intact mitochondria, the inhibitory effects of chloramphenicol, erythromycin, lincomycin, carbomycin and spiramycin on the incorporation of $[^{14}C]$leucine were measured. Rapid screening of ribosomes was possible by use of a modified fragment reaction for assay of peptidyl transferase activity. Effects of chloramphenicol, lincomycin and spiramycin can be measured directly. Erythromycin does not inhibit peptidyl transferase activity. Use was, therefore, made of its ability to displace chloramphenicol from erythromycin-sensitive, but not erythromycin-resistant ribosomes and thus relieve chloramphenicol inhibition of peptidyl transferase activity (4).

Fig. 2 summarizes the results of this approach for erythromycin-resistant mutant 514. The behaviour of the isolated mitochondria correlates well with *in vivo* data. The strain is, in addition, resistant to carbomycin and lincomycin (data not shown). This pattern of resistance is retained by purified mitochondrial ribosomes. This mutant, therefore, possesses altered mitochondrial ribosomes.

The properties of several other mutants are summarized in Table II. By and large, the pattern of response observed *in vivo* is retained by both mitochondria and ribosomes *in vitro*. All mutants probably possess altered mitochondrial ribosomes.

These results also underline further the differences between erythromycin-resistant mutants 354, 514 and 553. 514 and 354, besides being distinct genetically, can be further distinguished in that 514 is resistant to lincomycin, while 354 is not. 553, as might be predicted from its response *in vivo*, is only poorly resistant to erythromycin at both mitochondrial and ribosomal levels. The significance of a slight resistance of ribosomes from this mutant to spiramycin is at present not clear.

ARE MUTANT RIBOSOMES DEFECTIVE RIBOSOMES?

It might be expected that a change in any component of a complex organelle like a ribosome would be detrimental to function. This expectation is borne out in a number of cases (6). A side effect of mutations conferring resistance to erythromycin in *Escherichia coli* is a conformational change, which leads to negligible activity of the ribosomes under normal ionic conditions. The change is reversed and activity restored by high salt (7). In order to see whether the mutations affecting mitochondrial ribosomes resemble in any way those occurring in bacterial ery-

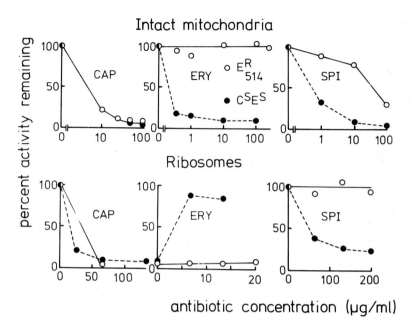

Fig. 2. Properties of erythromycin-resistant mutant 514. Data taken from ref.5.

TABLE II

A COMPARISON OF THE RESPONSE OF INTACT MITOCHONDRIA
AND ISOLATED RIBOSOMES TO ANTIBIOTICS
Data taken from ref. 5.

	Phenotype intact mitochondria/isolated ribosomes			
	CAP	ERY	LIN	SPI
$C^S E^S$	S/S	S/S	S/S	S/S
E^R_{514}	S/S	3R/3R	2R/2R	2R/2R
E^R_{354}	S/S	3R/3R	S/S	R/2R
E^R_{553}	S/S	R/R	S/S	S/R ?
C^R_{321}	2R/2R	R/?	R/R	R/R
C^R_{323}	3R/3R	R/?	R/R	R/R
S^R_{352}	R/R	R/R	?/S	3R/2R

279

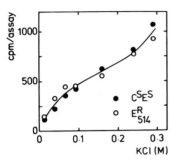

Fig. 3. Effects of salt on peptidyl transferase activity by ribosomes from E_{514}^R and wild type. Peptidyl transferase activity was measured by means of a modified fragment reaction as described in ref. 3.

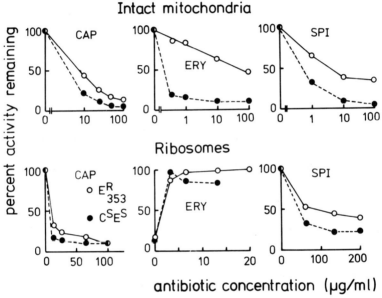

Fig. 4. Properties of erythromycin-resistant mutant 353. Characterisation was carried out as described in ref. 5.

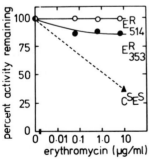

Fig. 5. Effect of erythromycin on protein synthesis directed by MS2 RNA. Assays were carried out as described in the legend to Table III.

thromycin-resistant mutants, E^R_{514} was examined for a possible anomalous dependence of peptidyl transferase activity on salt concentration. Fig. 3 demonstrates that such behaviour is absent and confirms previous indications (Table I) that these mutations are not deleterious to function.

THE FRAGMENT REACTION MAY NOT DETECT ALL MUTANT TYPES

Initial screening of mutants was facilitated by the availability of the fragment reaction as a simple assay for peptidyl transferase activity. Results with one erythromycin-resistant mutant suggest, however, that this technique may be limited in its ability to detect and characterise all types of mutant. The situation for this mutant, E^R_{353}, which maps at the same locus as highly-resistant ribosome mutant 514, is presented in Fig. 4. Although marginal cross-resistances to chloramphenicol and spiramycin remain, moderate resistance to erythromycin at the level of the intact mitochondria is not observed when purified ribosomes are examined. This is exactly the behaviour predicted of a mutant possessing an altered mitochondrial membrane. The occurrence at a single locus of mutations affecting both the mitochondrial membrane and ribosome would have important consequences for our ideas on the interaction between the two (8). This mutant was, therefore, submitted to further study, beginning with a check on the antibiotic response of the isolated ribosomes. Mitochondrial ribosomes are highly active in protein synthesis programmed by bacteriophage MS2 RNA and this activity is highly sensitive to a range of antibiotics (Table III). The response of mutant 353 in this system was, therefore, examined and is given in Fig. 5, together with the response of mutant 514 for comparison. By this criterion, mutant 353 is a moderately-resistant ribosomal mutant. It is fairly reasonable to suppose that the response of ribosomes in this system approaches most closely the situation obtaining *in vivo*, while the response in the fragment reaction reflects only the ability of erythromycin to compete with chlorampheni-

TABLE III

PROTEIN SYNTHESIS BY YEAST MITOCHONDRIAL RIBOSOMES
PROGRAMMED BY MS2 RNA: INHIBITION BY ANTIBIOTICS
Assays were carried out as described in ref. 9 at a
concentration of 6.5 mM Mg^{2+}, using $[^{14}C]$-(U)-leucine
(specific activity, 62 mCi/mmole) in place of $[^{14}C]$
phenylalanine. MS2 RNA was present at concentrations
between 100-150 µg per assay.

Antibiotic		Inhibition %
None		0
D-Chloramphenicol	(50 µg/ml)	91
Erythromycin	(10 µg/ml)	64
Spiramycin	(20 µg/ml)	59
Streptomycin	(20 µg/ml)	72
Spectinomycin	(13 µg/ml)	71

col for ribosome binding. We favour the interpretation that E_{353}^R is indeed a ribosomal mutant, but that
the mutation has in some way altered the (allosteric)
interaction between chloramphenicol and erythromycin
binding sites. If so, this mutant may be useful in
the study of the way in which two antibiotics interact with the ribosome. Cerná et al. (10) have recently reported the isolation of what may be a similar
mutant in E.coli.

DO MUTANTS WITH ALTERED MEMBRANE PERMEABILITY EXIST ?

None of the mutants examined by us so far have
turned out to display altered membrane permeability.
Do such mutants exist ? Compounds entering the mitochondrion may do so either via a specific translocator or by virtue of their general hydrophilic/hydro-
phobic properties. Impermeability may thus arise either from a mutation in a translocator or in a
"structural" membrane component, which alters the general permeability properties of the membrane. A
characteristic of this second type is that mutants
should be cross-resistant to a wide spectrum of compounds, unrelated in their action on mitochondria,
but similar in their solubility properties. Although

both types of mutation could confer resistance to antibiotics, it is unlikely that specific translocators
exist for many antibiotics and the second mechanism
probably predominates. Direct experimental support
for either type of mutant is so far lacking. Linnane
and co-workers (refs 8, 11) have, however, described
a number of mutants which they have identified as
membrane-permeability mutants on the basis of the
following indirect criteria:

1. Protein synthesis by mitochondria isolated from
such mutants is partially or completely sensitive to
antibiotics. *Rationale:* Permeability of the mitochondrial membrane is altered by damage incurred during
isolation.

2. Adaptation of anaerobically-grown cells to oxygen is sensitive to antibiotics. *Rationale:* The permeability properties of mitochondrial membranes are
drastically altered under anaerobic conditions.

Neither criterion is particularly compelling. In
the absence of full data on growth response, the level of sensitivity in mitochondria, or inability to
adapt to oxygen at high antibiotic concentrations is
difficult to evaluate. The low resistance of E_{553}^R,
for example, can be predicted directly from its behaviour *in vivo*. By the above criteria, however, E_{553}^R
would be a good candidate for a membrane mutant.

Further, it is not clear how the process of mitochondrial isolation should confer sensitivity on
such mutants. Analogy with bacteria (12) would suggest that impermeability results from an increased
ability of a membrane component to bind increased
amounts of the antibiotic. Comparison of protein synthesis of wild-type and resistant strains should still
reveal differences in response due to the protective
effect of the extra binding capacity of membranes.

Of the mutants isolated by Bunn *et al.* (8), one
fulfills both these criteria and is cross-resistant
to a wide range of antibiotics,including mikamycin,
tetracycline and oligomycin (13). The genetics of
this mutant, originally thought to be mitochondrial,
have since been found to be complex (14). Its pheno-

TABLE IV

PROPERTIES OF TWO MUTANTS RESISTANT TO TETRACYCLINE
Based on unpublished observations of H.W. Van der Glas
and L.A. Grivell. Tetracycline-resistant mutants and
the wild-type strain from which they were derived were kindly placed at our disposal by Dr. D.Wilkie.

Property	33-36	10-414
Genetics	Non-Mendelian(?)	Mendelian
Resistant to TC in $N_2 \rightarrow O_2$ transitions?	Yes	Yes
Response of isolated mitochondria	sensitive	sensitive
Response of purified mitochondrial ribo-somes	sensitive	?
Location of altered component	?	?

type may, therefore, result from multiple mutations.

Other mutants, resistant to erythromycin or spiramycin (11), resemble genetically mutants isolated by
Netter *et al.* (2). Their classification was on the
basis of criterion 1). These mutants could represent
ribosomal mutants of low or intermediate resistance,
but this should be verified by direct examination of
isolated mitochondrial ribosomes.

Apart from theoretical considerations, these criteria also appear inadequate in practice. Table IV
summarizes results obtained with two mutants resistant to tetracycline, isolated by Hughes and Wilkie
(15). Both mutants retain resistance during the $N_2 \rightarrow O_2$
transition, a property presumed to be characteristic
of ribosomal mutants by Bunn *et al.* (8). Mitochondria
isolated from the two strains are entirely sensitive
to tetracycline and are indistinguishable from the
wild type. Mitochondrial ribosomes from mutant 33-36
are, as might be expected, also highly tetracycline-
sensitive. The location of the component altered in
these mutants is still unknown, but in our opinion is
unlikely to be in either the mitochondrial membrane

or ribosome.

Whether general permeability mutants will ever be found remains to be seen. Considering the many difficulties involved in their characterisation, the search may well be a long one.

THE COMPONENT CHANGED IN MUTANT RIBOSOMES

Erythromycin resistance in bacteria is associated with an altered protein of the large ribosomal subunit, detectable either by polyacrylamide gel electrophoresis or by chromatography on carboxymethylcellulose (16, 17). So far, however, despite use of acrylamide gel electrophoresis in urea at pH 4.5 to effect separations of the proteins of the large ribosomal subunit into up to 28 components, no differences in protein composition have been detected for any of the several mutants examined. Separations performed at pH 8.7, or at pH 4.5 with differentially-labelled material (Fig. 6) also failed to reveal differences (5). This negative result leaves open the following possibilities:

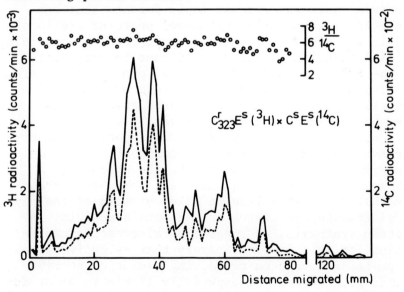

Fig. 6. Gel analysis of a differentially-labelled mixture of 50S proteins from wild type and C_{323}^R. From ref. 5. _____, ^{14}C; ----, ^3H.

1. Resistance is associated with a change in charge of a ribosomal protein, but techniques used so far are incapable of its detection.

2. A ribosomal protein is altered, but in each case the amino acid substitution is one which does not lead to a change of charge. Of the 549 possible mutations affecting the 61 amino acid coding triplets, 376 will be of this type. Too few mutants have been examined to exclude this possibility on statistical grounds, but this alternative is unattractive because similar mutants in bacteria are only poorly resistant (7).

3. Resistance results from an alteration, not in a protein, but in rRNA.

Of the three possibilities, the last is to us the most attractive. It localises the change in a ribosomal component known already to be specified by mtDNA (18) and fits with labelling experiments which suggest that all proteins of the mitochondrial ribosome are synthesized on cell-sap ribosomes (19). In the absence of evidence for mRNA export from mitochondria, such experiments imply further that the genes for mitochondrial ribosomal proteins are located in the nucleus. It should, therefore, be possible to find Mendelian mutants, resistant to antibiotics, in which a mitochondrial ribosomal protein is altered. Preliminary attempts to isolate such mutants (L.A. Grivell and J. Deutsch, unpublished observations) have been without success, however.

Beale *et al.* (20) have recently reported that erythromycin resistance in mitochondria of *Paramecium aurelia* could be correlated with the absence of a ribosomal protein. It is, however, somewhat improbable that a ribosome lacking a key protein of the peptidyl transferase centre should retain activity in protein synthesis. We suggest an alternative explanation for this finding, namely that selective losses of proteins have occurred during isolation of ribosomes. Such losses are especially liable to occur during the handling of small quantities of material and are accentuated by the strong tendency of mitochondrial proteins to aggregate (J.G. Huisman and L.A. Grivell, unpublished observations).

PROGNOSIS

rRNA mutations conferring antibiotic resistance in bacteria are rare. Indeed, the two cases reported to date (21, 22), are probably the result of a change in the properties of an RNA methylase rather than in the primary sequence of rRNA. Mitochondria are unusual, however, in the possession of only a single cistron for each rRNA (18). A mutation in this cistron will thus affect all rRNA made. The possibility that antibiotic resistance in these mutants is due to a change in the primary segment of rRNA, therefore, offers a unique opportunity to correlate a particular base sequence with a function.

ACKNOWLEDGEMENTS

I am indebted to Professor Piet Borst for stimulating discussions and much constructive criticism. Preliminary characterisation of several of the mutants described above was carried out during tenure of an EMBO Short Term Fellowship in the Centre de Génétique Moléculaire du C.N.R.S., Gif-sur-Yvette, France. This work was supported (in part) by a grant from The Netherlands Foundation for Chemical Research (S.O.N.) with financial aid from The Netherlands Organization for the Advancement of Pure Research (Z.W.O.) to P. Borst.

REFERENCES

1. P.P. Slonimski, G.Perrodin, M.Bolotin, D.Coen, J. Deutsch, B.Dujon, P.Netter and E.Petrochilo, Biochemie, in press.
2. P.Netter, E.Petrochilo, P.P.Slonimski, M.Bolotin, D.Coen, J.Deutsch and B.Dujon, Genetics, in press.
3. L.A. Grivell, L.Reijnders and H.De Vries, FEBS Letters 16, 159 (1971).
4. H.Teraoka, Biochim.Biophys.Acta 213, 535 (1970).
5. L.A.Grivell, P.Netter, P.Borst and P.P.Slonimski, Biochim.Biophys.Acta 312, 358 (1973).

6. D.Schlessinger and D.Apirion, Ann.Rev.Microbiol. 23, 387 (1969).
7. K.Tanaka, M.Tamaki, T.Itoh, E.Otaka and S.Osawa, Mol.Gen.Gen. 114, 23 (1971).
8. C.L.Bunn, C.H.Mitchell, H.B.Lukins and A.W.Linnane, Proc.Natl.Acad.Sci.U.S. 67, 1233 (1970).
9. L.A.Grivell, L.Reijnders and P.Borst, Biochim. Biophys.Acta 247, 91 (1971).
10. J.Cerná, J.Jonák and I.Rychlík, Biochim.Biophys. Acta 240, 109 (1971).
11. M.K.Trembath, C.L.Bunn, H.B.Lukins and A.W.Linnane, Mol.Gen.Gen. 121, 35 (1973).
12. C.R.MacKenzie and D.C.Jordan, Biochem.Biophys.Res. Commun. 40, 1008 (1970).
13. A.W.Linnane and J.M.Haslam, in: Current Topics in Cellular Regulation (B.L.Horecker and E.R. Stadtman, Eds.), Vol. 2, Academic Press, New York, pp. 101 (1970).
14. A.W.Linnane, J.M. Haslam, H.B.Lukins and P.Nagley Ann.Rev.Microbiol. 26, 163 (1972).
15. A.R.Hughes and D.Wilkie, Heredity 28, 117 (1972).
16. E.Otaka, H.Teraoka, M.Tamaki and S.Osawa, J.Mol. Biol. 48, 499 (1970).
17. I.Smith, C.Goldthwaite and D.Dubnau, Cold Spring Harbor Symp.Quant.Biol. 34, 85 (1969).
18. L.Reijnders, C.M.Kleisen, L.A.Grivell and P.Borst, Biochim.Biophys.Acta 272, 396 (1972).
19. G.S.P.Groot, this volume.
20. G.H.Beale, J.K.C.Knowles and A.Tait, Nature 235, 396 (1972).
21. C.J.Lai and B.Weisblum, Proc.Natl.Acad.Sci.U.S. 68, 856 (1971).
22. T.L.Helser, J.E.Davies and J.E.Dahlberg, Nature New Biol. 233, 12 (1971).

EVIDENCE OF INVOLVEMENT OF MITOCHONDRIAL POLYSOMES AND MESSENGER RNA IN SYNTHESIS OF ORGANELLE PROTEINS

Carole S. Cooper* and Charlotte J. Avers

Department of Biological Sciences, Douglass Campus - Rutgers University, New Brunswick, New Jersey 08903, U.S.A.

In the last few years much attention has been focussed on exploring the components and capacity of the mitochondrial genome (1). The mitochondrion may also differentiate, a phenomenon most clearly observed in organisms such as yeast, which undergo respiratory adaptation. The role of the mitochondrial genome in these developmental changes is complicated to analyze because the mitochondrion relies heavily on the nucleocytoplasmic system for much of its structural and enzymatic components, and perhaps for its regulation as well. We therefore developped an experimental design to study the contribution of the mitochondrial system to respiratory adaptation without inhibiting the nucleocytoplasmic compartment and, conceivably, altering the natural sequence of events.

Using intact mitochondria isolated from derepressing spheroplasts of yeast (2), we have been able to demonstrate that protein synthesis occurs on mitopolysomes and that changes in polysome activity may accompany respiratory adaptation. The mitopolysomes are programmed at least in part by mRNA synthesized from mtDNA. The mitochondrial mRNA molecules contain tracts of poly-A, which marks an essential difference between mitochondrial and prokaryotic protein synthesis (3).

*Present address: Biology Department, Massachusetts Institute of Technology, Cambridge, Massachusetts 02139, U.S.A.

MATERIALS AND METHODS

Cell growth conditions and isolation of mitochondria. Experimental details were as reported previously (4). Cells of *Saccharomyces cerevisiae*, strain iso-N, were grown in medium containing 10 % glucose and harvested for conversion to spheroplasts in 10 min using glusulase (Endo Labs). The spheroplasts then were allowed to derepress for 90-120 min in medium containing 2 % ethanol; then chilled, lysed, and nuclei sedimented at 3500 x g for 10 min. Mitochondria were washed three times in TMK buffer containing 0.5 M sorbitol, followed each time by sedimentation at 9000 x g for 15 min. Isolations were carried out at 0-5° C under sterile conditions.

Labeling of mitochondria and isolation of mitopolysomes. Mitochondria were suspended, at a concentration of 1 mg protein/ml, in 0.05 M Bicine, pH 7.4; 0.1 M KCl; 0.01 M $MgCl_2$; 0.01 M succinate; 0.005 M phosphoenol pyruvate; 0,04 M spermidine; 0.005 M proline; 0.002 M ATP; 0.01 M KH_2PO_4; 0.5 M sorbitol; and 2 enzyme units/ml pyruvate kinase. The suspension was incubated, with shaking, at 30° C. The reaction was terminated with 10 % trichloracetic acid when total mitochondrial incorporation was examined, or by chilling to 0-5° C when polysomes were to be isolated. In the latter case the mitochondria were recovered by centrifugation at 9000 x g, then suspended and lysed in TMK containing 2 % Triton X-100. Membrane fragments were removed by sedimentation at 27,000 x g for 20 min, and the supernatant was layered onto gradients of 15-40 % (w/v) ribonuclease-free sucrose (Mann Research Corp.) in TMK. For experiments analyzing the rRNA constituents, the polysome region of the gradient was pooled and polysomes sedimented in the Spinco Type 65 rotor for 1.5 h at 60,000 rpm and RNA was extracted from the pellets by dissolving in 0.01 M Tris, pH 7.5; 0.01 M NaCl; 0.001 M EDTA; 1 % SDS. For studying [3H] labeled polysomal RNA the polysome fractions of the gradient were pooled without further sedimentation and RNA was extracted by the hot phenol--SDS method (5).

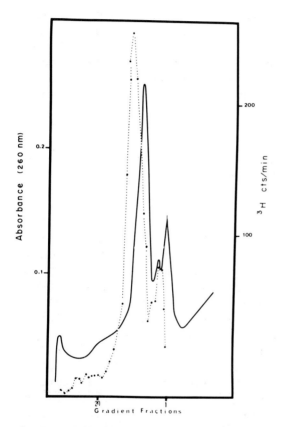

Fig. 1. Cosedimentation of purified ribosomes from mitochondria and cytoplasm. 2.7 A_{260} units of unlabeled mitochondrial ribosomes (obtained from centrifugation of mitochondrial lysate at 210,000 x g for 1 h) were layered onto a sucrose density gradient along with 0.02 A_{260} units of radioactive cytoplasmic ribosomes, labeled with [^{3}H]uracil to contain 2000 cpm. The gradient (4.6 ml, linear 15–30 % sucrose in 0.02 M Tris, pH 7.4; 0.005 M $MgSO_4$) was centrifuged in the Spinco SW50 rotor for 90 min, 46,000 rpm, 0–5°. Radioactivity:......; A_{260}:——.

Preparation of cytoplasmic ribosomes. Cytoribosomes were obtained from strain iso-N grown in medium containing 10 % glucose, as previously reported (4).

Binding polysomal RNA to poly-U sepharose. A 5 x 0.6 cm column was prepared of sepharose covalently linked to poly-U, according to the method of Firtel *et al.* (6). Polysomal RNA was layered on the column in "binding buffer" (0.01 M Tris, pH 7.4; 0.12 M NaCl; 0.5 % SDS) and elution of the column was continued in this buffer at 18° C until fractions had no measurable radioactivity or absorbance at 260 nm. Then the column was eluted with 12 ml of "eluting buffer" (0.01 M Tris, pH 7.4; 0.5 % SDS) at 18° C. The temperature finally was raised to 45° C and the column eluted with 50 % elution buffer plus 50 % formamide. RNA was precipitated from fractions with 2.5 volumes of ethanol.

RESULTS

Mitoribosomes can be isolated free of cytoribosome contamination. The success of the experimental design greatly on recovering mitochondria free of contamination by nuclei or cytoribosomes. Various independent lines of evidence showed the purity of our isolated mitochondrial preparations, as follows: (a) nuclei were not detected in electron micrographs; (b) RNA synthesis was sensitive to rifampicin, 250 µg/ml; (c) mitoribosome monomers sedimented more slowly than cytoribosomes, at an estimated sedimentation coefficient of 75S when compared with 80S yeast cytoribosomes and 70S *E.coli* ribosomes (Fig. 1); (d) RNA extracted from either mitoribosomes or mitopolysomes sedimented more slowly in sucrose gradients (Fig. 2) and migrated more slowly in polyacrylamide gels (Fig. 3) than the analogous RNA from cytoribosomes, as also reported by others (7,8); and (e) erythromycin (2 mg/ml) inhibited 60 % of the amino acid incorporation, while the cytoribosome inhibitor cycloheximide (100 µg/ml) was completely ineffective in blocking amino acid incorporation (Table I).

The clarity by which their characteristics could be

292

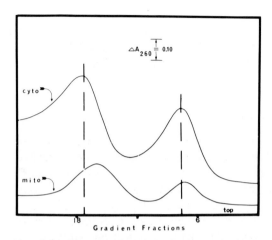

Fig. 2. Sedimentation of ribosomal RNAs derived from cytoplasmic ribosomes or mitochondrial polysomes. RNA prepared from pellets of mitopolysomes or cytoribosomes was layered onto identical sucrose gradients (5-20 % sucrose in TMK), centrifuged in the Spinco SW25.1 rotor at 24,500 rpm for 16 h at 0-5°. The absorbency profile for rRNA extracted from mitochondrial monoribosomes was identical to the polysomal RNA profile presented.

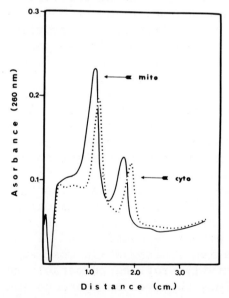

Fig. 3. Electrophoretic analysis of cytoplasmic ribosomal RNA and RNA extracted from mitopolysomes. RNA from the two sources was layered onto identical 5 cm gels of 2.4 % acrylamide, 0.12 % bis-acrylamide. Electrophoresis was accomplished at 4.5 mA/tube for 90 min at 20°.

TABLE I.

ANTIBIOTIC SENSITIVITY OF AMINO ACID INCORPORATION IN ISOLATED MITOCHONDRIA.
Mitochondria were incubated for 10 min with 5 µCi/ml [^{14}C] amino acids.

Additions to medium	% inhibition relative to control
None (control)	----
+ 2 mg/ml erythromycin	59.8 %
+ 100 µg/ml cycloheximide	0 %

resolved was another line of evidence that the mitochondria were essentially uncontaminated. The precise
level of contamination was assessed by a reconstruction experiment, performed by adding an excess of ex-
traneous, [^{3}H]labeled cytoribosomes to lysed spheroplasts and purifying the mitochondria from the mix u-
sing standard procedures. The mitoribosomes showed
less than 1.5 % level of contamination.

The incorporation of amino acids increased linearly
with time throughout the incubation intervals in all
the experiments, indicating the continuing activity
of the mitochondrial system. Contamination by bacteria
never exceeded 100 cells per mg mitochondrial protein
(based on plating aliquots of the incubation suspension on nutrient agar), far below the concentrations
which would affect amino acid incorporation.

Characteristics of mitochondrial polysomes. When
isolated mitochondria were incubated for a short time
with [^{3}H]amino acids, acid-insoluble radioactivity became associated with mitochondrial polysomes (Fig. 4).
To demonstrate that these products were nascent polypeptides rather than ribosomal proteins, puromycin
was added to the incubation mixture for 8 min leading
to the release of about 40 % of the polysome-associat-
ed radioactivity (Fig. 5). As has been noticed in o-
ther systems (9,10), this effect depended on NH$_4$Cl in
place of KCl in the incubation medium. While most experiments yielded radioactivity profiles comparable to
those in Fig. 4, we sometimes observed an enormous increase in amino acid incorporation by polysomes composed of 7 to 10 ribosomes. On other occasions amino acid
incorporation occurred primarily in dimer and trimer
polysomes, even though larger polysomes also appeared

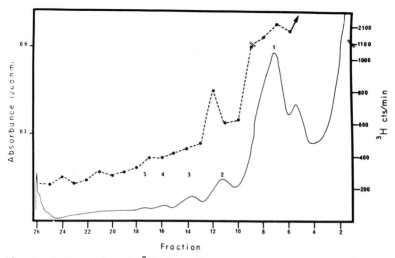

Fig. 4. Incorporation of [³H]amino acids by isolated mitochondria into peptides associated with mitochondrial ribosomes and polysomes. Mitochondria were incubated in incorporation medium with 20 µCi/ml [³H]amino acids for 7 min. Polysomes were prepared as described in Methods; gradient fractions were collected and precipitated by cold 10 % trichloroacetic acid and counted in a toluene-Triton-X-100 scintillation cocktail. Radioactivity: ------; A_{260}: ——— . Centrifugation was performed in the SW27 rotor for 3.5 h at 27,000 rpm.

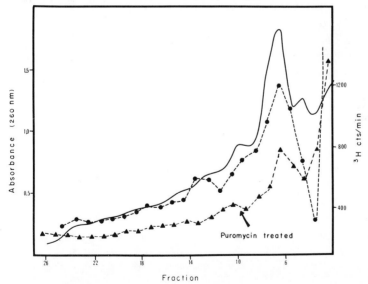

Fig. 5. Effect of puromycin on incorporation of [³H]amino acids by polysomes in isolated mitochondria. Mitochondria were incubated in incorporation medium, modified by the substitution of 0.1 M NH₄Cl for KCl, with 10 µCi/ml [³H]amino acids. After 5 min puromycin (0.5 mM) was added to one half of the mitochondria. Incubation of both samples was continued for an additional 8 min. Subsequent procedures are the same as detailed in the legend to Fig. 4. Radioactivity in puromycin treated sample: Δ---Δ; radioactivity in untreated sample: o---o; A_{260} of untreated sample: ———.

295

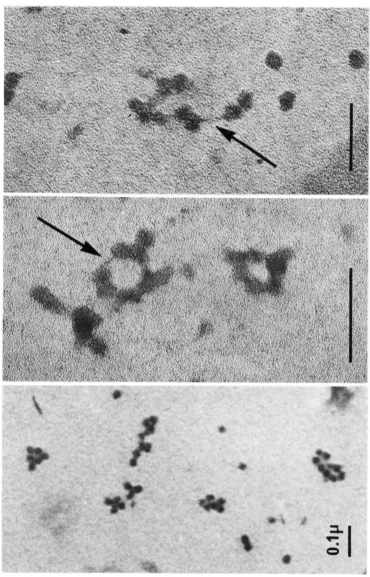

Fig. 6. Electron microscopy of mitochondrial polysomes. The polysome region of the sucrose gradient was collected and drops of the suspension were placed onto a clean, plastic Petri dish. Nickel grids (from Fullam, 300 mesh, coated with parlodian and stabilized with a thin layer of carbon) were placed onto the drops and allowed to remain for 30 min at 0–5°. The grids were then rinsed successively in 5 changes of 0.1 M sucrose in 10 % formalin and 3 changes of 0.4 % Photoflo (Kodak), then stained in 0.05 M uranyl acetate, 0.05 M HCl, freshly diluted 1:1000 with 90 % ethanol. Arrows indicate presumptive mRNA strand connecting the ribosomes.

in the absorbance profile of the gradient. These variations are currently under study.

Electron microscope examination of mitopolysomes taken from sucrose gradient fractions revealed ribosomes aggregated into coiled clusters. Occasionally the ribosomes were separated slightly and a strand of material, presumably mRNA, was exposed (Fig. 6). We previously had shown that mitopolysomes treated with ribonuclease sedimented as monomers in a sucrose gradient (4). Additional evidence that an RNA molecule was responsible for the structural integrity of the mitopolysomes was provided by collecting the polysome region of a gradient and treating it with ribonuclease then examining the fraction with the electron microscope. Subsequent scoring of the polysome sizes demonstrated that this treatment had induced breakdown of most of the polysomes, bringing about a concomitant increase in the number of monomers and small polysomes (Fig. 7).

RNA synthesis in isolated mitochondria. RNA synthesis, measured by incorporation of [^{3}H]uracil into isolated mitochondria, was sensitive to rifampicin (Table I), an antibiotic reported to affect prokaryotic-type RNA polymerase (11). We then sought to test the activity of the mitochondrial RNA polymerase as a function of respiratory adaptation. Mitochondria obtained from derepressing spheroplasts were 21 % more active in synthesizing RNA than mitochondria from spheroplasts kept in catabolite repression (Table II).

The utilization of some of the mitochondrially synthesized RNA as mRNA was indicated when rapidly-labeled RNA (synthesized in isolated mitochondria during a 5 min incubation with [^{3}H]uracil) was found with mitopolysomes (Fig. 8). There was no significant radioactivity in rRNA or tRNA extracted from the mitopolysomes after such a short period of labeling (data not shown). However, 95 % of the radioactivity in phenol-extracted polysomal RNA bound to a column of poly-U-sepharose under high salt conditions (Table III). The radioactivity was removed by elution in low salt plus 50 % formamide at 50° C. This behavior has been shown

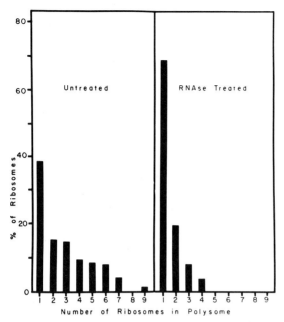

Fig. 7. Sensitivity of polysome structure to ribonuclease. The region of a sucrose gradient containing mitopolysomes was collected and divided into two portions. One portion was left untreated and the other incubated with 20 µg/ml pancreatic ribonuclease at 0-5° for 20 min. Samples were then fixed by addition of 10 % formalin, 0.1 M sucrose. Subsequent procedures as described in Fig. 6. The data presented in the graph were calculated by examining photographs of 1000 mitoribosomes and scoring the number of ribosomes in polysomes of different sizes.

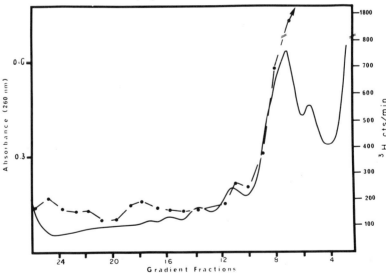

Fig. 8. Association of rapidly-labeled RNA with mitochondrial ribosomes and polysomes. Isolated mitochondria were incubated in incorporation medium with 20 µCi/ml [³H]uracil for 5 min. Subsequent procedures as in Fig. 4. Radioactivity:.---.; A₂₆₀: ──── .

298

TABLE II.

RNA SYNTHESIS IN ISOLATED MITOCHONDRIA.
A. Inhibition of [^{3}H]uracil incorporation into whole mitochondria by rifampicin. Mitochondria were incubated with 10 µCi/ml [^{3}H]uracil for 10 min.

Additions to medium	% inhibition relative to control
None (control)	--
+ 250 µg/ml rifampicin	28 %

B. Comparison of [^{3}H]uracil incorporation by isolated mitochondria obtained from derepressed or repressed spheroplasts. Mitochondria were incubated with 10 µCi/ml [^{3}H]uracil for 5 min.

Respiratory state of spheroplasts from which mitochondria were isolated	c.p.m. per mg mit. protein	% difference
Derepressed	8,558	--
Repressed	7,089	21 %

TABLE III.

BINDING OF POLYSOMAL RNA TO POLY-U-SEPHAROSE.
[^{3}H]labeled RNA was extracted from mitopolysomes and put over a column of poly-U-sepharose. Table information was calculated by pooling total cpm and A$_{260}$ from samples eluted under the following conditions. Details in Methods.

Eluting Conditions	% A$_{260}$ eluted	cpm eluted	% cpm eluted
"Binding buffer" and "Eluting buffer" at 18°	100 %	70	4 %
"Eluting buffer" and 50 % "eluting buffer" + 50 % formamide, at 50°	---	1509	96 %

by Singer *et al.* (12) and by control experiments with [^{14}C]poly-A (Kaufman, unpub. data) to be characteristic of the presence in the RNA of tracts of poly-A which form hydrogen bonds with the poly-U on the column. In recent experiments the RNA from mitopolysomes has been hybridized to [^{14}C]poly-dT, and has been demonstrated to protect the [^{14}C]label from digestion with single-strand-nuclease. These results show that the rapidly-labeled, polysomal RNA contains poly-A sequences. Moreover, since these RNA species were labeled in purified, isolated mitochondria they must originate as a transcript from mtDNA.

DISCUSSION

The experiments reported here demonstrate that protein synthesis in yeast mitochondria occurs on mitopolysomes which are directed by mRNA transcribed from mtDNA.

Mitopolysomes have been reported in yeast (4,13), *Euglena* (14), and HeLa cells (15). We have used several lines of evidence to characterize the mitopolysomes: (a) sedimentation as large units in sucrose density gradients, (b) composition of rRNA, (c) direct visualization by electron microscopy as ribosomes connected by a strand of RNA, (d) sensitivity of ribosome association to ribonuclease, (e) function in protein synthesis. The functional criterion was demonstrated by amino acid incorporation and subsequent release of nascent polypeptides from the polysomes when treated with the tRNA analogue puromycin.

Protein synthesis in the isolated mitochondria was found to be completely resistant to cycloheximide, a potent inhibitor of cytoplasmic ribosomes, while sensitive to erythromycin, an inhibitor of prokaryotic-type ribosomes. We have also shown that mitoribosomes can be purified away from radio-labeled cytoribosomes in a reconstruction experiment, and that the rRNA constituents isolated from the mitoribosomes from discrete bands in gradients or gels, with no visible "shoulder" contributed by cytoribosomes. These sets of data affirm that the components of the mitochondrial system can be isolated free from contamination by the cytoplasmic system.

Evidence that mtDNA transcribes mRNAs has previously been indirectly implied by experiments showing non-Mendelian mutations affecting mitochondrial structure or function or antibiotic sensitivity (16,17), and by inhibition of mitochondrial protein synthesis by drugs or by treatments which block RNA synthesis (18-20). Recently, Perlman *et al.* (21) labeled HeLa cells in the presence of camptothecin, an inhibitor of nuclear RNA synthesis, and found poly-A-containing RNA in the mitochondria. Our present results clearly demonstrate that native mitochondrial mRNA is used to program protein synthesis on mitopolysomes. These RNA molecules are synthesized in isolated mitochondria, completely free of nucleocytoplasmic contamination. They fulfill several criteria expected for messenger molecules: (a) they are labeled more rapidly than rRNA or tRNA, (b) they are found in association with polysomes, and (c) they have binding and hybridization characteristics which indicate the presence of poly-A sequences. Poly-A sequences have been identified as an integral portion of all eukaryotic mRNA molecules, with the exception of histone mRNAs (22).

The experimental system we have described takes advantage of the phenomenon of facultative respiration in yeast. The increase in RNA synthesis in mitochondria obtained from yeast spheroplasts undergoing respiratory adaptation is one indication that the isolated organelle retains the capacity for differentiation induced while in the cellular environment. We also have observed a varying level of amino acid incorporation in certain sizes of mitopolysomes. Although these data are preliminary, we speculate that these changes in polysome activity represent an important event in respiratory adaptation. Critical analysis of polysomes may thus prove to be a sensitive approach to studying mitochondrial differentiation. The lack of cytoplasmic contamination in this system would also facilitate the future isolation of pure mitochondrial components, such as mitopolysomes or mitochondrial mRNA, for *in vitro* studies to identify the proteins encoded in mtDNA. In addition, this experimental system may be exploited to test the effects on mitochondria of cytoplasmic extracts taken from cells in different metabolic or respiratory states.

ACKNOWLEDGEMENTS

This study was supported in part by Contract No. AT(11-1)-3516 from the U.S. Atomic Energy Commission. We are indebted to Dr. Hans-Peter Hoffmann for the electron microscopy portions of this study.

REFERENCES

1. M. Ashwell and T.S. Work, Ann.Rev.Biochem. 39, 251 (1970).
2. W.K. Neal, H.P. Hoffmann, C.J.Avers and C.A. Price, Biochem.Biophys.Res.Commun. 38, 414 (1970).
3. R.B. Perry, D.E. Kelley and J. LaTorre, Biochem. Biophys.Res.Commun. 48, 1593 (1972).
4. W.J. Stegeman, C.S. Cooper and C.J. Avers, Biochem.Biophys.Res.Commun. 39, 69 (1970).
5. S. Penman, J.Mol.Biol. 17, 117 (1966).
6. R.A. Firtel, A. Jacobson and H.F. Lodish, Nature New Biol. 239, 225 (1972).
7. I.T. Forrester, P. Nagley and A.W. Linnane, FEBS Letters 11, 59, (1970).
8. L.A. Grivell, L. Reijnders and P. Borst, Europ.J. Biochem. 19, 64 (1971).
9. C-C. Chuah and I.T. Oliver, Biochem. 10, 2990 (1971).
10. K. Moldave, F. Ibuki, P. Rae, M. Schneir, L. Skogerson and R.P. Sutter, in: Regulatory Mechanisms for Protein Synthesis in Mammalian Cells, A. San-Pietro, M.R. Lamborg and F.T. Kenney, Eds., Academic Press, New York, 191 (1968).
11. W. Wehrli, F. Knusel, K. Schmid and M. Staehlin, Proc. Natl.Acad.Sci.U.S.A. 61, 667 (1968).
12. R.H. Singer and S. Penman, Nature 240, 100 (1972).
13. P.V. Vignais, B.J. Stevens, J. Huet and J. André, J.Cell Biol. 54, 468 (1972).
14. N.G. Avadhani, M.J. Lynch and D.E. Buetow, Exptl. Cell Res. 69, 226 (1972).
15. D. Ojala and G. Attardi, J.Mol.Biol. 65, 273 (1972).
16. B. Ephrussi and P. Slonimski, Nature 176, 1207 (1955).
17. A.W. Linnane, G.W. Saunders, E.B. Gingold and H.B. Lukins, Proc.Natl.Acad.Sci.U.S.A. 59, 903 (1968).
18. J.G. Gamble and R.H. McCluer, J.Mol.Biol. 53, 557 (1970).
19. G.S.P. Groot, W. Rouslin and G. Schatz, J.Biol. Chem. 247, 1735 (1972).
20. H.R. Mahler and K. Dawidowicz, Proc.Natl.Acad.Sci.

U.S.A. 70, 111 (1973).

21. S. Perlman, H.T. Abelson and S. Penman, Proc.Natl. Acad.Sci.U.S.A. 70, 350 (1973).

22. M. Adesnik, M. Salditt, W. Thomas and J.E. Darnell, J.Mol.Biol. 71, 21 (1972).

Abbreviations used: mtDNA, mitochondrial DNA; mRNA, messenger RNA; rRNA, ribosomal RNA; tRNA, transfer RNA; poly-A, polyadenylic acid; poly-U, polyuridylic acid; SDS, sodium dodecyl sulfate; EDTA, ethylenediamine tetraacetic acid; TMK buffer, 0.01 M Tris, pH 7.5, 0.01 M KCl, 0.01 M $MgCl_2$.

MITOCHONDRIAL POLYSOMES FROM *NEUROSPORA CRASSA*.

E. Agsteribbe, R. Datema and A.M. Kroon

Laboratory of Physiological Chemistry, State University, Bloemsingel 10, Groningen, The Netherlands.

Mitochondria are equipped with a very active protein synthesizing system. Nevertheless the polysomal organization of the ribosomes has not been demonstrated conclusively. An exception may be made for *Euglena gracilis*. For this organism mitochondrial polysomes have been identified unambiguously by Avadhani and Buetow (1,2). The presence of mitochondrial polysomes in *Neurospora crassa* has been proposed by Küntzel and Noll (3) and by Neupert *et al.* (4). In these studies the main evidence for the polysomal organization of the mitochondrial ribosomes came from the distribution pattern of labelled nascent peptides after centrifugation of the ribosomes through sucrose gradients.

We have tried to isolate mitochondrial polysomes from *Neurospora crassa* by centrifuging ribosomal preparations through a layer of highly concentrated sucrose. In preparations consisting of ribosomes with the sedimentation value of 73S, a value considered characteristic for mitochondrial ribosomes of *Neurospora*, no polysomes could be detected. Moreover, sometimes two particle fractions with sedimentation values of 80S and 73S respectively, were obtained. It is the aim of this paper to show that the problem of the isolation of two types of ribosomal particles and the difficulties in finding mitochondrial polysomes, are related to each other.

80S AND 73S RIBOSOMES FROM MITOCHONDRIA OF *NEUROSPORA CRASSA*.

From mitochondria isolated in EDTA-containing buffer ribosomes were obtained with a sedimentation value

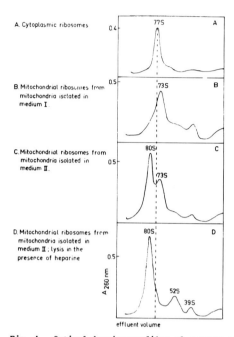

Fig. 1. Optical density profiles of cytoplasmic and mitochondrial ribosomes of *Neurospora crassa* after sedimentation through sucrose gradients. Cytoplasmic ribosomes were isolated as described by Küntzel (7). Mitochondrial ribosomes were isolated as described previously (5). Medium I consisted of 0.44 M sucrose, 100 mM NH_4Cl, 10 mM $MgCl_2$ and 10 mM Tris-HCl, pH 7.8. Mitochondria were lysed with 1/10 volume of 20 % Triton X-100 in a medium containing 100 mM NH_4Cl, 10 mM $MgCl_2$ and 10 mM Tris-HCl, pH 7.8. In the experiment presented in Fig. 1D heparin was added to the lysis medium in a concentration of 0.5 mg/ml. Sedimentation analysis was performed on isokinetic sucrose gradients in the SW27-1 tubes of the Spinco SW27-rotor. Sucrose concentration at the top of the gradient was 15 % (w/v). Salt concentrations in the gradient were 100 mM NH_4Cl, 10 mM $MgCl_2$ and 10 mM Tris-HCl, pH 7.8. Centrifugation was for 15 h at 22,000 rpm at 4° C.

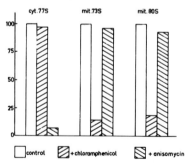

Fig. 2. Peptidyltransferase activity of cytoplasmic and mitochondrial ribosomes from *Neurospora crassa*. Peptidyltransferase activity was measured in the fragment reaction as described previously (5). 100 % activities for different preparations of all three types of ribosomes ranged from 4000 to 7000 dpm per incubation (1 OD_{260} nm ribosomes present).

306

of 73S as compared to a value of 77S for cytoplasmic ribosomes (Fig. 1A and B). Mitochondria isolated in Mg^{2+}-containing buffer gave ribosomal preparations showing two peaks - *viz.* at the 80S and the 73S position - in isokinetic sucrose gradients (Fig. 1C). An obvious interpretation of these results is that the latter preparations are contaminated with cytoplasmic ribosomes. However, there were two facts that make the correctness of this interpretation doubtful. In the first place the sedimentation value obtained for the faster sedimenting particles was 80-82S and this is significantly different from the value for cytoplasmic ribosomes. Secondly it was possible to obtain ribosomal preparations consisting of only 80S ribosomes by adding the ribonuclease inhibitor heparin to the medium in which the mitochondria were lysed (Fig. 1D). Our preliminary conclusion was, therefore, that the 80S particles were of mitochondrial origin.

CHARACTERIZATION OF THE 80S AND 73S MITOCHONDRIAL RIBOSOMES.

Additional evidence for the mitochondrial origin of the 80S particles came from the assay of the peptidyl transferase activity. Peptidyl transferase activities of ribosomes from bacteria and nucleated cells show opposite sensitivity to chloramphenicol and anisomycin as could be measured in the fragment reaction. Mitochondrial ribosomes behave in this respect like bacterial ribosomes, thus offering a possibility to pinpoint either the mitochondria or the cytoplasm as the source of ribosomes (5). As shown in Fig. 2 the peptidyl transferase of both the 80S and the 73S ribosomes is sensitive to chloramphenicol and not to anisomycin; the opposite holds for the 77S cytoplasmic ribosomes.

Usually mitochondrial ribosomes from ascomycetes are discriminated from cytoplasmic ribosomes on the basis of physical criteria, *e.g.* sedimentation values of ribosomes, ribosomal subunits and ribosomal RNAs. A number of physical properties are brought together in Table I. Values obtained for the 80S and 73S par-

TABLE I

PHYSICAL CHARACTERIZATION OF CYTOPLASMIC AND MITOCHONDRIAL RIBOSOMES FROM
NEUROSPORA CRASSA.
Sedimentation analysis of ribosomes and ribosomal subunits was performed on isokinetic sucrose gradients as described in the legend to Fig. 1. Sedimentation
analysis of ribosomal RNAs was performed on isokinetic sucrose gradients in the
SW27-tubes of the Spinco SW27-rotor. Sucrose concentration at the top of the gradient was 5 % (w/v). Salt concentrations in the gradient were 20 mM Tris-HCl, pH
8.1, 0.1 mM EDTA and 2 mM $MgCl_2$. Centrifugation was for 15 h at 24,500 rpm at 4°
C. The degree of dissociation of ribosomes into subunits was measured after dialysis of the ribosomes for 2.5 h against a medium consisting of 100 mM NH_4Cl, 1
mM $MgCl_2$ and 10 mM Tris-HCl, pH 7.8. Hybridization experiments were performed as
described by Schäfer and Küntzel (10).

Characteristics	Type of ribosomes		
	cyt. 77S	mit. 73S	mit. 80S
S-values of ribosomal subunits	60 ; 39	52 ; 39	52 ; 39
S-values of ribosomal RNA	25 ; 18	24 ; 17	24 ; 17
S_E-values of ribosomal RNA	25 ; 17	22 ; 15	22 ; 15
Degree of dissociation into subunits in 1 mM Mg^{2+}	ca 20 %	ca 80 %	ca 80 %
Hybridization with mit DNA	–	2.5 %*	2.45 %

* value obtained by Schäfer and Küntzel (10).

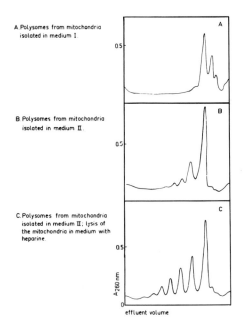

Fig. 3. Optical density profiles of mitochondrial polysomes from *Neurospora crassa*
after sedimentation through sucrose gradients. The composition of media I and II is
described in the legend to Fig. 1. Ribosomal preparations enriched in polysomes were obtained by centrifuging the ribosomes from the clarified mitochondrial lysate
through a layer of 3 ml 2 M sucrose in 100 mM NH_4Cl, 10 mM $MgCl_2$ and 20 mM Tris-HCl,
pH 7.8 in tubes of the Ti50 Spinco rotor for 4 h at 50,000 rpm at 2° C. Sedimentation analysis on isokinetic sucrose gradients was performed as described in the legend to Fig. 1, except that centrifugation lasted 15 h at 13,500 rpm.

ticles appeared to be almost identical and to differ
from those of the cytoplasmic ribosomes. These data
substantiate the conclusion that both the 80S and 73S
ribosomes are mitochondrial particles.

MITOCHONDRIAL POLYSOMES; INVOLVEMENT OF 73S- AND 80S-RIBOSOMES.

The question that arose now with respect to the two
types of mitochondrial ribosomes, was: are we dealing
with some kind of an isolation artifact or does there
exist a heterogeneous population of ribosomes within
the the mitochondria *in vivo* ? The latter possibility
seemed unlikely because ribosomal preparations could
be ontained exclusively consisting either of 80S- or
of 73S-particles. Thinking in terms of an isolation
artifact, more information was required about the
functional integrety of both particles. Ribosomes actively engaged in protein synthesis will most likely
be organized in polysomes. The properties of the monomeric ribosomes present in polysomes may be considered,
therefore, to reflect the properties of the native ribosomes most closely. Since, as stated already in the
introduction, no polysomes could be detected in the
73S ribosomal preparations, we reasoned that it was
worthwhile to search for polysomes in the 80S ribosomal preparations. Ribosomal preparations enriched in
polysomes were obtained by centrifuging ribosomal preparations through a high-sucrose layer. The results
are presented in Fig. 3. Ribosomes from mitochondria
isolated in medium I, did not show any polysomes (Fig.
3A). Under these conditions only 73S-ribosomes were
found. Conditions favouring the presence of 80S-ribo-
somes, *i.e.* isolation of mitochondria in medium II,
gave rise to a distinct polysomal profile on sucrose
gradients (Fig. 3B). The most pronounced polysomal profile was obtained from mitochondria lysed in the presence of heparine, under conditions where only 80S-
particles were found (Fig. 3C). Up to 40 % of the ribosomes could be obtained as polysomes, indicating
that a large part of the ribosomes was participating
in protein synthesis. The polysomes could be degraded

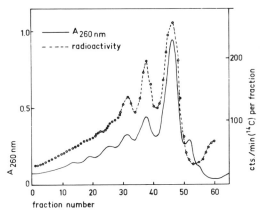

Fig. 4. Optical density profiles of polysomes and radioactivity distribution of nascent peptides from mitochondria of *Neurospora crassa* after sedimentation through isokinetic sucrose gradients. The incubation medium for *in vitro* labeling of mitochondrial nascent peptides contained:100 mM sucrose, 50 mM tricine, 50 mM KCl, 12 mM MgCl$_2$, 1 mM EDTA, 4 mM proline, 2 mM of the amino acids not present in the [^{14}C] amino acid mixture, 2 mM ATP, 10 mM phosphoenol-pyruvate, 2 µg/ml oligomycin, 50 µg/ml cycloheximide, 0.1 mg/ml pyruvate kinase, 1 mg/ml bovine serum albumin and about 6 mg/ml mitochondrial protein, final pH 7.6. The mixture was preincubated for 5 min at 30° C. The incubation was started by adding 0.2 ml of a [^{14}C]amino acid mixture containing ala, arg, asp, glu, gly, his, leu, ile, lys, met, phe, pro, ser, thr, tyr and val (specific activity 57 mCi/mAtom) to 4.8 ml of the incubation mixture. Incubation took place for 30 min at 30° C and was stopped by cooling in ice. Mitochondrial polysomes were isolated as described in the legend to Fig. 3.

TABLE II

IN VITRO PROTEIN SYNTHESIS BY MITOCHONDRIAL POLYSOMES FROM *NEUROSPORA CRASSA*.
The incubation mixture contained 40 mM KCl, 25 mM MgCl$_2$, 10 mM NH$_4$Cl, 30 mM Tris-HCl, pH 7.6, 1 mM dithiothreitol, 10 mM phosphoenolpyruvate, 1 mM ATP, 0.6 mM GTP, 3 mM spermidine, 2 mM of the amino acids not present in [^{14}C]amino acid mixture, 200 µg/ml heparine, 60 µg/ml pyruvate kinase, 3 mg/ml transfer RNA, 0.4 mg/ml polysomes. Incubations were carried out in a total volume of 0.125 ml, including 0.025 ml *E.coli* S100. *E.coli* S100 was prepared as described by Nirenberg (11). Incubation was started by adding 0.5 µC of a [^{14}C]amino acid mixture (specific activity, 57 mCi/mAtom; see legend to Fig. 4), and was continued for 30 min at 30° C. 100 % activities ranged from 3000 to 4000 dpm/0.125 ml.

Additions or omissions	[^{14}C]amino acid incorporation
None	100 %
- polysomes	3-10 %
- *E.coli* S100	35-42 %
+ chloramphenicol (50 µg/ml)	15-20 %
+ pancreatic RNAse (16 µg/ml)	24 %

310

by pancreatic ribonuclease, the monomere sedimented at the 80S position.

CHARACTERIZATION OF MITOCHONDRIAL POLYSOMES.

The mitochondrial origin of the polysomes was investigated in two ways. Firstly by labelling of the nascent peptides. *In vitro* labelling of the nascent peptides gave a distribution of radioactivity in sucrose gradients, coinciding with the peaks of the optical density profile (Fig. 4). The labelling experiments were performed in the presence of cycloheximide, indicating the mitochondrial nature of the process. Mitochondrial polysomes were also active in cell-free protein synthesis (Table II). Amino acid incorporation was inhibited by chloramphenicol and also by pancreatic ribonuclease. Always a high rest-activity was found when *E. coli* S100 was omitted. This is ascribed to the presence of mitochondrial supernatant enzymes in the polysomal preparations, which were not washed extensively.

MITOCHONDRIAL MESSENGER-RNA.

Since we were now able to obtain mitochondrial polysomes in preparative yield, we could start attempts to isolate messenger RNA. We followed two simple procedures for the isolation of non-ribosomal RNA. Firstly total polysomal RNA was isolated and run on sucrose gradients. From the gradients the 5S to 12S fraction was isolated, reasoning that messenger RNA would be present somewhere in this fraction. In the second approach polysomes were incubated with puromycin in the presence of ATP and GTP; a run-off system was thus induced as could be checked on sucrose gradients. After this incubation the ribosomes were pelleted and RNA was isolated from the supernatant. Both RNA preparations were tested for messenger activity in a cell-free system with *E. coli* ribosomes and *E. coli* supernatant enzymes as described by Nathans for MS_2-RNA (6). Both RNAs stimulated amino acid incorporation. For different preparations values ranged between a two- and five-fold stimulation. The composition of the two

311

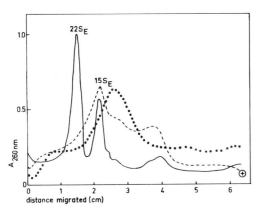

Fig. 5. Electropherograms of mitochondrial ribosomal RNA (———), 5S-12S fraction of mitochondrial ribosomal RNA (---), and "run-off" RNA from mitochondrial polysomes (o o o o) of *Neurospora crassa*. RNA was isolated by the phenol-extraction procedure of Kirby (8). Electrophoresis on polyacrylamide gels was performed as described by Loening (9).

non-ribosomal RNA preparations was investigated by electrophoresis on polyacrylamide gels (Fig. 5). The 5-12S RNA fraction was clearly contaminated with the small rRNA but also a faint peak was visible at the $9S_E$ position. The "run-off" RNA migrated as a broad peak, also at the $9S_E$ position.

CONCLUDING REMARKS

It has been shown that from mitochondria of *N. crassa* two types of ribosomes with sedimentation values of 73S and 80S can be obtained. Functional and physical characterization indicated that both particles are of mitochondrial origin. Evidently the 80S-particle is the native mitochondrial ribosome, because:
1. The 80S-particle is the monomer in polysomes obtained from mitochondria and is therefore actively involved in protein synthesis.
2. The difference in sedimentation value between the two particles is not reflected in the sedimentation values of ribosomal subunits and ribosomal RNAs. The effect of the ribonuclease inhibitor heparine, causing the absence of 73S-particles in mitochondrial ribosomal preparations, suggests that endogenous ribonucleases are responsible for the ocurrence of 73S.

ACKNOWLEDGEMENTS

The authors wish to thank Miss D. Oberman for technical assistance. This work was aided in part by the Netherlands Foundation for Chemical Research (S.O.N.) with financial aid from the Netherlands Organization for the Advancement of Pure Research (Z.W.O.).

REFERENCES

1. N.G. Avadhani and D.E. Buetow, Biochem.Biophys.Res. Commun. 46, 773 (1972).
2. N.G. Avadhani and D.E. Buetow, Biochem.J. 128, 353 (1972).
3. H. Küntzel and H. Noll, Nature 215, 1340 (1967).
4. W. Neupert, W. Sebald, A.J. Schwab, A. Pfaller, P. Massinger and Th. Bücher, Eur.J.Biochem. 10, 589 (1969).

5. H. de Vries, E. Agsteribbe and A.M. Kroon, Biochim. Biophys.Acta 246, 111 (1971).
6. D. Nathans, in "Methods in Enzymology", vol. XIIB, Ed. by L. Grossman and K. Moldave, Academic Press, New York, 787 (1969).
7. H. Küntzel, J.Mol.Biol. 40, 315 (1969).
8. K.S. Kirby, Biochem.J. 96, 266 (1965).
9. U.E. Loening, Biochem.J. 113, 131 (1969).
10. K.P. Schäfer and H. Küntzel, Biochem.Biophys.Res. Commun. 46, 1312 (1972).
11. M.W. Nirenberg, in "Methods in Enzymology", vol. VI, Ed. by S.P. Colowick and N.O. Kaplan, Academic Press, New York, 17 (1963).

NASCENT POLYPEPTIDE CHAINS ON MITOCHONDRIAL RIBOSOMES AND THEIR INTEGRATION INTO THE INNER MITOCHONDRIAL MEMBRANE

R. Michel and W. Neupert

Institut für Physiologische Chemie und Physikalische Biochemie der Universität, 8000 München 2, Goethestrasse 33, Federal Republic of Germany.

Little is known about the mechanism how mitochondrial translation products are transferred from their site of synthesis to their site of function, *i.e.* from the mitochondrial ribosomes into enzyme complexes of the inner mitochondrial membrane (1-4). The peculiar nature of the mitochondrial translation products may play a distinct role in this process. Therefore we have studied some properties of mitochondrial translation products before and after their integration into the membrane.

RESULTS.

Properties of nascent translation products on mitochondrial ribosomes. After labelling whole *Neurospora* cells with radioactive amino acids in the presence of cycloheximide (CHI), it is possible to isolate mitochondrial ribosomes in which the nascent peptide chains are radioactively labelled, whereas the ribosomal proteins are not. This is due to the fact, that the proteins of mitochondrial ribosomes are formed at the CHI sensitive cytoplasmic ribosomes (5-7). By this way, we have an experimental system in which the nascent translation products of mitochondria can easily be identified and analysed.

A) Mitochondrial ribosomes carrying nascent peptide chains.

Fig. 1 shows the result of density gradient centrifugation of mitochondrial ribosomes from cells exposed

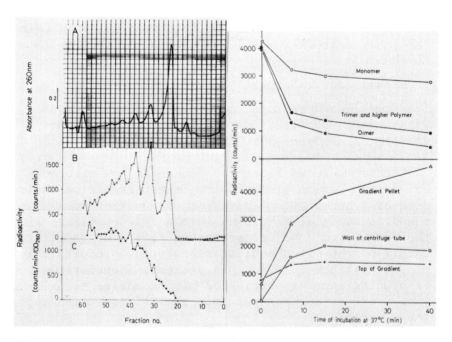

Fig. 1.   Sucrose density gradient centrifugation of mitochondrial ribosomes with radioactively labeled nascent peptide chains. A: Optical density pattern (260nm); B: Radioactivity pattern; C: Specific radioactivity related to absorbance at 260 nm. Cells were incubated with cycloheximide (100 μg/ml) for 2.5 min, [³H]leucine was added (1 μCi/ml) and after 1.5 min the cells were rapidly cooled to 0° C. Mitochondrial ribosomes were isolated and subjected to gradient centrifugation (9). Fraction 1 is top, and fraction 60 bottom of the gradient.

Fig. 2.   Effect of incubation of mitochondrial ribosomes in AMT buffer at 37° C on the distribution of radioactivity in the sucrose density gradient. Mitochondrial ribosomes were labeled as described in Fig. 1. They were kept at 37° C in AMT buffer (0.1 M NH₄Cl, 10 mM MgCl₂, 10 mM Tris-HCl, pH 7.5) for the time periods indicated in the figure. In the case of the radioactivity associated with the wall of the centrifuge tube the radioactivity in the control sample (8250 counts/min) was subtracted from all values.

316

to a pulse of radioactive leucine in the presence of CHI. In the optical density pattern (Fig. 1A) the peak of the monomeric ribosomes is prominent. A considerable amount of polymeric ribosomes (ca. 45 % of total ribosomes) is present. In the labelling pattern (Fig. 1B) monomeric, dimeric and higher polymeric ribosomes also can be distinguished. However, the majority of the radioactivity (ca. 85 %) is found associated with polymeric ribosomes. Accordingly, the specific radioactivity, related to A260nm (Fig. 1C) is about five times higher in the polymeric ribosomes. This difference in specific radioactivity of monomeric and polymeric ribosomes suggests, that monomers are not merely breakdown products of polymers. This is in contrast to cytoplasmic ribosomes, where monomers and polymers have the same specific radioactivity after short pulse labelling.

The amount of total radioactivity in the polymer region is subject to large alterations which depend on the preparation conditions of the ribosomes. However, the less the amount of radioactivity in the polymer region is, the more radioactivity appears in the pellet of the gradient. The yield of radioactivity at the monomer is rather constant. This leads us to suppose that the polymeric ribosomes have a high tendency to aggregate.

In order to test this, mitochondrial ribosomes were kept in Mg containing buffer at 37° C for different time periods and then subjected to gradient centrifugation. In Fig. 2 the radioactivity found in the different fractions of the gradient is shown. During the time period of incubation the radioactivity associated with the monomer remains constant after a slight initial decrease. In contrast, the radioactivity at the dimer and higher polymers decreases strongly. The radioactivity disappearing from this region does not appear at the top of the gradient, but can be traced in the pellet and at the wall of the centrifuge tube.

This selective tendency of polymeric ribosomes to aggregate and form heavy particles gives rise to the

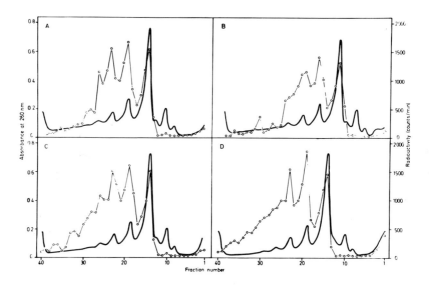

Fig. 3. Treatment of mitochondrial ribosomes with ribonuclease. Ribosomal suspensions in AMT buffer were kept at 0° C for 60 min with and without pancreatic ribonuclease, and then subjected to gradient centrifugation. o——o : radioactivity; ———: absorbance at 260 nm. A, control (without ribonuclease); B, 1 µg/ml ribonuclease; C, 4 µg/ml ribonuclease; D, 20 µg/ml ribonuclease.

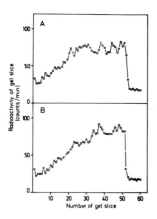

Fig. 4. Gel electrophoretic analysis of radioactively labeled nascent peptide chains associated with mitochondrial polymeric ribosomes. A, Ribosomes dissolved in 0.1 M Tris-HCl, 0.5 % SDS, pH 8, and kept for 1 h at 37° C; B, Ribosomes dissolved as in A, plus 20 µg/ml pancreatic ribonuclease.

318

question whether the polymeric ribosomes are real messenger-ribosome complexes or aggregates themselves. Treatment with ribonuclease is one way to test this. Cytoplasmic polymeric ribosomes are already converted to monomers to a considerable extent by incubation with 0.5 μg/ml ribonuclease for 60 min at 0° C. This conversion is complete at a concentration of 20 μg/ml ribonuclease. In contrast, the optical density pattern of mitochondrial ribosomes is not changed by incubation with corresponding ribonuclease concentrations (Fig. 3). Also no change is observed in the radioactivity pattern. This suggests that mitochondrial polymeric ribosomes are aggregates. In agreement with this is the observation that after treatment of isolated mitochondria with puromycin, aggregation of ribosomes is almost completely abolished, Therefore it appears that nascent chains are responsible for this aggregation. Similar observations and conclusions were made by Ojala and Attardi (8).

B) Properties of nascent translation products.

On the basis of these findings we are led to conclude that the peptide chains on mitochondrial ribosomes have certain peculiar properties. This is underlined by the observation that these peptides in the form of peptidyl-transfer-RNA cannot be separated from the ribosomal proteins by treatment with phenol, unless SDS is present. If SDS is removed from SDS solubilized ribosomes, the nascent peptides become insoluble.

In order to further elucidate these properties, it was examined whether the nascent peptide chains can be removed from the mitochondrial ribosomes as peptidyl-puromycin and as peptidyl-tRNA. To control the experimental setup, first cytoplasmic ribosomes were tested. In this case, treatment of ribosomes with puromycin, GTP and G-factor leads to the release of the nascent chains in a soluble form. Also, upon exposure of cytoplasmic ribosomes to EDTA, dissociation into subunits and release of peptidyl tRNA occurs. In contrast, when mitochondrial ribosomes are treated with puromycin, GTP and G-factor, the radioactive nascent

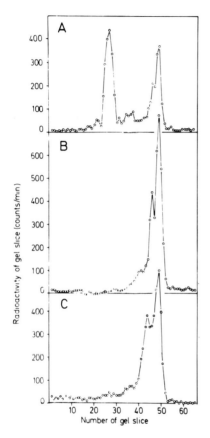

Fig. 5. Gel electrophoretic analysis of radioactively labeled nascent peptide chains associated with mitochondrial monomeric ribosomes. A, Ribosomes pretreated as described for Fig. 4A; B, Ribosomes dissolved in 0.1 M phosphate buffer, pH 11, and kept for 1 h at 37° C; C, Ribosomes pretreated as described for Fig. 4B.

chains disappear from monomers and polymers in the
gradient, however they are not released in a soluble
form, but are found at the wall and in the pellet of
the tube after centrifugation. Treatment with EDTA
dissociates the mitochondrial ribosomes but the radioactive chains remain associated with the large subunits and with ribosomal structures which cannot be
clearly identified on the basis of their sedimentation
behaviour. They probably represent aggregates of the
large subunit (9). It appears from these observations
that the nascent chains on mitochondrial ribosomes are
not watersoluble but aggregate in aqueous solutions.

For further analysis of the nascent chains the ribosomes were dissolved in SDS containing buffer and
subjected to gel electrophoresis in the presence of
SDS. This was done separately for poly- and monomeric
ribosomes. When nascent chains on cytoplasmic ribosomes were studied they showed a scattered distribution
of apparent molecular weights (AMWs). No significant
difference between chains of monomeric and polymeric
ribosomes is seen (9). In Fig. 4A gel electrophoresis
of radioactively labelled chains on mitochondrial polymeric ribosomes is shown. The AMWs also appear to
be quite spread here. Fig. 4B represents the distribution after treatment of the SDS solubilized polymeric ribosomes with ribonuclease. This was done to de-
grade peptidyl tRNA which might still be present after incubation of the ribosomes at pH 8 and 37° C for
one hour. The minor changes compared to Fig. 4A and
the observation that tRNA on this gel migrates corresponding to an AMW of ca. 15,000 suggest that hydrolysis of the peptide-tRNA bond has already occurred to
a large extent.

In Fig. 5A an electrophoretic separation of chains
associated with mitochondrial monomeric ribosomes is
presented. A peak with an AMW of 27,000 is prominent
in addition to a double peak with an AMW in the range
of 8,000 - 12,000. If the ribosomes are immediately
subjected to gel electrophoresis after having been
dissolved, mainly the first peak is present. Prolonged incubation of the ribosomes results in a decrease

of the first peak and in an increase of the second
one. At pH 6 this process is slower than at pH 8, and
at pH 11 (Fig. 5B) it is faster. Treatment of SDS solubilized ribosomes with ribonuclease causes the disappearance of the 27,000 AMW peak. At the same time
the double peak with the lower AMW increases (Fig. 5C).

Since determinations of AMWs are difficult or even
impossible in the low molecular weight range by gel
electrophoresis, gel chromatography was carried out.
The chains on the monomeric ribosomes are eluted from
the column (Sephadex G 200) together with the marker
cytochrome c. After digestion with trypsin smaller
products appear. These are eluted together with the
marker leucine. The chains at the polymeric ribosomes
show a very similar behaviour.

These results suggest that the 27,000 AMW peak represents peptidyl tRNA and that the peptides have AMWs
of 8,000 - 12,000. This is substantiated by the observation already mentioned that transfer RNA migrates
with an AMW of about 15,000.

On the basis of the results obtained by gel chromatography, the AMW of the polymer product is in the
same range as that of the monomer product. This can
only be reconciled with the electrophoretic data, if
we assume that on the gel the polymer chains aggregate either with each other or with the hydrophobic gel
material. This again would be in accordance with the
selective tendency of polymeric ribosomes to aggregate.

*Translation products after integration into the
membrane.* What do the translation products on the
monomeric ribosomes which display a rather uniform
AMW represent ? If they are completed translation products, then it should be possible to find them in the
mitochondrial membrane. To follow this question, *Neurospora* cells were labelled under the following conditions: CHI was added to the culture, after 2.5 min
radioactive leucine and after further 2 min a chase
of unlabelled leucine was given. After 45 min the
cells were cooled, mitochondrial membranes were prepared and subjected to gel electrophoresis. The dis-

322

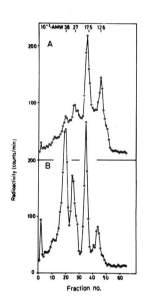

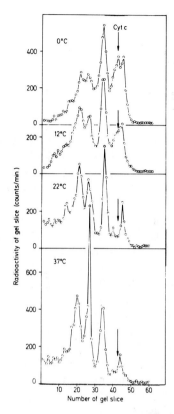

Fig. 6. Gel electrophoretic analysis of mitochondrial membranes after pulse labeling with radioactive leucine *in vivo* in the presence of cycloheximide. After 2.5 min preincubation of *Neurospora* cells with cycloheximide, a 2 min pulse of [³H]-leucine was given, followed by a chase of unlabeled leucine. In (A) the cells were immediately cooled after having given the chase, in (B) the chase lasted for 45 min at growth temperature (25° C).

Fig. 7. Gel electrophoretic analysis of mitochondrial membranes after pulse labeling with [³H]leucine *in vivo* in the presence of cycloheximide followed by a 1 h chase at different temperatures. The temperature during chase is given in the figure.

tribution of radioactivity on the gel is shown in Fig.
6B. Peaks are prominant at AMWs of 36,000, 27,000 and
18,000, and a small one at about 12,000. If all these
protein bands were original translation products, then
we would have to conclude that we do not recover completed translation products on the monomeric ribosomes.
However, if we stop the labelling procedure immediately after having added the chase, a quite different
labelling pattern of the membrane is observed (Fig.
6A). A large part of the radioactivity co-migrates
with cytochrome c or is even faster. The high molecular weight peaks are present only to a low extent,
the 18,000 AMW peak to a similar extent as after a
long chase. The background in the high molecular
weight region is very high and it looks like low molecular weight material is tailing. The total radioactivity in the membrane does not change, so it must be
conluded that a conversion of translation products
with lower AMWs to such with higher AMWs takes place
in the membrane. If chase times in between those shown
in Fig. 6 are investigated, a gradual shift of the
radioactivity is observed.

This process was studied in a further experiment.
Labelling was performed here as described for Fig. 6,
with the modification that after giving the chase,
the culture was divided into four equal portions and
these were rapidly adjusted to 0, 12, 22 and 37° C,
respectively. Incubation at these temperatures was
carried on for one hour. Fig. 7 shows the labelling
patterns of the mitochondrial membrane. A chase performed at 0° C does not change the labelling pattern
of the unchased cells (cf. Fig. 6A). At higher temperatures the conversion takes place. It is the faster
the higher the temperature is.

DISCUSSION.

1. The rather uniform apparent molecular weight of
the nascent chains at the mitochondrial monomeric ribosomes may have two possible reasons: a) It is an artifact, because nascent chains are cut down to a certain length; b) the monomers carry completed chains.

Possibility a) does not appear to be likely, since no indication for such a breakdown was found under the different conditions of isolation and separation of mitochondria and ribosomes. Furthermore, the observed conversion of translation products in the membrane could explain that no larger peptide chains are found at the mitochondrial ribosomes.

The presence of completed polypeptide chains at the monomeric ribosomes is related to the problem, how mitochondrial translation products reach the inner membrane. They are obviously so hydrophobic that it seems highly improbable that they are released from the ribosomes into the matrix in a soluble form. So we can speculate that the ribosome leaving the messenger-RNA does not immediately release the product and dissociate, but rather dissociates only after having transported the completed chain to the membrane. For discussion of the alternative mechanism that ribosomes are bound to the inner membrane, see ref. 9.

2. Experiments presented here suggest that the mitochondrial translation products found in enzyme complexes such as cytochrome aa_3, cytochrome b, and ATPase (10-13) are not original translation products, but are rather generated by conversion of peptides with lower apparent molecular weights. After pulse labelling, the mitochondrially synthesized subunits of cytochrome c oxidase appear in the enzyme protein with a time course similar to that described for the 36,000, 27,000 and 18,000 AMW peaks in the membrane (14). It is possible that for each subunit this conversion process is the rate limiting step.

3. It is concluded in this report that mitochondrial translation products exhibit a strong hydrophobic character. This is substantiated by amino acid analyses of subunits of cytochrome c oxidase formed inside the mitochondria (15). On the basis of these data we are led to suggest that in mitochondria a system of transcription and translation had to be maintained (if we follow the endosymbiont theory) or had to be created (if we follow some other theory (16), because the translation products are so hydrophobic that they can-

not be transported through the cytoplasm and the intercristae space. They rather have to be delivered to the inner membrane directly, *i.e.* from the matrix side.

ACKNOWLEDGEMENT.

This work was supported by the Deutsche Forschungsgemeinschaft, Schwerpunktprogramm "Biochemie der Morphogenese".

REFERENCES.

1. G. Schatz, in: "Membranes of Mitochondria and Chloroplasts", E. Racker Ed., Nostrand Reinhold Co., New York, p. 251 (1970).
2. H. Küntzel, Curr.Top.Microbiol.Immunol. $\underline{54}$, 94 (1971).
3. D.S. Beattie, Subcell.Biochem. $\underline{1}$, 1 (1971).
4. P. Borst, Ann.Rev.Biochem. $\underline{41}$, 333 (1972).
5. H. Küntzel, Nature (Lond.) $\underline{222}$, 142 (1969).
6. W. Neupert, W. Sebald, A. Schwab, A. Pfaller & Th. Bücher, Eur.J.Biochem. $\underline{10}$, 589 (1969).
7. P.M. Lizardi & D.J.L. Luck, J.Cell Biol. $\underline{54}$, 56 (1972).
8. D. Ojala & G. Attardi, J.Mol.Biol. $\underline{65}$, 273 (1972).
9. R. Michel & W. Neupert, Eur.J.Biochem. in press (1973).
10. H. Weiss, W. Sebald & Th. Bücher, Eur.J.Biochem. $\underline{22}$, 19 (1971).
11. T.L. Mason & G. Schatz, J.Biol.Chem. $\underline{248}$, 1355 (1973).
12. H. Weiss, Eur.J.Biochem. $\underline{30}$, 469 (1972).
13. A. Tzagoloff, M.S. Rubin & M.F. Sierra, Biochim. Biophys.Acta $\underline{301}$, 71 (1973).
14. A.J. Schwab, W. Sebald & H. Weiss, Eur.J.Biochem. $\underline{30}$, 511 (1972).
15. W. Sebald, W. Machleidt & A. Otto, Eur.J.Biochem. in press (1973).
16. R.A. Raff & H.R. Mahler, Science $\underline{177}$, 575 (1972).

ANALYSIS OF MITORIBOSOMES FROM *TETRAHYMENA* BY POLYACRYLAMIDE GEL ELECTROPHORESIS AND ELECTRON MICROSCOPY

Barbara J. Stevens, Jean-Jacques Curgy, Gérard Ledoigt and Jean André

Laboratoire de Biologie Cellulaire 4 Université de Paris XI, Orsay, France.

One of the fundamental features of mitochondrial ribosomes (mitoribosomes) from a variety of organisms is a sedimentation coefficient in sucrose gradients lower that that of the corresponding cytoribosomes (1,2,3). One exception has been found in the ciliate *Tetrahymena pyriformis* (4), where the cyto- and mitoribosomes exhibit the same sedimentation coefficient (80S). The two ribosome types, however, can be clearly distinguished by analysis of their heavy and light ribosomal RNA components and by buoyant density determinations in CsCl (4). A further unusual physicochemical property of this mitoribosome lies in its dissociation into subunits: in the presence of EDTA, the monosome peak is transformed into a single peak at 55S. Buoyant density determinations in CsCl can, however, distinguish two different dubunit particles having densities of 1.46 and 1.56 g/cm^3 (4).

The present work describes a further analysis of *Tetrahymena* mitoribosomes by polyacrylamide gel electrophoresis and electron microscopy, used correlatively. Mitoribosomes examined by these techniques exhibit distinctive properties, indicating a greater volume and a unique fine structure.

ABUNDANCE OF MITORIBOSOMES ACCORDING TO CULTURE CONDITION

Mitoribosomes *in situ* can be clearly visualized in thin sections of aldehyde-fixed material stained with lead citrate and examined in the electron microscope

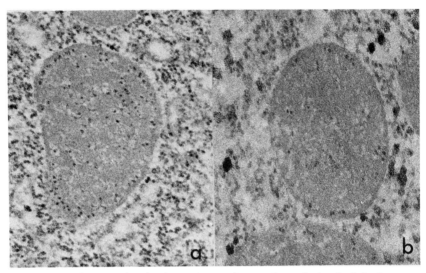

Fig. 1. Mitoribosomes *in situ* in representative sections of mitochondria from cells in exponential (a) and stationary (b) growth phases. Aldehyde fixed; lead stained. X 55,000.

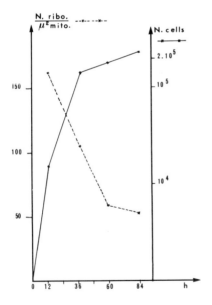

Fig. 2. Change in number of mitoribosomes during *Tetrahymena* growth cycle. Particles were counted and related to mitochondrial surface section.

328

(Fig. 1). A quantitative study of the number of mitoribosomes per unit mitochondrial surface at various
times during growth of a culture reveals a definite
relationship (Fig. 2). The number of particles observed in early exponential growth ($\sim160/\mu^2$) diminishes
sharply as the cultures reach late exponential growth
phase. A low value of $\sim50/\mu^2$ is found for cultures in
stationary growth. Since the mitochondrial population
within a single organism seems homogeneous with regard to the number of mitoribosomes, the presence of
these particles appears functionally related to cell
growth and to that of the mitochondrial population.

ANALYSIS OF PURIFIED MITORIBOSOMES

Mitoribosomes were extracted from purified mitochondria by lysis with Triton X-100 (2.5%) or deoxycholate (0.4 %) and the lysate was placed directly on
a linear sucrose gradient. A sedimentation coefficient
of 80S was found for both purified mito- and cytoribosomes, in agreement with the previous study by Chi &
Suyama (4). Electrophoretic analysis (5) of the mitoribosomal 80S peak gives a single, well-defined band
at 265 nm absorbance, which migrates only a short distance from the origin (Fig. 4). The electropherogram
of particles from a mitoribosomal pellet is similar
to that of the 80S peak particles; no polysome peaks
were present (Fig. 4). In comparison, the cytoribosome 80S fraction migrates more rapidly and produces a
single band clearly distinct from that of mitoribosomes. The two ribosome species, therefore, which are
indistinguishable in sucrose gradients, can be separated by polyacrylamide gel electrophoresis.

Examination of negatively-stained mitoribosomes in
the electron microscope reveals elongate profiles,
about 370 x 240 Å, which can be grouped into two main
classes (Fig. 3). The most frequent form exhibits an
electron-opaque spot situated at approximately equal
distance from the two ends of the profile. One end is
wider and bears a prominent lobe at one side; the opposite end is more narrow and tapered and shows a
notch at the rounded tip. The other principal form

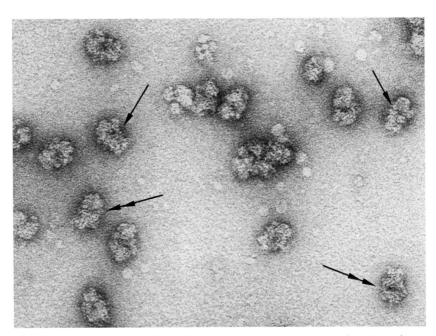

Fig. 3. General view of purified 80S mitoribosomes. Elongated profiles show either a dense, centrally positioned spot (single arrows) or a dense cleft, delimiting two nearly equal size subunits (double arrows). Unfixed particles were negatively stained with uranyl acetate. X 270,000.

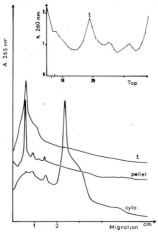

Fig. 4. Electropherogram of purified mitoribosomes. A fraction of the 80S mitoribosomal peak from a sucrose gradient (1 in inset) was placed on a 2.4 % acrylamide gel containing 40 mM Tris, 0.1 mM EDTA, 20 mM Na acetate and 2 mM Mg acetate at pH 7.8. Migration was for 4 h at 4° C. Tracings pellet and cyto correspond to migration of a mitoribosomal pellet and purified 80S cytoribosomes respectively.

330

does not show the opaque spot but contains a more or less evident electron-opaque cleft which separates the profile transversely into two nearly equal size subunits. The centrally positioned spot, the lobe at one extremity and the definitely elongate form of this mitoribosome are all unique morphological features not observed in their cytoplasmic counterpart.

MITORIBOSOMAL SUBUNITS

When the 55S subunit fraction of mitoribosomes dissociated in a sucrose gradient containing 1 mM EDTA was analyzed by gel electrophoresis, a single symmetrical peak was observed (Fig. 6). The mobility of the subunit band coincides with that of the large 60S subunit from cytoribosomes. Ribosomes dissociated directly on the gel produce a coincident band. Variation of certain electrophoretic conditions, such as gel concentration, pH, time of migration, was unable to demonstrate two subunit types. In the presence of increased monovalent cation concentration (40 or 80 mM), the peak is spread and displays shoulders, indicating some heterogeneity of the particles.

The 55S subunit particles appear in the electron microscope as rounded but irregular profiles with an average diameter of 255 Å (Fig. 5). No indication of two distinct classes of profiles was found and we are thus unable to detect two types of subunits. Some profiles show a dense region along one edge, suggestive of the notch seen in various cytoribosome large subunits (6,7,8). None of the regular features of monomeric ribosomes, *i.e.* a dense spot, an indication of two subunits or recurrent profiles of the several projections of the particle, are observed in subunit profiles, indicating that these particles do not represent mini-ribosomes.

COMPARISON OF MITORIBOSOMES WITH OTHER RIBOSOMES

From recent data on the electrophoretic behavior of a variety of ribosomes (manuscript in preparation), we have observed that the relationship between the sedimentation coefficient, and the electrophoretic mobili-

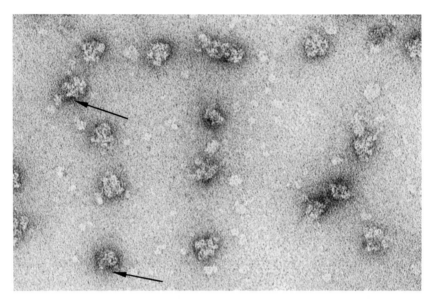

Fig. 5. General view of 55S mitoribosomal subunits. Profiles are rounded with irregular contours and cannot be classified into two different types. No monomeric ribosome profiles are observed. An electronopaque notch is noted in some (arrows). X 270,000.

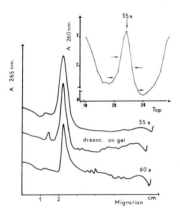

Fig. 6. Electropherogram of mitoribosomal 55S subunits. Various regions (arrows, inset) of the 55S peak from a sucrose gradient containing 1 mM EDTA were placed on a 2.6 % acrylamide gel containing 40 mM Tris, 0.1 mM EDTA, 20 mM Na acetate and no Mg acetate at pH 7.8. Migration was for 2 h at 4° C. Tracing dissoc. on gel corresponds to monomers treated with 2 mM EDTA prior to electrophoresis; 60 S corresponds to large cytoribosomal subunits.

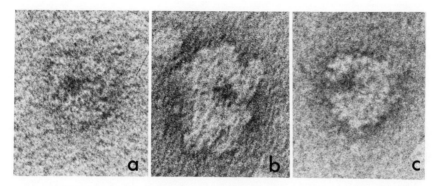

Fig. 7. Selected images of negatively stained *E.coli* (a), *Tetrahymena* mito- (b) and cyto- (c) ribosomes. X 850,000.

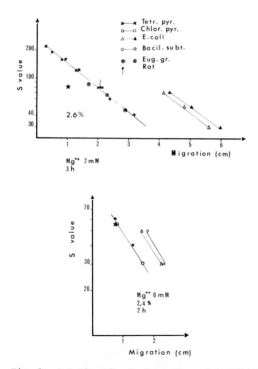

Fig. 8. Relationship of electrophoretic mobilities and sedimentation coefficients of various polysomes, monosomes and subunits. Eukaryotic particles on the one hand, prokaryote on the other, are aligned on two different lines with approximately the same slope. Mitoribosomal monomers and subunits are positioned at the stars.

ty of polysomes, monosomes and subunits is linear on semi-log coordinates for prokaryotic ribosomes on the one hand, and eukaryotic ribosomes on the other (Fig. 8). Thus, two major categories of ribosomes are revealed, corresponding to prokaryote and eukaryote origins.

The slow mobility of mitoribosomes places them off the alignment of eukaryotic particles but at a greater distance from prokaryotic particles (Fig. 8). The behaviour of mitoribosomal subunits tends to approach them to eukaryotic subunits.

The relative resistance of mitoribosomes to dissociation by low Mg^{++} ion concentration indicates a further resemblance to eukaryotic ribosomes. In sharp contrast to prokaryotic ribosomes, mitoribosomes are dissociated only in the presence of EDTA and the resulting subunits are stable in the absence of Mg^{++} ions.

A brief morphological comparison of mitoribosomes with other types serves to emphasize their uniqueness (Fig. 7). Features of the mitoribosome, such as the more elongated form, the centrally positioned spot and the characteristic lobe differ drastically from the typical structure of both pro- and eukaryotic type ribosomes (6,7,8,9,10).

The various properties observed for the mitoribosome of *Tetrahymena* tend to place it apart from both pro- and eukaryotic ribosomes, perhaps in a novel class of ribosomes. Support for such a class of ribosomes must come from further studies on other ciliate mitoribosomes.

ACKNOWLEDGEMENTS

This work was supported by the C.N.R.S. (E.R.A.174) and the D.R.M.E. (contract 72/778).

REFERENCES

1. P. Borst and L.A. Grivell, FEBS Letters 13, 73 (1971).
2. P. Borst, Ann.Rev.Biochem. 41, 333 (1972).
3. A.M. Kroon, E. Agsteribbe and H. de Vries, In:

The Mechanism of Protein Synthesis and its Regulation. (L. Bosch, Ed.), North-Holland Publishing Co., Amsterdam 539 (1972).

4. J.C.H. Chi and Y. Suyama, J.Mol.Biol. 53, 531 (1970).
5. U.E. Loening, Biochem.J. 102, 251 (1967).
6. Y. Nonomura, G. Blobel and D. Sabatini, J.Mol. Biol. 60, 303 (1971).
7. G. Lutsch, H. Bielka, K. Wahn and J. Stahl, Acta Biol.Med.Germ. 29, 851 (1972).
8. V.I. Bruskov and N.A. Kiselev, J.Mol.Biol. 37, 367 (1968).
9. M. Lubin, Proc.Natl.Acad.Sci.U.S.A. 61, 1454 (1969).
10. P.V. Vignais, B.J. Stevens, J. Huet and J. André, J.Cell Biol. 54, 468 (1972).

FINE STRUCTURE OF MITOCHONDRIAL RIBOSOMES OF LOCUST FLIGHT MUSCLE

W. Kleinow, W. Neupert and F. Miller

Institut für Physiologische Chemie und Physikalische Biochemie und Institut für Zellbiologie der Universität München, 8 München 2, Goethestrasse 33, German Federal Republic.

From the mitochondria of *Locusta mirgatoria* thoracic muscle, ribosomes were obtained which are clearly distinct from their cytosolic counterparts. They exhibit a sedimentation constant of 60S, dissociate into subunits of 40S and of 25S and contain RNA molecules of 0.52 and $0.28.10^6$ (1-4). In contrast, cytosolic ribosomes display sedimentation constants of 80S and dissociate into subunits with 60S and 40S and contain RNA molecules of 1.5 and $0.7.10^6$. It was shown that also the protein parts of both ribosome types are different (2).

SECTIONS OF MUSCLE TISSUE.

In sections of muscle tissue from *Locusta Migratoria* cytoplasmic ribosomes lie between myofibrils and are densely packed around the mitochondria (Fig. 1. They have a mean diameter of 180-200 Å. In the mitochondria, corresponding particles are smaller (mean diameter about 130 Å) and less abundant. They are scattered irregularly in the matrix and show no preferential arrangement along mitochondrial membranes. The difference in the diameters between cytosolic and mitochondrial ribosomes points to a volume difference of 3:1 which agrees well with the difference of the molecular weights of the corresponding RNA-molecules. Similar dimensions for intramitochondrial ribosomal particles have also been observed for vertebrate tissues (5). After fixation in potassium permanganate or

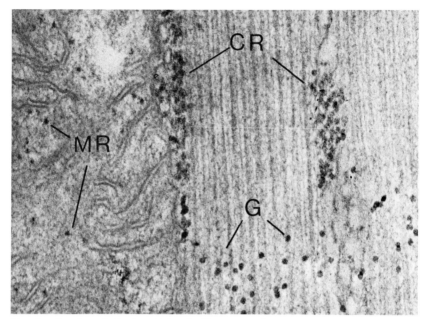

Fig. 1. Section of the dorsal longitudinal thoracic muscle from a locust 2 days after imaginal moult. Fixation in 3 % Glutaraldehyde and 2 % osmium tetroxide; section stained with uranyl acetate and lead citrate; x 90,000. CR: cytosolic ribosomes; G: glycogen particles; MR: mitochondrial ribosomes.

digestion with RNAse cytosolic and mitochondrial ribosomes are not visible and can be discriminated by this behaviour from glycogen particles. Glycogen particles are located in the I-band of myofilaments and measure about 230 Å (Fig. 1).

ISOLATED RIBOSOMES FROM LOCUST MUSCLE.

In a previous communication (2) it was shown that undissociated mitochondrial ribosomes from *Locusta* are heavily contaminated by a protein particle, provisionally called "Mx". In order to characterize isolated mitochondrial ribosomes, this protein impurity has to be removed. This was accomplished by freezing and thawing of the mitochondrial lysate at elevated concentrations of ammonium chloride. As shown in Fig. 2 this treatment leads to a partial dissociation of the ribosomes (most of the small subunits being lost during sedimentation of the particles). Parallel to the dissociation, however, the amount of the protein contamination in the residual monosomes is diminished. This is apparent from the decreased absorbancy at 280 nm.

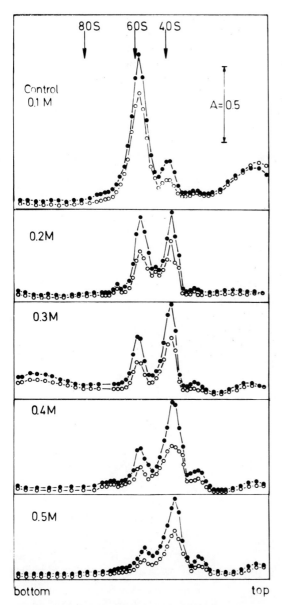

Fig. 2. Gradient profiles of mitochondrial ribosomes from Locusta migratoria after freezing and thawing the mitochondrial lysate in different concentrations of ammonium chloride. Preparation of ribosomes and gradient centrifugation was performed as described (2) except that the molarity of ammonium chloride in the mitochondrial lysate was changed as indicated in the graphs and that the lysate was frozen for 12-18 h at $-80°C$.

At a concentration of 0.3 M ammonium chloride contamination with "Mx" is neglegible and on the other hand enough undissociated mitochondrial ribosomes are left to perform electron microscopic studies.

Section of cytoplasmic ribosomal pellets and of pellets of purified mitochondrial ribosomes show a homogeneous distribution of particles without contamination by other cell constituents (Fig. 3A and B). In

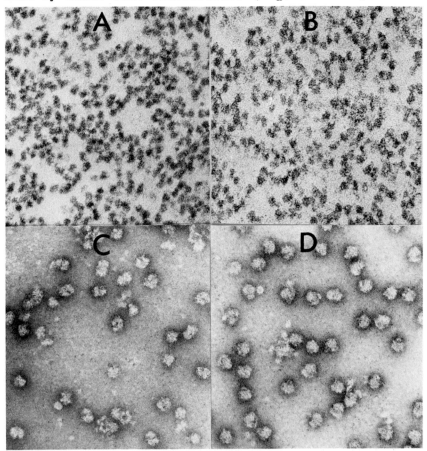

Fig. 3. Fields of positively and negatively stained mitochondrial and cytosolic ribosomes from Locusta thoracic muscle. A and B: positively stained sections of ribosomal pellets. Fixation and staining as in Fig. 1; x 120,000. A: purified mitochondrial ribosomes; B: cytosolic ribosomes. C and D: unfixed ribosomal particles after negative staining; x 150,000. C: mitochondrial ribosomes collected from the 60S region of a sucrose gradient as shown in Fig. 2 (0.3 M); D: cytosolic 80S ribosomes.

these sections cytosolic ribosomes exhibit a mean diameter of 200 Å, whereas mitochondrial ribosomes measure about 180 Å.

For negative staining of undissociated mitochondrial ribosomes, the "Mx"-free 60S peak of a sucrose
gradient containing mitochondrial ribosomes prepared
by freezing in 0.3 M ammonium chloride, is pooled and
dialysed against a buffer containing 1 M methanol (6)
2 mM Mg^{2+} and 30 mM Tris; pH 7.6. Cytosolic ribosomes
are collected from the 80S peak of a corresponding sucrose gradient and processed in the same way. The optical density of the preparations was adjusted to 0.2-
0.4 A_{260}-units/ml and the negative staining was performed according to Nonomura *et al.* (7).

In Fig. 3C and D fields of mitochondrial and cytosolic ribosomes are shown side by side. Both are homogenous preparations containing particles of homogenous
size. Frequently in both types of particles the cleft
can be seen which separates large and small subunit.
More similarities become evident if the particles are
properly orientated. For comparison we have selected
(Fig. 4) pictures which correspond to the pictures
which have been described by Nonomura *et al.* (7) as
"frontal view" and "lateral view" in the case of rat
liver cytosolic ribosomes. In the "frontal view" the
ribosomes show an oval outline, the small subunit extending along the whole width of the ribosome. Frequently in this profile a dense spot is to be observed
off center on the cleft between small and large subunit (Fig. 4A). "Lateral views" (Fig. 4B) are more rarely seen, especially in the case of mitochondrial ribosomes from *Locusta*. They are represented by kidney
shaped profiles with the small subunit sitting asym-
metrically and seen from their small end. Other profiles are present in the fields, however, and unambiguous interpretation is difficult and must await the
results of tilting experiments in progress.

In any case the galeries show that cytosolic and
mitochondrial ribosomes from *Locusta* display in principle the same profiles. So the nature of the 60S particle from the locust flight muscle mitochondria as a

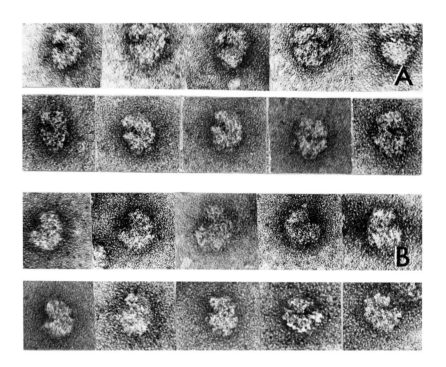

Fig. 4. Selected images of negatively stained mitochondrial and cytosolic ribosomes; x 400,000. A: frontal views, upper row cytosolic ribosomes, lower row mitochondrial ribosomes. B: lateral views, upper row cytosolic ribosomes, lower row mitochondrial ribosomes.

complete ribosome is substantiated by the negative staining experiments. This confirms and extends earlier attempts to demonstrate by negative staining the ribosomal nature of 60S particles from mitochondria of other higher organisms (8,9).

Beside obvious similarities between cytosolic and mitochondrial ribosomes from *Locusta* there are also indications of different morphological features. For example, as mentioned before, "lateral views" are more seldom observed in the case of mitochondrial ribosomes than in the preparations of cytosolic ribosomes. Also the pictures of the frontal views show differences. In no case a small subunit partition can be identified with certainty in the mitochondrial ribosomes. A small subunit partition, however, is clearly seen in cytosolic ribosomes. Most of the pictures of mitochondrial ribosomes show also the cleft between the subunits only in one half of the particle.

The dimensions of the mitochondrial ribosomes after

negative staining are smaller than the dimensions of the cytosolic ribosomes (Table I). This holds especially for the total heigth and for the width along the cleft between the subunits. Both types of ribosomes appear as slightly prolate elipsoids, the cytosolic ribosomes measuring about 300 x 260 x 240 Å , Whereas the mitochondrial ribosomes measure 270 x 210 x 210 Å. It is obvious that after isolation of the ribosomes, the difference in the dimensions between cytosolic and mitochondrial ribosomes is always lesser than between extra- and intramitochondrial ribosomal particles in tissue sections. The reason for this discrepancy is not clear. It may well be due to an artifact produced by one of the preparation techniques or to a different behaviour of the mitochondrial ribosomes after isolation from the cell environment.

SUBUNITS OF MITOCHONDRIAL RIBOSOMES.

Demonstration of the isolated subunits from the mitochondrial ribosomes was possible after dissociation

TABLE I

DIMENSIONS OF NEGATIVELY STAINED RIBOSOMES FROM *LOCUSTA MIGRATORIA* THORACIC MUSCLE.
Given are the dimensions in Å units ± S.D. 50 measurements were averaged for each value, only 20 in the case of lateral views of mitochondrial ribosomes.

	Cyto-Ribosomes	Mito-Ribosomes
Frontal view		
total height	294 ± 18	271 ± 18
height of large subunit	185 ± 16	159 ± 16
height of small subunit	106 ± 15	103 ± 13
width along the cleft between the subunits	245 ± 21	210 ± 18
Lateral view		
total height	294 ± 22	268 ± 25
height of large subunit	191 ± 16	157 ± 23
height of small subunit	104 ± 16	110 ± 19
width of large subunit	257 ± 16	215 ± 28
width of small subunit	145 ± 19	140 ± 28

in 4.0 mM EDTA and subsequent fixation of the subunits according to Subramanian (10). The subunits were separated by density gradient centrifugation and the particles from the 40S peak and from the 25S peak of the gradient were collected separately and processed as described for the undissociated ribosomes. Fields

of negatively stained large and small subunits of the
mitochondrial ribosomes from *Locusta* are shown side by

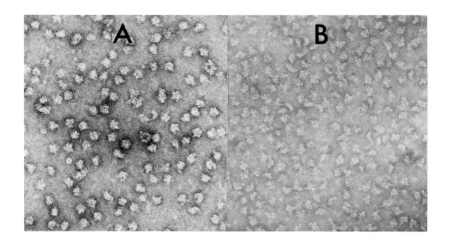

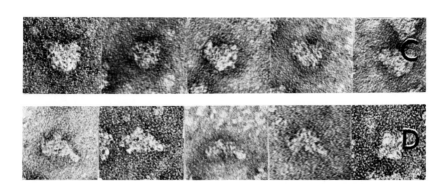

Fig. 5. Subunits of mitochondrial ribosomes from *Locusta migratoria*. The ribosomes
were partly dissociated by incubation at 4° C in 4.0 mM EDTA and without magnesium.
The suspension was fixed with glutaraldehyde according to (10) and was placed immi-
diately after fixation on a sucrose gradient. The peaks in the gradient corresponding
to the particles sedimenting with 40S and 25S were collected and processed as described in the text. A: general view of a field of large (40S) subunits; x 112,000.
B: general view of a field of small (25S) subunits; x 112,000. C: selected images
of large subunits; x 400,000. D: selected images of small subunits; x 400,000.

side in Fig. 5A and B. Clearly each of the fields contains a homogeneous particle fraction and evidently the morphological features of the particles obtained from the 40S peak are different from the particles in the field obtained from the 25S peak. The pictures of the large subunits (Fig. 5C) coincides well with those described for ribosomes of other origin (7,11,12). They show a rounded or semicircular profile. One side of the particles seems to be flattened or concave, Frequently on this concave side a dense spot is observed, which may correspond to a shallow groove on the surface of the particle. A knoblike projection as described for other mitochondrial large subunits (11) could not be identified with certainty.

The small subunit shows elongated profiles as is to be expected from the pictures of small subunits from other ribosomal types. However, in contrast to small subunits from rat liver cytosol (7) or from $E.coli$ (12,13) the profiles are not rounded and the small subunit partition is absent or at least difficult to demonstrate. Instead angular profiles prevail: tetragonal, triangular or slightly concave forms are the most abundant. In this respect the morphological features of the 25S small subunit from the 60S mitochondrial ribosome of locusts coincide possibly with the features of the 36S small subunit from the 72S mitochondrial ribosome of yeast (11).

ACKNOWLEDGEMENTS.

This work was supported by the "Deutsche Forschungsgemeinschaft". We wish to thank Prof. Th. Bücher for steady support and Miss H. Rothe for excellent technical assistance.

REFERENCES.

1. W. Kleinow, W. Neupert and Th. Bücher, FEBS Letters 12, 129 (1971).
2. W. Kleinow and W. Neupert, FEBS Letters 15, 359 (1971).
3. W. Kleinow and W. Neupert, Z.Physiol.Chem. 351, 1205 (1970).

4. W. Kleinow, W. Neupert and F. Miller, Verh.Dtsch. Zool.Gesellsch. 66, 87 (1973).
5. J. André and V. Marinozzi, J.Microsc.(Paris) 4, 615 (1965).
6. V.D. Vasiliev, FEBS Letters 14, 203 (1971).
7. Y. Nonomura, G. Blobel and D. Sabatini, J.Mol. Biol. 60, 303 (1971).
8. T.W. O'Brien and G.F. Kalf, J.Biol.Chem. 242, 2180 (1967).
9. C. Aaij, N. Nanninga and P. Borst, Biochim.Bio-phys.Acta 227, 140 (1972).
10. A.R. Subramanian, Biochemistry 11, 2710 (1972).
11. P.V. Vignais, B.J. Stevens, J. Huet and J. André, J.Cell Biol. 54, 468 (1972).
12. M.R. Wabl, P.J. Barends and N. Nanninga, Cytobiol. 7, 1 (1973).
13. M.R. Wabl, H.G. Doberer, S. Höglund and L. Ljung, Cytobiologie 7, 111 (1973).

THE STRUCTURE, COMPOSITION AND FUNCTION OF 55S MITOCHONDRIAL RIBOSOMES

T.W. O'Brien, N.D. Denslow and G.R. Martin

Department of Biochemistry, J.Hillis Miller Health Center University of Florida, Gainesville, Florida 32601, U.S.A.

In earlier studies of mammalian mitochondrial ribosomes it was apparent that these ribosomes differed from others in at least 3 major respects: their low sedimentation coefficient, their low RNA content and low buoyant density (1,2). The sedimentation coefficient of mitochondrial ribosomes is uniformly low, 55-57S in all mammals we have studied (3). The RNA content of these ribosomes appears to be only about half the amount in other ribosomes, based both on chemical measurements (1) and upon estimates of the RNA content from buoyant density determinations (2,4). Their low RNA content or disproportionately high protein content could account for their sedimentation properties and low buoyant density. Alternatively, the distinctive physical-chemical and functional properties of this ribosome may derive from a novel ribosome structure, perhaps including structural lipoprotein components (4) or other material of a "membrane" nature. It is the aim of this paper to clarify the major structural and functional properties of 55S mammalian mitochondrial ribosomes permitting us to define better their relation to other ribosomes.

FUNCTIONAL PROPERTIES OF 55S RIBOSOMES

Isolated 55S monoribosomes incorporate [3H]phenylalanine in a poly U-directed system (5) when incubated in the presence of homologous factors (Table I), showing that the isolated bovine 55S ribosomes is active in protein synthesis. As is true for other ribosomes,

347

TABLE I

INCORPORATION OF [^{3}H]PHE BY MITOCHONDRIAL 55S RIBOSOMES.
Mitochondrial ribosomes, prepared from bovine liver as described in Fig. 4, and
E.coli ribosomes, extracted by sonication and washed in buffer A, were incubated
for 30 min with poly-A, 65 μg of E.coli tRNA and 100 μg (protein) of mitochondrial supernatant fraction using the system described by Swanson and Dawid (5).

Ribosomes	[^{3}H]phe incorporation pmoles/mg RNA/30 min
Mitochondrial 55S	34
E.coli 70S	48
E.coli 70S, minus factors	3.5
None	5.6

peptidyl transferase activity is localized to the large (39S) subunit of 55S ribosomes; the small (28S) subribosomal particle is inactive (Table II). In this reaction, where peptidyl transferase activity is independent on added factors, 55S ribosomes show an activity comparable to that of 70S ribosomes. These findings indicate that the functional unit of protein synthesis in mammalian mitochondria is a 55S particle.

Active intramitochondrial ribosomes appear to be
associated with the inner mitochondrial membrane. A
fraction of mitochondrial ribosomes resists extraction
by treatment of mitochondria with non-ionic detergents
under relatively mild extraction conditions (Fig. 1).
These ribosomes can be extracted from a membrane residue fraction by treatment with Na deoxycholate (DOC;
Fig. 1b). If the mitochondria are incubated with [^{3}H]
leucine to pulse label active ribosomes, the membrane-
associated ribosomes (Fig. 1b) exhibit almost twice
the specific radioactivity of those solubilized by

TABLE II

PEPTIDYL TRANSFERASE ACTIVITY OF BOVINE LIVER MITOCHONDRIAL RIBOSOMES.
Puromycin was reacted with N-Ac-[^{3}H]-leu tRNA (10,000 cpm) from E.coli in a modified fragment reaction, after Monro (6), to measure the peptidyl transferase activity of ribosomes and subunits. The reaction was carried out for 15 min at 22°
in the presence of approximately 1 A$_{260}$ unit of ribosomes in a volume of 0.15 ml.

Ribosomes	N-Ac-[^{3}H]-leu tRNA-puromycin formed, cpm/mg ribosomes
Mitochondrial 55S	40,000
Mitochondrial 39S subunit	17,800
Mitochondrial 28S subunit	84
E.coli 70S	44,500
None	57

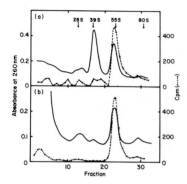

Fig. 1.   Sedimentation analysis of mitochondrial ribosomes extracted (a) by treatment of 140 mg (protein) of rabbit liver mitochondria with Triton X-100 (1 %) in buffer A (10 mM Mg^{2+}, 0.05 M NH_4Cl, 0.05 M KCl, 5 mM Tris, pH 7.5). After detergent treatment, the extract was centrifuged at 60,000 x g for 10 min to sediment a membrane residue fraction. (b) Additional mitochondrial ribosomes were extracted by resuspending the membrane residue in buffer A containing 0.5 % DOC. Before extraction of the ribosomes, the 6-times washed mitochondria were incubated for 5 min with [^{3}H]leucine as described (3). Centrifugation of linear 10-30 % sucrose gradients in buffer A was for 13.5 h at 20,000 rpm in the Beckman SW27 rotor.

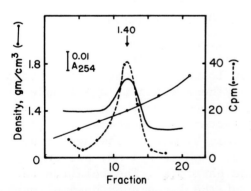

Fig. 2.   Buoyant density analysis of active mitochondrial ribosomes. Mitochondrial ribosomes from rabbit liver were pulse labelled with [^{3}H]leucine and extracted with Triton X-100 and DOC using the conditions of Fig. 1. 55S ribosomes recovered from sucrose gradients were fixed with 5 % formaldehyde and centrifuged to equilibrium in a cesium chloride gradient containing 1 % formaldehyde.

gentle lysis with Triton X-100 alone (Fig. 1a). In o-
ther systems (7,8) association of ribosomes with membranes is known to involve at least two sorts of interactions: a primary, salt-labile interaction of the
large subribosomal particle with the membrane, and a
puromycin-labile binding, interpreted to represent intimate physical association of the nascent polypeptide
of active ribosomes with the membrane. Puromycin
treatment (50 µg/ml) of mitochondria under conditions
of active protein synthesis is without effect on the
overall yield of ribosomes extracted by Triton X-100
alone, showing that the binding of active mitochondrial
membranes is not solely via a puromycin-labile linkage.

   These findings, and observations in other laboratories (8,9) emphasize the functional and physical association of mitochondrial ribosomes with membranes, and
raise the possibility that the low buoyant density of
55S mitochondrial ribosomes (Fig. 2) is an artifact
arising from the retention of membrane fragments by
mitochondrial ribosomes. This possibility, and the implications concerning the composition and structure
of 55S ribosomes will be treated below.

SUBUNIT STRUCTURE OF 55S RIBOSOMES

   Runoff or puromycin-treated 55S ribosomes of mammalian mitochondria dissociate in media of Mg^{2+}/μ (ratio of Mg^{2+} concentration to ionic strength of buffer)
values less than 0.1 (3). These subunits are stable
over a wide range of ionic conditions, generally be-
having like subunits of bacterial ribosomes. In 0.5 M
salt they move progressively slower as the Mg^{2+} concentration is reduced from 5 mM to 0.1 mM, presumably
due to "unfolding" of the particles. The relative sedimentation coefficient of the large subribosomal particle decreases to about 35S over this range, while
that of the small subunit decreases from 28S to about
25S under these conditions. This behavior is strikingly different from that of extramitochondrial subribosomal particles, which show marked discontinuities in
their sedimentation rates over the same range of Mg^{2+}

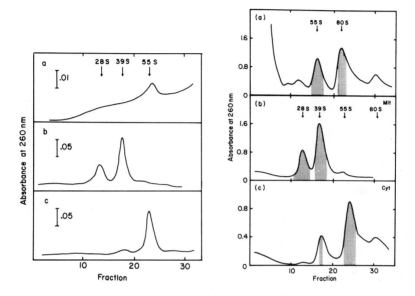

Fig. 3. Reversible dissociation of 55S mitochondrial ribosomes. 55S monoribosomes prepared (Fig. 4) in buffer C (20 mM Mg^{2+}, 0.05 M KCl, 0.05 M NH_4Cl, 1 mM dithiothreitol, 5 mM Tris, pH 7.5) were split into aliquots for (a) dialysis and sedimentation analysis in buffer C, (b) dialysis and sedimentation analysis in buffer D (5 mM Mg^{2+}, 0.8 M KCl, 0.05 M NH_4Cl, 1 mM dithiothreitol and 5 mM Tris, pH 7.5), and (c) dialysis in buffer D (to dissociate the ribosomes as in (b), followed by dialysis in buffer C to promote reassociation, and sedimentation analysis in buffer C.

Fig. 4. Sedimentation analysis of bovine mitochondrial and extramitochondrial subribosomal particles. 55S mitochondrial ribosomes and 80S extramitochondrial ribosomes (a) prepared by treatment of partially purified bovine liver mitochondria in buffer B (20 mM Mg^{2+}, 0.1 M KCl, 5 mM 2-mercaptoethanol, and 20 mM triethanolamine, pH 7.5) with Triton (2 %) and DOC (0.5 %), as described (3). 55S ribosomes and 80S ribosomes were pooled and analyzed in high salt buffer Z (5 mM Mg^{2+}, 0.5 M KCl, 5 mM 2-mercaptoethanol, 20 mM triethanolamine, pH 7.5). Sedimentation analysis of (b) 55S subribosomal particles and (c) 80S subribosomal particles in buffer Z. Centrifugation was for 5 h at 27,000 rpm in (a), and for 13.5 h at 20,000 rpm in (b) and (c).

concentrations.

Mitochondrial subribosomal particles derived by dissociation of 55S monoribosomes in high salt (Fig. 3b) will reassociate under favorable conditions (Fig. 3c).

Analysis of subribosomal particles in media of high ionic strength removes loosely bound proteins. For comparative purposes, the 80S extramitochondrial ribosomes used in this study were obtained from lysates of partially purified mitochondria (Fig. 4a). It is important to note that major, consistent differences in buoyant density (ρ) are observed between 80S ribosomes ($\rho = 1.54$) and mitochondrial ribosomes ($\rho = 1.40$ to 1.42) extracted and analyzed under identical conditions. Treatment of mitochondrial subribosomal particles with high salt buffers that cause destabilization and stripping of ribosomal proteins from extramitochondrial ribosomes, produces relatively minor changes in their buoyant density. The presumed limit densities (before marked destabilization) of mitochondrial subribosomal particles are about 1.44 and 1.45 g/cm^3 for the large and small subribosomal particles, respectively.

PROTEIN CONTENT OF MITOCHONDRIAL RIBOSOMES

To study the protein content of mitochondrial salt-washed subribosomal particles, bovine 55S ribosomes were prepared (Fig. 4a) and analyzed (Fig. 4b), along with 80S ribosomes (Fig. 4c) for comparative purposes, in sucrose gradients containing high salt buffer Z. Under these conditions, the mitochondrial small (Fig. 5a) and large (Fig. 5c) subribosomal particles have buoyant densities of 1.44 and 1.43 g/cm^3, respectively, while the cytoplasmic small (Fig. 5b) and large (Fig. 5d) subribosomal particles have densities of 1.54 and 1.59, respectively. It should be pointed out that the small peak of material of $\rho = 1.43$ accompanying the cytoplasmic small subribosomal particle (Fig. 5b) represents large (39S) mitochondrial subribosomal particles originally present in dimers of 55S ribosomes in the 80S region of Fig. 4a.

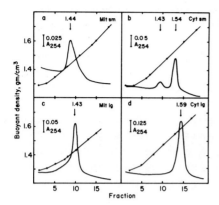

Fig. 5. Buoyant density analysis of bovine mitochondrial and cytoplasmic sub-ribosomal particles. Subribosomal particles, pooled as indicated in Fig. 4b and c, were fixed with 5 % formaldehyde and centrifuged to equilibrium in CsCl gradients containing 0.3 % formaldehyde.

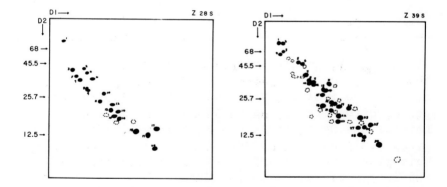

Fig. 6. Analysis of proteins extracted (Urea-LiCl) from the small, salt-washed subribosomal particle (Fig. 4b) by 2-D electrophoresis in polyacrylamide gels. Electrophoresis in the first dimension was conducted in 7.5% gels (3% bis-acryl-amide) at pH 4.5 in the presence of 8 M urea. Electrophoresis in the second dimension was performed in a slab of 10% polyacrylamide (3% bis-acrylamide) in the presence of 0.5% SDS and 5 M urea. Standard proteins of known molecular weight were run as molecular weight markers.

Fig. 7. Analysis of proteins from the large, salt-washed subribosomal particle from bovine mitochondria (Fig. 4b) by 2-D polyacrylamide gel electrophoresis. Conditions as in Fig. 6.

The buoyant density values for mitochondrial salt-washed subribosomal particles can be used (10), with the known molecular weight values for the RNA of the 55S mitochondrial ribosome (12,13), to estimate the protein content of these particles from a calibration curve based on the buoyant density of 1.28 g/cm^3 determined (10) for proteins of the mitochondrial ribosome. In this manner, we predict a protein content of 0.56×10^6 daltons for the salt-washed small mitochondrial subribosomal particle and 1.08×10^6 daltons for the corresponding large subribosomal particle (Table III) on the basis of their buoyant density in CsCl.

TABLE III

BUOYANT DENSITY AND PROTEIN CONTENT OF MITOCHONDRIAL RIBOSOMAL AND SALT-WASHED SUB-RIBOSOMAL PARTICLES.

Particle	ρ	RNA %	MW	Daltons x 10^{-6} Composition Predicted	Observed
55S (T)	1.417	30.2	2.75	Protein 1.92	
28S (Z)	1.44	35	.86	RNA (.3)	
				Protein .56	.59 - .64
39S (Z)	1.43	33	1.61	Protein 1.08	.99 - 1.57
				RNA (.53)	

To provide an alternative estimate of the protein content of the salt-washed subribosomal particles, we arrayed the isolated proteins by 2-D polyacrylamide gel electrophoresis, using procedures outlined by Leister and Dawid (14) to visualize proteins of the cytoplasmic and mitochondrial ribosomes of *Xenopus*. This procedure is a variation of methods employing SDS in the second dimension run so that a molecular weight estimate can be made for each of the resolved proteins. In this manner, we visualize 20 major and 3 minor proteins in the small salt-washed subribosomal particle (Fig. 6), ranging in size from approximately 10,000 to 79,000 daltons, with an average of 29,400 daltons. Similarly, we resolve 30 major and 20 minor components of the large, salt-washed subribosomal particle (Fig. 7), ranging in size from approximately 10,500 daltons to 81,000 daltons, with an average of 33,000 daltons.

If we assume the major components visualized by 2-D gel electrophoresis represent unit copies of ribosomal proteins, then we observe (Table III) 0.59×10^6 daltons of protein in the small subribosomal particle, a value in close agreement with that of 0.56×10^6 daltons, predicted on the basis of its buoyant density. The agreement is almost as good for the protein content of the large subribosomal particle. Of a total protein content of 1.08×10^6 daltons predicted on the basis of its buoyant density, we can account for 0.99×10^6 daltons by the major components alone.

If the unusually low buoyant density values for these particle results from their content of low density materials such as membrane fragments or lipoproteins, we would predict an inordinately high protein content for the particle. Because the values predicted on the basis of buoyant density are borne out by the 2-D gel analysis of the individual proteins, we conclude that low buoyant density is an inherent property of 55S ribosomes, resulting primarily from a protein content higher than that of other ribosomes, and not from lipoprotein components, adventitiously bound proteins or membrane fragments. Thus, the 55S ribosomes of animal mitochondria are distinguished from other ribosomes of the prokaryotic type, including mitochondrial ribosomes of primitive eukaryotes, on the basis of characteristic physical-chemical, as well as functional differences.

ACKNOWLEDGEMENT

This work was supported by the U.S.P.H.S., National Institute of General Medical Sciences Research Grant GM-15438. We gratefully acknowledge the technical assistance of Mark Critoph, Warren Clark and Diane Jernigan.

REFERENCES

1. T.W. O'Brien and G.F. Kalf, J.Biol.Chem. 246, 2180 (1967).
2. S. Perlman and S. Penman, Nature 227, 133 (1970).

3. T.W. O'Brien, J.Biol.Chem. 246, 3409 (1971).
4. T.W. O'Brien, Fed.Proc. 31, 456 (1972).
5. R.F. Swanson and I.B. Dawid, Proc.Natl.Acad.Sci. U.S.A. 66, 117 (1970).
6. R.E. Monro, Methods in Enzymol. XX, part C, 472 (1971).
7. M.R. Adelman, D.D. Sabatini and G.Blobel, J.Cell Biol. 56, 206 (1973).
8. Y. Kuriyama and D.J.L. Luck, this volume
9. N.R. Towers, G.M. Kellerman, J.K. Raison and A.W. Linnane, Biochim.Biophys.Acta 299, 153 (1973).
10. M.G. Hamilton, Methods in Enzymol. XX, part C, 512 (1971).
11. T.W. O'Brien, Manuscript in preparation.
12. I.B. Dawid and J.W. Chase, J.Mol.Biol. 63, 217 (1972).
13. D. Robberson, Y. Aloni and G. Attardi, J.Mol.Biol. 60, 473 (1971).
14. D.E. Leister and I.B. Dawid, Fed.Proc. 32, 647 (1973).

PHYSICOCHEMICAL AND FUNCTIONAL CHARACTERIZATION OF THE 55S RIBOSOMES FROM RAT-LIVER MITOCHONDRIA

H. de Vries and A.M. Kroon

Laboratory of Physiological Chemistry, State University, Bloemsingel 10, Groningen, The Netherlands.

Since the first description of 55S ribosomes from mammalian mitochondria in 1967 (1) an impressive number of studies on these peculiar ribosomes ("miniribosomes") has appeared (for review, see ref. 2). Although many properties have been settled, there are still some gaps in the functional characterization of the isolated ribosomes, and also the question how "mini" these ribosomes really are has not been answered satisfactorily in our opinion. In this paper we want to concentrate on the dimensions of the 55S ribosome from rat liver as investigated by electrophoretic analysis and on its activity in protein-synthetic reactions.

PREPARATIVE ISOLATION OF MITOCHONDRIAL RIBOSOMES.

We isolated ribosomes from the mitochondria of 40 rat livers by lysis with 1.3 % Triton X-100 followed by pelleting the crude ribosomes through a 1-M sucrose layer. This crude fraction was then spun on an isokinetic sucrose gradient. When the mitochondria were washed twice with 0.25 M sucrose, 1 mM EDTA, pH 7.4 before lysis, besides the main peak of 56S a substantial amount of 82S particles was found. However, when the mitochondria were treated by the method of Loewenstein *et al.* (3) the 82S peak disappeared whereas the yield of 56S particles was not affected. This method, involving a short treatment of the mitochondria with a low digitonin concentration, was designed to remove lysosomal contamination from the mitochondria, and used by us to eliminate a side reaction interfering

357

with the peptidyl transferase assay, as will be discussed below.

The disappearance of the 82S peak strongly suggests that these particles are merely adhering cytoplasmic ribosomes and not, as we supposed previously (4), mitochondrial ribosomes. The assignment of a mitochondrial origin to these particles was mainly based on chloramphenicol-sensitivity in the peptidyl transferase reaction. Careful reexamination, however, showed a substantial contamination of the 82S fraction with 55S ribosomes and possibly also with disomes of 55S. Moreover, the fact that 55S ribosomes are much more active in the peptidyl transferase reaction than cytoplasmic ribosomes (see below) resulted in an overestimation of the mitochondrial portion in this fraction. The dissociation behaviour of the 82S particles and the electrophoretic mobilities of their RNAs further confirmed the cytoplasmic nature of these ribosomes.

For the experiments described here we only used ribosomes from digitonin treated mitochondria. For the determination of physicochemical properties the 55S ribosomes were further purified on a second sucrose gradient. After this purification the ribosomes showed spectral ratios of 1.70 (A_{260}/A_{280}) and 1.20 (A_{260}/A_{230}).

PHYSICOCHEMICAL PROPERTIES.

Buoyant density of 55S ribosomes. Fig. 1 shows the buoyant densities in CsCl of glutaraldehyde-fixed 55S and cytoplasmic 80S ribosomes. Whereas the latter have a buoyant density of 1.55, 55S particles have a density of only 1.43. This low density accords to 37 % RNA (5), which is much lower than found for more orthodox ribosomes. This is in reasonable agreement with analytical determinations of RNA and protein content. It shoudl be pointed out that we checked for the presence of phospholipids in the 55S particles but that no sigh of these membrane components could be detected. Also from the spectral data it could be concluded that membrane contamination was not likely. Therefore, the low density is not caused by adhering membrane frag-

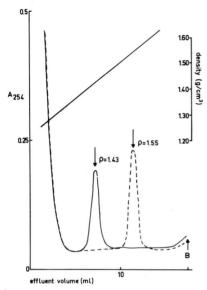

Fig. 1. Buoyant density of 55S mitochondrial and 80S cytoplasmic ribosomes. ——: 55S ribosomes; — — —: 80S ribosomes. Ribosomes were fixed with glutaraldehyde and analyzed on a linear CsCl gradient in a SW27.1 rotor. The whole procedure was exactly as described by Baltimore and Huang (20).

ments.

Electrophoretic behaviour of 55S ribosomes. For *E. coli* ribosomes it has been found by Talens *et al.* (6) that the smaller a particle is, the faster it moves to the anode in a polyacrylamide gel. We have used their method for exploring the differences in size between 55S ribosomes and *E.coli* 70S and ratliver 80S ribosomes. Table I shows that indeed 80S migrate slower than 70S ribosomes, but that, surprisingly, 55S ribosomes move even slower than 80S ribosomes. The reasons for

TABLE I.

ELECTROPHORESIS OF RIBOSOMES IN 2.2 % POLYACRYLAMIDE GELS.
The electrophoresis was performed according to Talens *et al.* (16), except that preswelling was omitted. The buffer used was 74 mM ammonium acetate, 5 mM MgCl$_2$, 50 mM Tris-acetate, pH 7.8. Electrophoresis was for 3 h at 12 mA/gel at 4° C. Gels were scanned at 260 nm on a Gilford 2400-S spectrophotometer with linear transport system.

Type of ribosomes	Distance migrated to the anode (mm)
55S mitochondrial	12.5
80S cytoplasmic	22.5
70S *E.coli*	29.0

for this unexpected behaviour can be: the volume of
55S ribosomes is larger than that of 80S ribosomes, or
their charge/mass ratio is very low, or both factors
play a role. We have tested these possibilities and
found that a large volume and a low charge/mass ratio
are responsible for the slow migration of the 55S ribosomes. Table II gives the migration of ribosomes in

TABLE II.

ELECTROPHORESIS OF RIBOSOMES IN 0.4 % AGAROSE GELS.
The electrophoresis was performed according to Talens *et al.* (7). The buffer
was 60 mM ammonium acetate, 4 mM $MgCl_2$, 6 mM β-mercaptoethanol, 10 mM Tris-
acetate, pH 7.6. Electrophoresis was for 2.5 h at 9 mA/gel at 4° C. Gel scanning was as described in Table I.

Type of ribosomes	Distance migrated to the anode (mm)
55S mitochondrial	20.0
80S cytoplasmic	29.3
70S *E.coli*	48.0

a 0.4 % agarose gel. In this type of gel sieving effects are virtually absent and only the charge/mass
ratio determines the velocity. It is evident that 70S
ribosomes have the highest velocity, 80S ribosomes run
slow and 55S ribosomes still slower. This fits with
the RNA contents of these three types of ribosomes:
the more RNA they contain, the higher their charge/
mass ratio will be.

Regarding the volume of the 55S ribosomes, it has
been found (7) that smaller ribosomes penetrate further into a polyacrylamide gradient than larger ones.
In such a gradient the pore size becomes smaller with
increasing acrylamide concentration and ribosomes will
run to the anode until no further penetration is pos-

TABLE III.

ELECTROPHORESIS OF RIBOSOMES IN 3-8 % POLYACRYLAMIDE GRADIENT GELS.
Preparation of the gels and electrophoresis were as described by Talens *et
al.* (7). The buffer was the same as in Table II. Electrophoresis was in 8 cm
gels at 3.6 mA/gel at 4° C, with continuous circulation of the buffer via a
peristaltic pump. After 7 days equilibrium was reached. Scanning of the gels
was as described in Table I.

Type of ribosomes	Distance migrated to the anode (mm) to reach equilibrium
55S mitochondrial	40.7
80S cytoplasmic	28.5
70S *E.coli*	54.7 and 64.5

360

sible. Table III shows the equilibrium positions reached after 7 days by 55S, 70S and 80S ribosomes in 3-8 % polyacrylamide gels. *E. coli* ribosomes migrated farthest to the anode. The phenomenon of splitting of 70S ribosomes into two fractions has also been found by Talens *et al.* (7). 80S ribosomes got stuck rather close to the origin and 55S ribosomes arrived at a position intermediate between 70S and 80S ribosomes. Therefore, it seems highly probable that the volume of 55S ribosomes is also intermediate between those of 70S and 80S ribosomes. That this conclusion is not unreasonable also follows from rough calculations, taking into account that the percentage of RNA in the 55S ribosomes (37 %) gives it a molecular weight equal to that of *E. coli* ribosomes (2.7×10^6 Daltons). The very low buoyant density must then lead to a volume even higher than that of the bacterial ribosome. So, the "miniribosome" is not as mini as it seemed in view of its S-value and its small RNAs, but actually has a mass about equal to that of the *E. coli* ribosome and even a larger volume. Independently, also Sacchi *et al.* (8) and O'Brien *et al.* (9) discounted the "miniribosome" model, based on buoyant density measurements (8) and on analysis of the separated ribosomal proteins (9).

FUNCTIONAL PROPERTIES.

The 55S ribosomes were characterized functionally by poly-U directed polyphenylalanine synthesis and by

TABLE IV.

POLY(U) DIRECTED PHENYLALANINE POLYMERIZATION BY 55S RIBOSOMES.
Phenylalanine polymerization was carried out as described by Hosokawa *et al.* (18) with 15 mM $MgCl_2$ and with 5 μM of [^{14}C]phenylalanine (514 mC/mmole). The reaction was stopped by the addition of 0.1 ml 10 mM cold phenylalanine and 3 ml 5 % trichloroacetic acid. As a carrier about 2 mg of mitochondrial protein was added. Processing of the precipitates and counting of radioactivity has been described earlier (19). An *E. coli* supernatant fraction was used. The reaction took place for 15 min at 37° C. The values are means of 9 different experiments. Activities are corrected for blanks without ribosomes.

Addition or omission	none	−poly(U)	+chloramphenicol (120 μg/ml)	+cycloheximide (80 μg/ml)	−S100 fraction
Activity (%)	100*	43	49	71	2

*100 % value: 25 pmoles/mg RNA per 15 min.

the peptidyltransferase reaction under the conditions of the "fragment reaction".

Table IV shows the results of polyphenylalanine synthesis by 55S ribosomes with an *E.coli* supernatant. It is evident that these ribosomes display poly(U)-directed protein synthesis, which is chloramphenicol-sensitive and also slightly inhibited by cycloheximide. The activity, although in the same order of magnitude as found for *Tetrahymena* mitochondrial ribosomes (10) and much higher than for *Xenopus laevis* mitochondrial ribosomes (11), is very low as compared to *E.coli* ribosomes, which under the same conditions, polymerized about 2.5 nmoles per mg RNA in 15 min. We did not obtain higher activities by prior treatments like run-off incubation or dissociation of the ribosomes, by incubating the mitochondria with puromycin as described by Grivell *et al.* (12) or by raising the MgCl$_2$ concentration to 25 mM and adding 3 mM spermidine (*cf.* ref. 13). The most probable explanation for this low activity seemed to us that the *E.coli* supernatant does not quite fit the 55S ribosomes. However, Greco *et al.* (14) were able to get an activity of about 800 pmoles per mg RNA in 15 min. Since their incubation system is essentially the same as ours, the reason for our failure to obtain a similar high activity is not clear.

In the peptidyl transferase reaction, prokaryotic and organelle ribosomes are inhibited by chloramphenicol and not by anisomycin, whereas eukaryotic ribosomes are anisomycin-sensitive and chloramphenicol-insensitive (*cf.* ref. 2). Table V shows the peptidyl transferase activities of 55S ribosomes and of *E.coli* and rat liver ribosomes. The 55S particles contain a very active peptidyl transferase, about as active on RNA basis as that of *E.coli* ribosomes. Furthermore, their mitochondrial origin is evident from the strong inhibition by chloramphenicol and the lack of inhibitions by anisomycin. We like to point out that the digitonin treatment of the mitochondria was introduced to reduce the high blanks without puromycin we originally encountered in our preparations (15). We suspected lysosomal hydrolytic enzymes to be responsible for

TABLE V.

PEPTIDYL TRANSFERASE ACTIVITIES OF 55S RIBOSOMES AND OF *E.COLI* AND RAT-LIVER
CYTOPLASMIC RIBOSOMES; INFLUENCE OF CHLORAMPHENICOL AND ANISOMYCIN.
Peptidyl transferase was measured as described previously (15). Time of incubation: 1 h. Corrections were made for blanks without ribosomes. The reaction was performed with 25 (mitochondrial) or 50 (bacterial or cytoplasmic)
μg ribosomal RNA per tube. 22,000 dpm of acetyl-[^{3}H]leucyl-tRNA (specific
activity 10 C/mmole) was added.

Addition or omission	Type of ribosomes		
	mitochondrial 55S	*E.coli*	rat-liver cytoplasmic
none	100 %	100 %	100 %
+chloramphenicol (67 μg/ml)	16.3	6	99
+anisomycin (27 μg/ml)	89.6	95	14
-puromycin	4.6	2	2
number of experiments	7	6	3
100 % value (dpm)	3,715	8,000	5,000

this effect, causing splitting off of acetyl-leucine
or other small ethylacetate-soluble molecules from
acetyl-leucyl-tRNA. The Loewenstein treatment indeed
caused a remarkable diminishing of the blanks without
puromycin, in contrast to the EDTA-washing of mitochondria. Therefore, these results on the antibiotic
sensitivities of mitochondrial ribosomes in the peptidyl transferase reaction essentially confirm our earlier observations (15), but since the very high blanks
without puromycin are absent now, the results are much
more reliable and clear-cut. Moreover, the low protein-synthetic activity of our 55S preparations evidently is not caused by contamination with degrading
lysosomal enzymes.

CONCLUDING REMARKS.

From the experiments summarized above the mitochondrial 55S ribisome emerges as a remarkable particle:
in spite of its low sedimentation coefficient and its
small RNAs it seems to have a molecular weight about
equal to that of the *E.coli* ribosome, whereas its volume is intermediate between those of bacterial and of
eukaryotic cytoplasmic ribosomes. The low buoyant density points to an unusually low RNA content only about
37 %. The low charge/mass ratio is most likely caused

by this low RNA content. That this particle sediments at such a slow rate as compared to its mass might well be a consequence of its low specific gravity and large volume.

Notwithstanding these peculiarities the ribosome functions actively in the peptidyl transferase reaction, with full inhibition by chloramphenicol. Moreover, all the evidence now available points to the 55S ribosome as the active animal mitochondrial ribosome, and in isolated mitochondria it shows a very high protein-synthetic activity if we express it on an RNA basis (30 pmoles or more leucine incorporated per µg of total mitochondrial RNA in 1 h; $cf.$ ref's. 16,17).

Finally, the question still remains how a ribosome with such a remarkable composition still manages to synthesize proteins. Much research will still be necessary to gain more insight into the functioning of this peculiar ribosome.

ACKNOWLEDGEMENTS.

We wish to thank Mrs. Rya van der Koogh for her excellent technical assistance and Prof. F.J. Loomeijer for encouragement. This work was supported in part by the Netherlands Foundation for Chemical Research (S.O.N.) with financial aid from the Netherlands Organization for the Advancement of Pure Research (Z.W.O.).

REFERENCES.

1. T.W. O'Brien and G.F. Kalf, J.Biol.Chem. <u>242</u>, 2172 (1967).
2. A.M. Kroon, E. Agsteribbe and H. de Vries, in:"The mechanism of Protein Synthesis and its Regulation" L.Bosch, Ed., North Holland, Amsterdam, 539 (1972).
3. J. Loewenstein, H.R. Scholte and E.M. Wit-Peeters, Biochim.Biophys.Acta <u>223</u>, 432 (1970).
4. H. de Vries, H.J.L. van der Koogh and A.M. Kroon, 8th FEBS Meeting, Amsterdam, abstr. 545 (1972).
5. R.P. Perry and D.E. Kelley, J.Mol.Biol. <u>16</u>, 255 (1966).
6. J. Talens, F. Kalousek and L. Bosch, FEBS Letters

$\underline{12}$, 4 (1970).

7. J. Talens, O.P. van Diggelen. M. Brongers, L.M. Popa and L. Bosch, Eur.J.Biochem. $\underline{37}$, 121 (1973).
8. A. Sacchi, F. Cerbone, P. Cammarano and U. Ferrini, Biochim.Biophys. Acta $\underline{308}$, 309 (1973).
9. T.W. O'Brien, D.N. Denslow and G.R. Martin, this volume.
10. N.E. Allen and Y. Suyama, Biochim,Biophys.Acta $\underline{259}$, 369 (1972).
11. R.F. Swanson, Biochemistry $\underline{12}$, 2142 (1973).
12. L.A. Grivell, L. Reijnders and P. Borst, Biohhim. Biophys.Acta $\underline{247}$, 91 (1971).
13. G. Schlanger, R. Sager and Z. Ramanis, Proc.Nat Acad.Sci. USA $\underline{69}$, 3551 (1972).
14. M. Greco, G. Pepe and C. Saccone, This volume
15. H. de Vries, E. Agsteribbe and A.M. Kroon, Biochim.Biophys.Acta $\underline{246}$, 111 (1971).
16. A.M. Kroon and H. de Vries, in:"Control of Organelle Development", P.L. Miller, Ed., Cambridge University Press, Cambridge, 181 (1970).
17. A.M. Kroon, Chimia $\underline{25}$, 114 (1971).
18. K. Hosokawa, R.K. Fujimura and M. Nomura, Proc. Nat.Acad.Sci.USA $\underline{55}$, 198 (1966).
19. A.M. Kroon and H. de Vries, in:"Autonomy and Biogenesis of Mitochondria and Chloroplasts", N.K. Boardman, A.W. Linnane and R.M. Smillie, Eds., North Holland, Amsterdam, p. 318 (1971).
20. D. Baltimore and A.S. Huang, Science $\underline{162}$, 572 (1968).

CHARACTERIZATION OF THE MONOMER FORM OF RAT LIVER MITOCHONDRIAL RIBOSOME AND ITS ACTIVITY IN POLY U-DIRECTED POLYPHENYLALANINE SYNTHESIS

M. Greco, G. Pepe and C. Saccone

Institute of Biological Chemistry, University of Bari, Bari, Italy.

Since the first paper of O'Brien and Kalf in 1967 (1) many authors (see ref. 2 for review) have reported the occurrence of particles having a sedimentation coefficient of about 55-60S as major ribosomal components of animal mitochondria. These studies led to the general assumption that all animal cells containing 5 μ mtDNA circles possess also ribosomes of small size called miniribosomes. However a number of experimental observations reported in the literature were not in complete agreement with this assumption. Many authors reported the occurrence in mitochondrial lysate of rodent tissues of larger ribosomal particles (70-80S) which do not originate from the contaminating cytoplasmic ribosomes (2,3,4,5). A number of hypotheses were advanced in order to explain these discrepancies. It was suggested that 55S ribosomes could represent a degraded form of the original larger ribosome or consist of a mixture of two types of subunits. Another possibility was that the larger particles could represent small polysomes or aggregated miniribosomes. De Vries *et al.* (6) have put forward that rat-liver mitochondria possess two populations of ribosomes each giving rise to different pairs of subunits. The weakest point for the evidence that miniribosomes represent the basic unit of the protein synthetic apparatus in mammalian mitochondria was the fact that often the 55-60S particles were characterized as real mitochondrial ribosomes on the basis of indirect criteria such as determining the size of the

particles from which nascent proteins were recovered
when mitochondria were incubated with radioactive ami-
noacids, of the ability of RNA extracted from these
particles to specifically hybridize with mitochondrial
DNA. The best direct criterion, however, for the characterization of a given particle as a ribosome remains its biological activity *in vitro*. Very recently
(7) we have succeeded in isolating mitochondrial ribosomes highly active in protein synthesis *in vitro* and
using these particles we have undertaken a study on
their characterization.

RESULTS AND DISCUSSION.

The procedure used to obtain from rat liver mitochondrial ribosomes highly active in poly U-directed
polyphenylalanine synthesis has been already described (7). Table I shows that two kinds of particles having a sedimentation coefficient of 55S and 80S respectively are able to carry out *in vitro* protein synthesis which is sensitive to chloramphenicol. However,
the 80S particles are clearly contaminated with cytoplasmic ribosomes since the protein synthesizing activity is inhibited also by anisomycin and cycloheximide other than by chloramphenicol. On the basis of
these data only, it is possible to conclude that 55S

TABLE I.

EFFECT OF CHLORAMPHENICOL, ANISOMYCIN AND CYCLOHEXIMIDE ON POLY(U)-DIRECTED
POLYPHENYLALANINE SYNTHESIS BY MITOCHONDRIAL, CYTOPLASMIC AND *E.COLI* RIBOSOMES.
Assay of activity was under the conditions described by Hosokawa *et al.*(15).
Ribosomes were present at a concentration equivalent to 0.65 A_{260} Units per assay. Incorporation of phenylalanine by the *E.coli* supernatant in the absence of
ribosomes was 0.3 (range 0.2 - 0.6) pmoles per assay. For other experimental
details see ref. 7.

System	[^{14}C]Phenylalanine incorporated (pmoles/mg RNA/15 min) Ribosomes			
	mit. 55S	mit. 80S	cytopl. 80S	*E.coli* 70S
Complete	1090	2210	508	3200
+ Chloramphenicol (60 µg/ml)	348	1450	558	1440
+ Anisomycin (180 µg/ml)	1090	1547	150	3104
+ Cycloheximide (180 µg/ml)	980	1480	130	3520

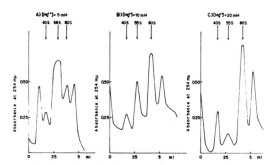

Fig. 1. Sedimentation pattern of mitochondrial ribosomes at different Mg^{2+} concentrations. Reprinted by permission from ref. 7.

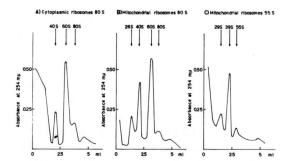

Fig. 2. Sedimentation pattern of EDTA-treated 80S cytoplasmic, 55S and 80S mitochondrial ribosomes. Reprinted by permission from ref. 7.

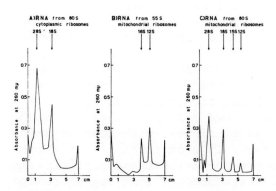

Fig. 3. Polyacrylamide gel electrophoresis of RNA extracted from cytoplasmic 80S ribosomes and mitochondrial 55S and 80S ribosomes. Reprinted by permission from ref. 7.

particles are able to synthesize protein *in vitro* and
therefore must be considered a complete ribosome or a
mixture of two subunits. On the other hand the 80S
particles could be considered either aggregates of 55S
ribosomes of a population of different mitochondrial
ribosomes. In order to clarify the nature of both
kinds of particles we have undertaken a study on their
characterization using different experimental approaches (7). The sedimentation pattern of 80S and 55S
particles in different ionic conditions was analyzed.
Fig. 1 demonstrates that raising the Mg^{2+} concentration causes an increase in the 80S peak of the gradient at the 55S peak's expense, whereas at a concentration of 5 mM Mg^{2+} the 55S peak is higher but broa-
der than under our standard conditions (10 mM Mg^{2+}).
This experiment indicates that probably the 80S peak
may contain aggregates of mitochondrial 55S monomers.
The dissociation experiments using 20 mM EDTA (Fig. 2)
in our opinion mean that the 39S and 29S are the mitochondrial subunits which originate from the 55S monosome which is also present, probably in an aggregated
form, in the 80S peak of the gradient in close association with contaminating cytoplasmic ribosomes. The
same conclusion can be reached by studying the nature
of the RNA extracted from mitochondrial particles
(Fig. 3): 16S and 12S RNA seem to be the specific mitochondrial ribosomal RNA's present in 55S as well as
in 80S particles. All these results suggest that 55S
particles represent the monomer form of mitochondrial
ribosomes in rat liver. This conclusion is now supported by studies at the electron microscope. The micrographs of mitochondrial ribosomal preparations show
that they are free from contamination by membranes.
Electron microscopy of negatively stained 55S mitochondrial preparations shows many particles having the
characteristics of real ribosomes, with rounded and
regular profiles. A dense line or sometimes an electron-opaque spot appears to divide these particles into unequal size subunits. Moreover some aggregation of
two ribosomes and some smaller particles, probably
isolated subunits, are observed in these preparations

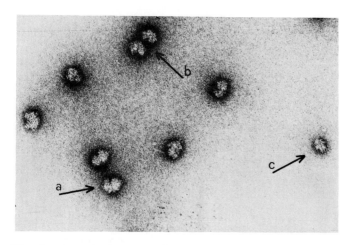

Fig. 4. General view of a field of 55S mitochondrial ribosomes: the most particles show a division in subunits (arrow a). In this preparation is also present an aggregate of two ribosomes (arrow b) and a smaller particle which may
be considered as an isolated subunit (arrow c). x 260 000. For technical details
of the negative staining and of the micrographs see ref. 14.

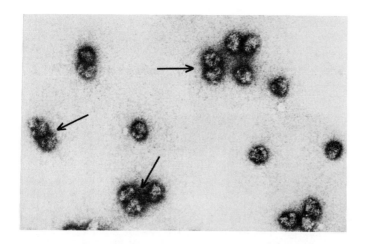

Fig. 5. General view of a field of 80S mitochondrial ribosomes in which are
evident many aggregates (dimers) of two ribosomes (arrows). x 260 000.

(Fig. 4). The 80S mitochondrial preparations (Fig. 5) show ribosomes which correspond with the 55S particles in form and dimension, but they seem to contain much more aggregates composed of two whole ribosomes, which may be considered "dimers" of 55S monomers. In Fig. 6 selected images of a 55S monosome and of a 80S dimer are represented. The dimensions of the 55S ribosome and of the monosomes of 80S dimers appear to be 304 x 252 $\overset{\circ}{A}$ and 308 x 255 $\overset{\circ}{A}$ respectively. These morphological results seem to support our opinion that the 55S particles are real mitochondrial monosomes which are probably present as "dimers" in the 80S peak. The dimensions of these particles are not in agreement with the values reported by Aaij *et al.* (8), but they appear to be in the range observed for other ribosomes. From these experiments it seems that the low sedimentation coefficient of the 55S particles is not due to their small size but probably depends on other intrinsic properties of animal mitochondrial ribosomes, such as on the size of ribosomal RNAs and on their high protein/RNA ratio. These conclusions are in agreement with the observations reported by De Vries and Kroon (9) and by Sacchi *et al.* (10).

Having characterized the 55S particle as the monomer form of the rat liver mitochondrial ribosome, in further experiments we have studied some properties of 55S supported *in vitro* protein synthesis. Fig. 7 shows the effect of Mg^{2+} and NH_4Cl on the poly U-directed synthesis of polyphenylalanine by purified mitochondrial ribosomes. Optimal Mg^{2+} concentration is about 15 mM and the reaction seems to be only slightly affected by raising the NH_4Cl concentration. This response is in contrast to that of yeast mitochondrial ribosomes which are strongly inhibited by concentrations above 25 mM as reported by Grivell *et al.* (11). Table II shows the effect of erythromycin on the activity of the 55S particles. It can be seen that protein synthesis is inhibited by erythromycin. The effect of macrolides on ribosomes from animal mitochondria is still controversial. Firkin and Linnane (12) have found that incorporation of aminoacid into proteins by

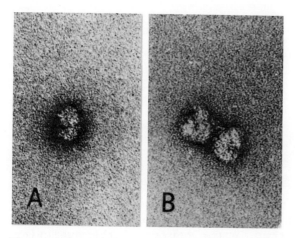

Fig. 6. Selected images of 55S monosomes (A) and of 80S "dimers" (B). x 260 000.

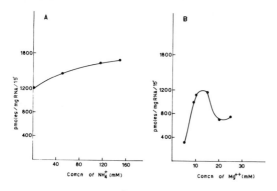

Fig. 7. Effect of Mg^{2+} and NH_4^+ on the poly(U)-directed synthesis of polyphe-nylalanine by purified mitochondrial 55S ribosomes. Experimental conditions as in ref. 7. The Mg^{2+} concentration in the experiments in A was 15 mM.

TABLE II.

EFFECT OF CHLORAMPHENICOL, ANISOMYCIN AND ERYTHROMYCIN ON POLY(U)-DIRECTED
POLYPHENYLALANINE SYNTHESIS BY 55S MITOCHONDRIAL RIBOSOMES.
100 % activity was 455 pmoles/mg RNA/15 min. Experimental conditions as in
Table I.

System	Activity %
Complete	100
+ Chloramphenicol (60 μg/ml)	30
+ Anisomycin (180 μg/ml)	100
+ Erythromycin (130 μg/ml)	55
+ Erythromycin (260 μg/ml)	20

mitochondria from mammalian species is insensitive to
erythromycin and lincomycin. These results were interpreted to suggest that in the course of the evolution
of the mammalian mitochondrial protein synthesizing
system an intrinsic change in the system has occurred
which reflects a change in cytoplasmic genetic information. Kroon and de Vries (13) on the other hand have reported that the insensitivity of the mitochondrial protein synthesizing system to macrolides is only due to the fact that the mitochondrial membranes
are impermeable to these drugs. Using swollen mitochondria protein synthesis is in fact strongly inhibited by erythromycin and other macrolides. Our results
confirm the interpretation of Kroon by demonstrating
that isolated mitochondrial ribosomes are themselves
susceptible to the action of the inhibitors.

We further tested the activity of mitochondrial ri-

TABLE III.

RIBOSOME SPECIFICITY OF SUPERNATANT PREPARATIONS.
The mitochondrial supernatant enzymes from rat liver were prepared according
to the method of Richter and Lipmann (16). For other experimental details see
ref. 7.

Ribosomes	Supernatant	[^{14}C]Phenylalanine incorporation	
		(pmoles/mg RNA/15 min)	%
E. coli	*E. coli*	1.050	100
	Mitochondria	82	8
Mitochondrial	Mitochondria	56	100
	E. coli	465	810

bosomes in the presence of homologous or heterologous supernatant enzymes (Table III). It can be seen that the mitochondrial ribosomes are much more active with *E. coli* supernatant enzymes than with mitochondrial enzymes, whereas the *E. coli* ribosomes are naturally much more active when tested with the homologous enzymes. This means that an endonuclease is probably present in rat liver mitochondrial extract. A better purification of mitochondrial polimerizing enzyme is therefore required.

In conclusion the results demonstrate that isolated 55S mitochondrial ribosomes are able to carry out a poly U-directed poly-phenylalanine synthesis which is strongly inhibited by chloramphenicol and by erythromycin. Physicochemical and morphological characterization of the particles present in lysates of rat-liver mitochondria reveals that 55S particles can be considered as real monosomes dissociating into 39S and 29S subunits and possessing peculiar RNA species sedimenting at 16S and 12S. On the basis of these studies we conclude that the 55S ribosome is the basic unit of the mitochondrial protein synthetic apparatus in rat-liver cells.

ACKNOWLEDGEMENTS.

The electron microscopy has been performed in the Laboratory of Cellular Biology of Orsay, France. We thank J. André and B. Stevens for their collaboration and helpful discussions. This work was supported in part by C.N.R. Italy.

REFERENCES.

1. T.W. O'Brien and G.F. Kalf, J.Biol.Chem 242, 2172 (1967).
2. P. Borst and L.A. Grivell, FEBS Letters 13, 73 (1971).
3. H. de Vries, E. Agsteribbe and A.M. Kroon, Biochim. Biophys.Acta 246, 111 (1971).
4. C. Saccone and M.N. Gadaleta, 8th Int.Congr.of Biochem., abstr. 185 (1970).
5. C. Saccone, M. Greco, C. de Giorgi and G. Pepe,

Boll.Soc.It.Biol.Sper. <u>48</u>, 243 (1972).

6. H. de Vries, H.J.L. van der Koogh and A.M. Kroon, Commun.Meeting Fed.Eur.Biochem.Soc. <u>8</u>, abstr., 545 (1972).

7. M.Greco, P.Cantatore, G. Pepe and C. Saccone, Eur.J.Biochem. <u>37</u>, 171 (1973).

8. C. Aaij, N. Nanninga and P. Borst, Biochim.Biopjys.Acta. <u>277</u>, 140 (1972).

9. H. de Vries and A.M. Kroon, this volume.

10. A. Sacchi, F. Cerbone, P. Cammarano and U. Ferrini, Biochim.Biophys.Acta <u>308</u>, 390 (1973).

11. L.A. Grivell, L. Reijnders and P. Borst, Biochim. Biophys.Acta <u>247</u>, 91 (1971).

12. F.C. Firkin and A.W. Linnane, FEBS Letters <u>2</u>, 330 (1969).

13. A.M. Kroon and H. de Vries, in:"Autonomy and Biogenesis of mitochondria and chloroplast", North Holland, 318 (1971).

14. P.V. Vignais, B.J. Stevens, J. Huet and J. André, J.Cell Biol. <u>54</u>, 468 (1972).

15. K. Hosokawa, R.K. Fujimura and M. Nomura, Proc. Nat.Acad.Sci. USA <u>55</u>, 198 (1966).

16. D. Richter and F. Lipman, Biochemistry <u>9</u>, 5065 (1970).

RNA FROM MITOCHONDRIAL RIBOSOMES OF *LOCUSTA MIGRATORIA*

Walter Kleinow

Institut für Physiologische Chemie und Physikalische Biochemie der Universität München, 8 München 2, Goethestrasse 33, Germany.

From mitochondria of *Locusta migratoria* two RNA species in the high molecular weight range have been extracted together with a 4S type RNA (1,2,3). It was suggested that these two species correspond to the RNA from mitochondrial ribosomes (2,3). However, it was not possible until now to extract high molecular weight RNA from isolated mitochondrial ribosomes or from their subunits. This has been achieved for the first time with mitochondrial 60S ribosomes from *Xenopus* (4). The reason for failing in the case of *Locusta* was found in a high specific activity of RNA splitting enzymes which are associated with mitochondrial ribosomes under our preparation conditions (5). After inhibiting this RNAse, high molecular weight RNA could be extracted from isolated mitochondrial ribosomes.

Mitochondria and mitochondrial ribosomes were prepared as described (6). In order to block RNAse activity, lysis of the mitochondria was performed in a medium containing either 55 mM $CaCl_2$ or 55 mM $MgCl_2$ (instead of 10 mM $MgCl_2$) and 10-20 µg proteinase K (7). Furthermore the medium contained 2,5 % Triton X-100, 100 mM ammonium chloride and 30 mM Tris-HCl, pH 7.6 as usual. RNA from undissociated ribosomes was extracted from the 90 min 144,000 xg pellet of the mitochondrial lysate by the diethyl pyrocarbonate method (8) as modified by Forrester *et al.* (9).

In other experiments the ribosomes were dissociated by centrifuging them in a magnesiumfree gradient

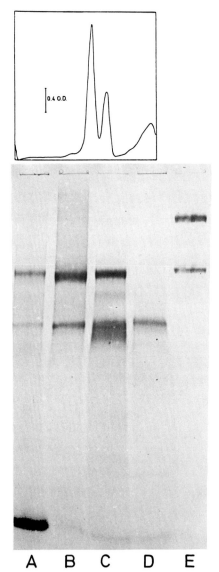

A   B   C   D   E

Fig. 1. Upper part: Profile (absorbancy at 260 nm) of a magnesium free gradient containing mitochondrial ribosomes. Sedimentation from right to left; 5 h 41,000 rpm., Spinco rotor SW41 Ti. The mitochondrial ribosomes were prepared as described in the text (lysis of mitochondria in 55 mM $CaCl_2$).

Lower part: gel electrophoretic patterns of RNA extracted from different mitochondrial fractions of *Locusta migratoria*. A: whole mitochondrial; B: undissociated mitochondrial ribosomes; C: RNA extracted from the large subunit of the gradient shown directly above; D: RNA extracted from the small subunit of the gradient shown directly above; E: RNA from *E. coli* ribosomes.

(6). The gradient fractions were collected into tubes
containing 10 µg proteinase K. From the peak fractions
(corresponding to 40S and 25S) the RNA was then extracted as mentioned above for the undissociated ribosomes. Gel electrophoresis was performed according
to Dingman and Peacock (10).

The gelelectrophoretic pattern of the RNA from undissociated mitochondrial ribosomes (Fig. 1 B) shows
two main bands which are present in a proportion of
2 : 1. Only traces of 4S RNA can be seen, which may
be attributed to residual tRNA. In some experiments a
very weak band in the 5S region was also observed. The
two main bands obtained from the undissociated mitochondrial ribosomes show the same electrophoretic mobility as the two slow moving bands from whole mitochondria. The molecular weights of these two species
were re-estimated to 520,000 and 280,000 (instead of
520,000 and 250,000). This was accomplished by comparing their electrophoretic mobility to RNA from *E. coli*
ribosomes (Fig. 1 E) on gels running at temperatures
between -2° C and +10° C (5).

On the other hand RNA from whole mitochondria (Fig.
1 A) contains a large amount of 4S RNA. A 5S RNA species never was observed (2). The 4S RNA obtained from
whole mitochondria makes up about 30-50 % of the extractable RNA. We have evidence that this 4S RNA is
not a degradation product of the high molecular weight
species. This large amount of 4S RNA could be explained by an incomplete extraction of ribosomal RNA from
whole mitochondria. Another possibility is that RNA
with a different function than tRNA is present in the
4S RNA fraction (for instance free poly(A); *cf*. ref's
11 and 12).

From the small subunit obtained from magnesium free
gradients (upper part in Fig. 1) only RNA of 280,000
could be extracted (Fig. 1 D). On the other hand the
RNA from the large subunit (Fig. 1 C) shows two bands.
One corresponds to a molecular weight of 520,000 and
presumably represents native RNA from the large ribosomal subunit. The second band corresponds to a molecular weight of about half of the first. This one is

probably a degradation product of the first species, as can be concluded from the rather broad appearance of the band.

These results make it seem rather improbable that the RNA species obtained from locust mitochondria are split pieces of RNA molecules of higher molecular weight.(2,3). It can be concluded that the RNA species with molecular weights of 520,000 and 280,000 are native RNAs derived from the large and small subunit of mitochondrial ribosomes from *Locusta migratoria*. The molecular weights estimated for these RNAs are in good agreement with data for RNA from mitochondrial 60S ribosomes obtained from other animal organisms and with other methods (13,14).

ACKNOWLEDGEMENTS

This work was supported by the Deutsche Forschungsgemeinschaft (Schwerpunktprogramm "Biochemie der Morphogenese"). The steady support by Professor Th. Bücher, the excellent technical assistance of Miss Heide Rothe and the advice of Mr. A. v. Rücker in the preparation of the manuscript is gratefully acknowledged.

REFERENCES.

1. W. Kleinow and W. Neupert, Z.Physiol.Chem. 350, 1166 (1969).
2. W. Kleinow and W. Neupert, Z.Physiol.Chem. 351, 1205 (1970).
3. W. Kleinow, W. Sebald, W. Neupert and Th. Bücher in:"Autonomy and biogenesis of mitochondria and chloroplasts", N.K. Boardman, A.W. Linnane and R.M. Smillie Eds., p. 140, North Holland, Amsterdam (1971).
4. R. Swanson and I.B. Dawid, Proc.Nat.Acad.Sci.US. 66, 117 (1970).
5. W. Kleinow, in preparation.
6. W. Kleinow and W. Neupert, FEBS Letters 15, 359 (1971).
7. U. Wiegers and H. Hilz, Biochem.Biophys.Res.Commun. 44, 513 (1971).

8. F. Solomosy, I. Fedorcsák, A. Gulyás, G.L. Farkas
   and L. Ehrenberg, Eur.J.Biochem. 5, 520 (1968).
9. I.T. Forrester, Ph. Nagley and A.W. Linnane, FEBS
   Letters 11, 59 (1970).
10. C.W. Dingman and A.C. Peacock, Biochemistry 7,
    659 (1968).
11. N.G. Avadhani, M. Kuon, P. VanDer Lign and R.I.
    Rutman, Biochem.Biophys.Res.Commun. 51, 1090 (1973).
12. G. Attardi, P. Constantino and D. Ojala, this
    volume
13. D.L. Robberson, Y. Aloni, G. Attardi and N. David-
    son, J.Mol.Biol. 60, 473 (1971).
14. I.B. Dawid and J.W. Chase, J.Mol.Biol. 63, 217
    (1972).

CODING DEGENERACY IN MITOCHONDRIA

Nancy Chiu, A.O.S. Chiu and Yoshitaka Suyama

Department of Biology, University of Pennsylvania, Philadelphia, Pa., U.S.A.

Mitochondria contain transfer RNAs that support protein synthesis occurring within the organelle. These tRNA species have been found to differ from the cytoplasmic tRNAs in terms of 1: their differential responses to mitochondrial and cytoplasmic tRNA synthetases (1); 2: chromatographic elution patterns (2, 3,4,6); 3: coding response (5,6) and 4: specific hybridization capacity to mitochondrial DNA (7,8,9). In any organism so far examined, there exist multiple isoaccepting species of tRNA in support of universally accepted coding degeneracy. In *Escherichia coli*, at least five isoaccepting leucyl tRNAs have been found which respond differentially toward different synonym codons (10,11,12). In mitochondria, however, rigorous evidence is lacking to support the existence of a similar coding degeneracy. In this paper we review evidence for the presence of rather extensive coding degeneracy in *Tetrahymena* mitochondria.

CYTOPLASMIC AND MITOCHONDRIAL LEUCYL tRNAs SHOW MULTIPLE SPECIES.

Recent development of the rapid and reproducible reversed-phase column chromatography (RPC 5) by Kelmers and Heatherly (13) enables us to resolve at least six leucyl tRNA species in whole-cell *Tetrahymena* tRNA when charged by the purified mitochondrial and cytoplasmic synthetases (Fig. 1). The mitochondrial enzyme can charge all six species, while the cytoplasmic enzyme preferentially charges species I, III, IV and a small amount of species V. The cytoplasmic enzyme exhibits no recognition toward species II and VI. These

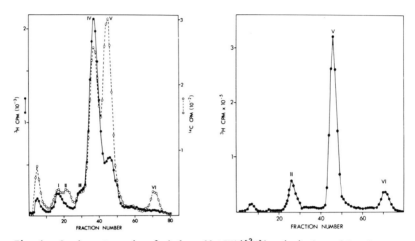

Fig. 1. Co-chromatography of whole-cell tRNA([³H]leucine) charged by the cytoplasmic enzyme and whole-cell tRNA([¹⁴C]leucine) by the mitochondrial enzyme.
The tRNAs^{leu} charged separately in 0.5 ml each reaction mixture and deproteinized were mixed, loaded onto a column and developed with a linear NaCl gradient
from 0.5 - 0.8 M. A total of 80 fractions (1.2 ml/fraction) was collected. Input
count ratio of [³H]:[¹⁴C]was 4:1. The first peak before the 10th fraction represents wash-out counts.

Fig. 2. Chromatography elution pattern of mitochondrial tRNA charged by the mitochondrial enzyme. Reaction mixture in 0.216 ml contained 0.16 mg deacylated
tRNA, 20 μCi [³H]leucine, 0.4 mg mitochondrial enzyme. Other conditions were same as in Fig. 1.

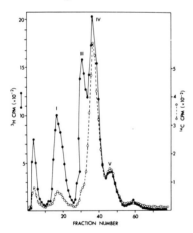

Fig. 3. Co-chromatography of cytoplasmic tRNA ([³H]leucine) charged by the cytoplasmic enzyme and whole-cell tRNA ([¹⁴C]leucine) charged by the cytoplasmic enzyme. A reaction mixture contained in 0.216 ml: 0.39 mg (c)tRNA, 20 μCi [³H]leu-
cine, 0.4 mg cytoplasmic enzyme. The other reaction mixture in 0.5 ml: 0.4 mg
whole-cell tRNA, 1 μCi [¹⁴C]leucine, 1 mg cytoplasmic enzyme. Charged tRNAs (leucine) were separately deproteinized and mixed in an input count ratio of [³H]:
[¹⁴C]= 11:1.

data suggest that both cytoplasmic and mitochondrial fractions contain multiple isoaccepting leucyl tRNA species, and that species I, III and IV probably reside in the cytoplasm, whereas species II and VI and possibly V reside in mitochondria.

In attempts to verify the specific localizations of these tRNA species within the cell, tRNAs were separately purified from the cytoplasm or mitochondrial fraction and chromatographed after charging with homologous enzymes. The results (Fig. 2) show that mitochondrial tRNA [tRNA^{leu}] charged by the mitochondrial leucyl tRNA synthetase develops three isoaccepting species II, V and VI, species V being the predominant one. Homologous charging (Fig. 3) of cytoplasmic tRNA by the cytoplasmic enzyme produces radioactive peaks at species I, III and IV positions as expected and the presence of species V is also apparent. These results generally agree with the notion of localization specificities of these tRNAs and their specific responses to their homologous enzymes. However, the presence of a small but recognizable amount of species V in cytoplasmic tRNA contradicts this notion. A cross-charging of cytoplasmic tRNA with the mitochondrial synthetase reveals a charging pattern similar to the pattern of whole-cell tRNA as charged by the cytoplasmic enzyme and no differential enhancement of peak V was observed. This suggests that peak V in the cytoplasm may not be identical to the mitochondrial species V.

The present data establish that mitochondria contain at least three isoaccepting leucyl tRNA species, two (II and VI) of which show no response to the cytoplasmic leucyl tRNA synthetase. Species V of mitochondria can be charged, albeit in a very reduced efficiency with the heterologous enzyme. This cross-charging is not due to the mitochondrial enzyme contaminating the cytoplasmic enzyme preparation, since these two enzymes have been purified to near homogeneity through hydroxyapatite and DEAE-cellulose chromatography and isoelectric focusing, all of which differentiate these enzymes (14).

A major problem exists in identifying isoaccepting

tRNA species. It is necessary to demonstrate that they are not artifacts of preparation. Structural transformation from one form to another (15,16) or induced nicks (17) resulting from nucleases in the cell should be considered as possible sources of artifacts. With our tRNAs, various conditions such as varying concentrations of magnesium and sodium ions, heating and cooling in water and dialysis do not alter the present chromatographic patterns. This and other evidence provided below prove that these tRNAs indeed represent three separate molecules.

THREE MITOCHONDRIAL LEUCYL tRNA SPECIES ARE TRANSCRIBED FROM MITOCHONDRIA.

The discovery of three isoaccepting species of mitochondrial leucyl tRNA prompted us to examine their structural relationship with one another and their transcriptive origins. The individual tRNA species acylated with [^{3}H]leucine were separated from RPC fractions. Rechromatography of these species demonstrates that they are not contaminated with one another. It can be shown that all three species bind to mitochondrial DNA fixed on nitrocellulose filters (Fig. 4). No significant hybridization, however, can be detected with nuclear DNA filters. All tRNAs together show no additivity of hybridization; instead three tRNAs compete for hybridization. Although these results suggest that three mitochondrial leucyl tRNAs contain many similar base sequence regions, species II appears to have a different base composition from the other two species.

THREE MITOCHONDRIAL LEUCYL tRNA ISOACCEPTING SPECIES ARE TRANSCRIPTIVE PRODUCTS OF DIFFERENT DNA SEGMENTS.

Since the *Tetrahymena* nuclear DNA has a genomic size 1000 times greater than the mitochondrial DNA, as determined by renaturation kinetics, the lack of binding of mitochondrial [^{3}H]aminoacyl tRNA to it at the available radioactivity level does not constitute good evidence for the lack of a small number of cistrons in the nucleus. Similarly its binding to mitochondrial

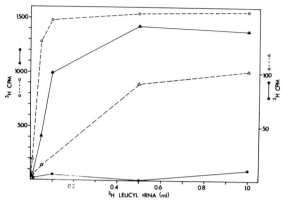

Fig. 4. Hybridization of charged mitochondrial tRNA with DNA. Mitochondrial tRNA was charged with [³H]leucine and *E.coli* enzyme. After deproteinization, charged tRNA^leu was chromatographed, and fractions corresponding to each tRNA^leu species (II, V or VI) were separately pooled, and hybridized by the method of Weiss *et al.* (23). mDNA (14 µg) and nDNA (50 µg) on filter. △---△: species II (631 cpm/10 µl solution) with mDNA; ●---●: species II (631 cpm/10 µl solution) with nDNA; O---O: species V (3,697 cpm/10 µl solution) with mDNA; ▲---▲: species VI (465 cpm/10 µl solution) with mDNA. Counts on blank filters varied with an average 32 cpm/filter and were subtracted from counts on DNA filters.

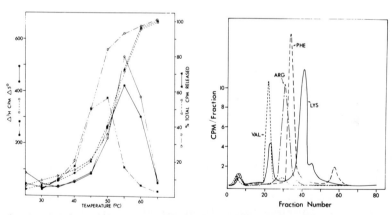

Fig. 5. Melting profiles of tRNA/DNA hybrids. Hybrid filters were heated successively for 7 min at each temperature in vials each containing 1 ml 50 % formamide, 2X SSC, 0.05 M ammonium acetate, pH 5.0. Every temperature increment was insured by measuring directly with a thermometer in vial. After removing the filter, 10 ml Triton X-100: spectrafluor mixture (1:2) were added to each vial and radioactivity was counted. Melting profiles of hybrid: species V (△---△); species VI (▲---▲); species II (□---□); Differential melting profiles: species V (O---O); species VI (●--●); species II (■---■).

Fig. 6. Reversed phase column chromatography elution patterns of mitochondrial tRNAs charged with homologous synthetases. Separate RPC-5 elution patterns are superimposed on one another. The scale of ordinate axis is chosen arbitrarily. Cpm at the highest peaks are: 2,100 (arg), 5,470 (lys), 3,100 (phe), and 10,300 (val).

DNA does not necessarlily mean that true hybrids are formed. Melting profiles and temperatures of the hybrids can be valuable measures at least to assess the nature of hybridization reaction in the latter case.

As shown in Fig. 5, melting profiles of these hybrids between three leucyl tRNAs and mitochondrial DNA were relatively sharp with high melting temperatures for two tRNAs (V,VI) at 52° and 51°, respectively. The hybrid with species II melts at a low temperature, 44°. Since an average Tm of *Tetrahymena* mitochondrial DNA in these conditions is estimated to be 44°, it appears reasonable to assume that true and stable hybrids were formed between these mitochondrial tRNAs and DNA, and that mitochondrial DNA contains at least one cistron for each of these tRNA species. The differences in melting temperatures for these hybrids suggest that three isoacceptors have different base compositions and hence are transcribed from different regions of mitochondrial DNA which consist of at least three discrete melting regions.

DO MITOCHONDRIAL DNA CONTAIN CISTRONS FOR MORE THAN 20 tRNAs?

As will be seen later, multiple tRNA species for other amino acids are also found in the mitochondria of *Tetrahymena*. Previously Epler, (4) obtained indications for the presence of multiple chromatographic peaks for *Neurospora* mitochondrial tRNAs for the following amino acids: Glu, His, Ile, Ser, Thr, Tyr, Val, Arg, Asp, Lys, Leu, and Phe. Although these data should be treated with caution, it is possible that mitochondria contain more than 20 tRNAs that are transcribed from mitochondrial DNA. As one of us (Y.S.) has shown before (18,19), hybridization with bulk mitochondrial tRNA did not produce sufficient cistron numbers to account for this expectation. It appears that there are two reasons for this contradiction: 1. tRNA is a collection of heterogeneous molecules in terms of base composition and structure. A single hybridization condition does not provide an optimum condition for all tRNA molecules to complete hybridiza-

tion; 2. ribosomal RNA often used as a competitor for
tRNA hybridization has some degree of sequence homology with tRNA molecules and thus competes with tRNA in
the hybridization reaction.

The wide Tm's difference, as observed between individual tRNA/DNA hybrids in the present case, supports
the necessity of examining point 1 carefully. One
needs rather extensive temperature and time dependence
studies to deduce any conclusion as to tRNA cistron
numbers. Concerning point 2, it has been shown in *E.
coli* that rRNA abolishes as much as 40% of the hybrids
formed between tRNA labeled specifically at the terminal adenosine and DNA (19). Extending this information
and other available data, it is likely that any hybridization data of tRNA with DNA would underestimate the
cistron numbers. The degree of uncertainty naturally
depends on the extent of heterogeneity in molecular
compositions of tRNA molecules and their sequence re-
lations to other RNAs and DNA. The only means to as-
sure ourselves of obtaining correct cistron numbers is
to test individually acylated tRNA molecules for hybridization.

ISOACCEPTING LEUCYL tRNA SPECIES RESPOND TO DIFFERENT SYNONYM COCONS.

The ultimate demonstration for the biological significance of isoaccepting tRNA species is to show that
they are able to recognize specific codons in messenger RNA. A simple method of identifying their codon
responses is to test their binding reactions to ribosomes in response to trinucleotide codons (10,11,12).
With *E. coli* leucyl tRNA species, 5 different codons
have been assigned: UUG, CUU, CUC, CUA, CUG, CUC (10-12).
One additional codon UUA has been assigned to leucine
but tRNA corresponding to this codon has never been
found. In *E. coli*, tRNAs Leu_1 and Leu_2 corresponding
to CUA and CUC respectively, are major species. Distributions of these leucine codons have not been studied in any other organism than *E. coli*. Regarding the
coding specificities of mitochondrial leucyl tRNAs,
several questions may be asked: 1. Is the universality

of codon applicable to the mitochondrial system?
2. Does codon frequency in mitochondria differ from
that in *E. coli* or *Tetrahymena*? 3. Do patterns of synonym codon recognition follow those predicted from
the Wobble hypothesis (20)?

Results of binding of individual tRNAs to ribosomes
with trinucleotide codons are summarized in Table I.

TABLE I.

CODON RECOGNITION PATTERNS OF LEUCYL tRNAs FROM *TETRAHYMENA* MITOCHONDRIA AND
CYTOPLASM, AND *E. COLI*.

Codon specificities for *E. coli* tRNA^{leu} and the present presentation are after
Caskey *et al.* (25). Major species are shown by double circles. The joined symbols represent synonym codons recognized by one tRNA species. Codon responses
of *Tetrahymena* mitochondrial and cytoplasmic tRNA^{leu} were determined by binding
of each tRNA to *E. coli* ribosomes with trinucleotide codons (24), poly (U$_2$C), and
poly (U$_7$G).

CODONS	E. coli	T. pyriformis	
		Mitochondrial	Cytoplasmic
UUA UUG CUU CUC CUA CUG	• Leu$_{4,5}$ ◉ Leu$_2$ • Leu$_3$ ◉ Leu$_1$	•VI ◉V •II	• I ◉ IV

From extensive studies on coding patterns in *E. coli*,
liver and yeast (10), it has been assumed that universality exists. It is not surprising that this can be
indeed extended to the mitochondrial system. The coding
unit in mitochondria appears to be also triplet. Usa-
ges of different codons might be dictated by or manifested in different sets of isoaccepting species corresponding to different synonym codons. As is seen in
Table I, the major mitochondrial species V responds to
CUA, which also shows some binding to CUC and CUU codons. This coding pattern, although consistent with
the Wobble theory, is not found in *E. coli*. Species VI
responds weakly to CUU codon but does not bind to CUC
or CUA. In many ways, this species is similar to Leu$_2$
of *E. coli*. However, since the anticodon of Leu$_2$ has

been found to be GAG (21), its corresponding codon
should more likely be CUC. Species II binds very weakly to CUG and also to CUU and CUA. This coding pattern
is illegal or ambiguous as far as the Wobble theory is
concerned. Mitochondrial leucyl tRNAs do not respond
to UUG or UUA, while Leu$_4$ and Leu$_5$ of *E. coli* were
assigned to UUG. In comparison, only two cytoplasmic
species have been tested for coding responses. The major cytoplasmic species (IV) binds to CUU and less to
CUC and CUG. It does not respond to poly (U$_7$G) and
hence its codon is likely CUU or CUC. Species I responds only to UUG like Leu$_4$ and Leu$_5$ of *E. coli*.

These data suggest that codon frequency and patterns
of codon recognition in mitochondria are quite different from those in *E. coli* and *Tetrahymena*. As to the
validity of the Wobble theory, the present data do not
provide any more information than already put forth
so elegantly by Ninio (22). Codon specificities, cellular locations and charging responses of six leucyl
tRNA species in *Tetrahymena* are diagrammatically summarized in Table II.

TABLE II.

DIAGRAMMATIC PRESENTATION OF CODON-, LOCALIZATION- AND ACYLATION SPECIFICITIES
OF SIX LEUCYL tRNA SPECIES FROM *TETRAHYMENA*.

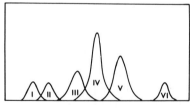

CODON	UUG	CUG	N.D.	CUU	CUA	CUU
tRNA Sources						
Mitochondria		+			+	+
Cytoplasm	+		+	+	?	
ACYLATION by LEUCYL tRNA SYNTHETASE from						
Mitochondria	+	+	+	+	+	+
Cytoplasm	+	−	+	+	±	−
E. Coli	±	+	±	±	+	+

MULTIPLICITY OF tRNAs IS ALSO FOUND WITH LYS-, PHE- AND VAL tRNAs.

To examine the extent of coding degeneracy in mitochondria, we have further examined RPC patterns of four other mitochondrial tRNAs (Arg, Lys, Phe, and Val). These tRNAs are found highly active in mitochondria, and different number of synonym codons have previously been assigned to these amino acids, *e.g.*, 6 codons to Arg, 4 to Val and two to Lys or Phe. In *E. coli*, rat liver and yeast, two to four different tRNA isoacceptors have been found for Arg, whereas in *Tetrahymena* mitochondrial tRNA, only one arginyl tRNA species can be detected. RPC-5 of cytoplasmic tRNA reveals two arginine isoaccepting species, one major and one minor peak, neither of which corresponds to the mitochondrial arginyl tRNA species. As seen in Fig. 6, valyl-, phenyl- and lysyl-tRNAs all produced 2 peaks in RPC. These peaks do not coincide with the peaks obtained with cytoplasmic tRNAs for corresponding amino acids. Although further evidence is necessary, it appears reasonable to conclude that these RPC peaks really represent isoacceptors. If so, a rather extensive coding degeneracy is expected to be present in mitochondria.

Multiplicity of tRNA species corresponding to different synonym codons have been implicated in regulation of protein synthesis (10). It is certainly interesting to know why so many tRNAs are necessary in mitochondria for such a small genome. Is protein synthesis in mitochondria subject to as much regulation as in *E. coli* cells ? The elucidation of the roles of these tRNAs in mitochondria should provide further understanding of mitochondrial biogenesis.

ACKNOWLEDGEMENTS.

This work was supported by U.S. Atomic Energy Commission contract AT (11-1)-3083. Nancy Chiu was supported by an NIH Training Grant administered by the Molecular Biology Group of the University of Pennsylvania.

REFERENCES.

1. W.E. Barnett, D.H. Brown and J.L. Epler, Proc.Nat. Acad.Sci.USA 57, 1775 (1967).
2. Y. Suyama and J. Eyer, Biochem.Biophys.Res.Commun. 28, 746 (1967).
3. C.A. Buck and M.M.K. Nass, J.Mol.Biol. 41, 67 (1969).
4. J.L. Epler, Biochemistry 8, 2285 (1969).
5. J.L. Epler and W.E. Barnett, Biochem.Biophys.Res. Commun. 28, 328 (1967).
6. E. Bernstein, Thesis, Princeton University (1972).
7. M.M.K. Nass and C.A. Buck, J.Mol.Biol. 54, 187 (1970).
8. A. Halbreich and M. Rabinowitz, Proc.Nat.Acad.Sci. USA 68, 294 (1971).
9. J. Casey, M. Cohen, M. Rabinowitz, H. Fukuhara and G.S. Getz, J.Mol.Biol. 63, 431 (1972).
10. M. Nirenberg, T. Caskey, R. Marshall, R. Brima-combe, D. Kellogg, B. Doctor, D. Hatfield, J. Le-vin, F. Rottman, S. Pestka, M. Wilcox and F. An-derson, Cold Spring Harbor Quant.Biol. 31, 11 (1966).
11. J. Kan, M.W. Nirenberg, N. Sueoka, J.Mol.Biol. 52, 179 (1970).
12. D. Söll, E. Ohtsuka, D.S. Jones, R. Lohrmann, H. Hayatsu, S. Nishimura and H.G. Khorana, Proc. Nat. Acad.Sci. USA 54, 1378 (1965).
13. A.D. Kelmers and D.E. Heatherly, Anal.Biochem. 44, 486 (1971).
14. A.O.S. Chiu, Thesis, University of Pennsylvania (1973).
15. T. Ishida, D. Snyder and N. Sueoka, J.Biol.Chem. 246, 5965 (1971).
16. T. Lindahl, A. Adams and J.R. Fresco, J.Biol.Chem. 242, 3129 (1967).
17. S. Nishimura and G.D. Novelli, Biochim.Biophys. Acta 80, 574 (1964).
18. Y. Suyama, Biochemistry 9, 2829 (1967).
19. Y. Suyama, Atti del Seminario di Studi Biologici, E. Quagliariello, Ed., Adriatica, Bari, IV, 83 (1969).

20. F.H.C. Crick, J.Mol.Biol. 19, 548 (1966).
21. H-U. Blank and D. Söll, Biochem.Biophys.Res.Commun. 43, 1192 (1971).
22. J. Ninio, Prog.in Nucleic Acid Res. 13, 301 (1973).
23. S.B. Weiss, W.T. Hsu, J.W. Foft,Proc.Nat.Acad.Sci. USA 61, 114 (1968).
24. A. Kaji, in:"Methods in Enzymology", L. Grossman and K. Moldave, Eds., XII, 692 (1968).
25. C.T. Caskey, A. Beaudet and M. Nirenberg, J.Mol. Biol. 37, 99 (1968).

ON THE SENSITIVITY OF MAMMALIAN MITOCHONDRIAL PROTEIN SYNTHESIS TO INHIBITION BY THE MACROLIDE ANTIBIOTICS

A.M. Kroon, A.J. Arendzen and H. de Vries

Laboratory of Physiological Chemistry, State University, Bloemsingel 10, Groningen, The Netherlands.

Of the macrolide antibiotics erythromycin is frequently used as a therapeutic agent in medicine. It inhibits bacterial growth through its effect on bacterial protein synthesis; it interferes with the bacterial ribosomes. With respect to this group of antibiotics it has been convincingly shown that they also inhibit mitochondrial protein synthesis in yeast and that the mitochondrial ribosomes are the target for this inhibition (for review, see ref. 1). Mammalian mitochondria appear to be insensitive to erythromycin. We have previously explained this insensitivity by the impermeability of the mitochondrial membranes to this antibiotic (2). This explanation, however, has recently been seriously questioned. Other workers state that the mammalian mitochondrial ribosome as such is insensitive to erythromycin (3,4). From a pharmacological point of view it is highly important to know which of the two alternative explanations is correct, because the latter predicts that mitochondrial biogenesis will not be affected under any circumstance, whereas the former keeps open the possibility that mitochondrial synthetic processes may be impaired by the antibiotic in cases of altered or aberrant permeability of the mitochondrial membranes.

Effects of different macrolides on mitochondrial biogenesis in BHK 21 cells. We have shown recently that amino acid incorporation in mitochondria isolated from rat liver or BHK 21 cells is inhibited by tylosin tartrate, spiramycin and carbomycin (5). These three antibiotics belong to the macrolides but are disting-

uished from *e.g.* erythromycin by a larger lactone
ring. Their mode of action is thought to be only gra-
dually and not essentially different from the macrolides with the smaller lactone ring (6). With respect
to mitochondrial protein synthesis *in vitro* carbomycin
is the strongest inhibitor, next spiramycin and then
tylosin tartrate, the I50's being about 4, 10 and 40
µg/ml, respectively. These observations supported our
working hypothesis that mitochondrial ribosomes as
such are sensitive to the macrolides and that the apparent resistance to erythromycin has its cause in the
impermeability of the mitochondrial membranes. To extend these studies to effects on whole cells we looked for the effects of the macrolides mentioned on the
formation of cytochrome *c* oxidase and on the cycloheximide-resistant amino acid incorporation in BHK 21
cells. The results are given in Table I. It can be
seen that only carbomycin affects these processes in
the intact cells. With respect to erythromycin we had
previously established the absence of inhibition. This
led us to add erythromycin routinely to the culture
medium of the cells in an amount of 25 µg/ml. For this

TABLE I.

EFFECT OF DIFFERENT MACROLIDES ON THE FORMATION OF CYTOCHROME *C* OXIDASE AND ON
THE INCORPORATION OF [^{14}C]LEUCINE INTO MITOCHONDRIAL PROTEIN IN WHOLE BHK-21 CELLS.
Cytochrome *c* oxidase activity was measured in the cells of one single culture.
The 100 % value expressed as K(min^{-1}) per mg protein varied from 12 to 16 in the
different experiments due to differences in the age of the cultures at the time
of collection of the cells and to the fact that different cytochrome ᵌ preparations were used in the various experiments. In the incorporation experiments with
whole cells the 100 % values varied from 15 to 18 pmoles [^{14}C]leucine incorporated per mg total cell protein in a 45 min period of incubation. For further
experimental details see ref. 5 from which these data are taken.

Additions	Cytochrome *c* oxidase activity (% of control)		[^{14}C]leucine incorporation (% of control)	
	mean ± SE	n	mean ± SE	n
none	100 ± 4	10	100 ± 2	12
erythromycin (200 ᵤg/ml)	–		105 ± 3	8
tylosin tartrate (200 ᵤg/ml)	98 ± 4	19	103 ± 3	12
spiramycin (100 ᵤg/ml)	108 ± 9	8	102 ± 4	12
carbomycin (10 ᵤg/ml)	50 ± 4	8	46 ± 3	6

reason erythromycin was present in the experiments in which the effects of the other macrolides on cytochrome c oxidase formation were tested. We mentioned above that tylosin tartrate and spiramycin do inhibit amino acid incorporation by isolated mitochondria, whereas Table I shows that there is no effect on the cycloheximide-resistant incorporation in intact cells. We tend to explain this apparent discrepancy by assuming that the cell membrane is impermeable to these two macrolides. At present we are not able to decide whether the cell membrane is also impermeable to erythromycin.

Effects of different macrolides on mitochondrial ribosomes from rat liver. In order to obtain a final answer on the question whether or not mitochondrial ribosomes as such are sensitive to the different macrolides one should preferably test the antibiotics in a reconstituted system for protein synthesis with the purified mitochondrial ribosomes. However, we have not yet been able to reconstitute such a system. We therefore investigated the effects of the various macrolides on the peptidyl transferase activity of the isolated 55S ribosomes from rat-liver mitochondria. The results are presented in Table II. Erythromycin itself has no inhibitory effect on the peptidyl transferase but is able to reverse inhibition by chloramphenicol (7,8). The table shows that erythromycin indeed reversed the chloramphenicol inhibition of 55S ribosomes, and that the other three macrolides all inhibited the reaction. In neither case, however, inhibition or reversal was complete and relatively high concentrations were necessary. To erythromycin the 55S ribosomes are less sensitive than isolated yeast mitochondrial ribosomes (*cf.* ref. 8). However, the same holds for the mitochondrial ribosomes from *Neurospora crassa* (5), an ascomycete like yeast, and the sharp barrier, drawn by Towers *et al.* (3), between antibiotic sensitivities of animal and ascomycete mitochondrial protein synthesis, seems to be non-existent. It appears, then, that the rat-liver mitochondrial ribosomes are sensitive to erythromycin and the other macrolides tested. In agree-

TABLE II.

EFFECT OF MACROLIDE ANTIBIOTICS ON THE PEPTIDYL TRANSFERASE ACTIVITY OF 55-S RIBOSOMES.

For experimental details, see ref. 5.

Additions		Activity ± S.E. (% of control value)*			Number of experiments
chloramphenicol 67 µg/ml (CAP)		14.5	±	2.3	10
CAP + erythromycin	133 µg/ml**	30.1	±	7.5	5
CAP + erythromycin	333 µg/ml**	33.2	±	1.8	5
CAP + erythromycin	1000 µg/ml**	43.5	±	5.0	3
tylosin tartrate	100 µg/ml	90.0			2
tylosin tartrate	250 µg/ml	86.1			2
tylosin tartrate	600 µg/ml	66.7			2
spiramycin	200 µg/ml	57.6			2
spiramycin	600 µg/ml	52.7			2
carbomycin	30 µg/ml	87.8	±	3.4	5
carbomycin	100 µg/ml	78.0	±	4.0	4
carbomycin	300 µg/ml	49.7	±	5.2	5

* 100 % activity ranged from 2,500 to 5,000 dpm.

** Values without chloramphenicol were between 100 % and 112 %.

ment with this observations, Greco *et al.* (9) report inhibition by erythromycin in a system for polyphenyl-alanine synthesis by rat-liver mitochondrial ribosomes programmed with poly(U).

The concentration of carbomycin by isolated mitochondria. In view of the considerable difference in the concentrations of carbomycin giving complete inhibition of amino acid incorporation by mitochondria at the one hand and those giving only partial inhibition of the 55S-ribosomal peptidyl transferase, we considered it necessary to explore the possibility that mitochondria concentrate carbomycin. If so, one should expect the concentration of carbomycin in the medium of the amino acid incorporation to decrease during the incubation. Since this decrease may be expected to be only in the order of micrograms, one

398

needs a sensitive and reliable method to measure small
quantities of carbomycin. Since we were not aware of
any process more sensitive to inhibition by carbomycin
than mitochondrial protein synthesis itself, we tackl-
ed this problem as follows. Rat-liver mitochondria were incubated at a concentration of 10 mg mitochondrial
protein per ml in a medium suitable for amino acid incorporation with omission of radioactive leucine. Carbomycin was added in concentrations between 1 and 5
µg/ml in the various preincubations. After 10 min at
30° C the incubates were cooled to 0° C and the mitochondria spun down. The supernatants were added to
fresh mitochondria together with [^{14}C]leucine and re-
incubated 45 min at 30°. The inhibition by carbomycin
obtained in the second incubation was related to a
standard inhibition curve. From the difference between
the inhibition measured and that expected on the basis
of the concentration present in the first incubation
medium the carbomycin concentration in the second incubation could be calculated. The experimental results
are given in Table III and show clearly that about 60%
of the carbomycin added to the first incubation is not
recovered in the second incubation. From this value
one can compute that carbomycin is concentrated about
30-fold on the basis of total mitochondrial volume and
about 75-fold on the basis of mitochondrial water space (*cf*. ref. 5). That these results are not due to the
fact that carbomycin is inactivated by the mitochon-

TABLE III.

THE CONCENTRATION OF CARBOMYCIN BY ISOLATED RAT-LIVER MITOCHONDRIA AS MEASURED
BY THE INHIBITION OF [^{14}C]LEUCINE INCORPORATION BY A CARBOMYCIN CONTAINING ME-
DIUM IN TWO SUCCESSIVE INCUBATIONS WITH FRESH MITOCHONDRIA.
The values given (in per cent) are the means of 3 different experiments. The
100 % values ranged from 30-52 pmoles/mg protein in a 45 min incubation. For
further details, see text and ref. 5.

Additions	[^{14}C]leucine incorporation % of control		calculated carbomycin concentration in second incubation
	expected	obtained	
none	100	100	–
1 µg carbomycin/ml	75	88	0.4 µg/ml
2 µg carbomycin/ml	58	77	0.9 µg/ml
5 µg carbomycin/ml	42	65	1.9 µg/ml

TABLE IV.

EFFECT OF PREINCUBATION OF MITOCHONDRIA WITH CARBOMYCIN ON ITS INHIBITORY ACTI-
VITY.
All experimental conditions were the same as reported elsewhere (5) except that
the mitochondria were preincubated in a medium containing <u>all</u> the necessary components for the incorporation of amino acids *minus* [^{14}C]leucine. The latter was
added to start the incubation after the time periods indicated in the first column.

Preincubation time	Incubation time	[^{14}C]leucine incorporation (pmoles/mg/45 min)		
		no additions	+ 2 µg/ml carbomycin	+ 5 µg/ml carbomycin
5 min	45 min	25	15	8
20 min	45 min	24	10	6
35 min	45 min	21	6	2

dria or otherwise in the course of the incubation can
be concluded from the experiment given in Table IV. It
can be seen that the percentual inhibition of amino
acid incorporation by carbomycin is not decreased if
the [^{14}C]leucine is added after different periods of
preincubation. Rather, the contrary is the case; the
percentual inhibition increases with the length of
the preincubation time suggesting that the process of
concentration of carbomycin within the mitochondria is
not yet completed after 10 min, the period used in the
experiments of Table III.

As yet it is difficult to decide with certainty
that the phenomenon described is not due merely to adsorption of the carbomycin to the mitochondrial membranes. However, there is no doubt that carbomycin
acts within the mitochondria. In view of the data presented one should realize that within the mitochondria
this action might come about at much higher concentrations of the drug than those expected on the basis of
the micrograms added per ml.

*Is there more than one binding site for macrolides
in isolated mitochondria ?* It has been postulated by
Towers *et al.* (4) that erythromycin as well as carbomycin bind to the mitochondrial ribosomes if added to
isolated mitochondria. In their concept of macrolide
action, binding of both antibiotics occurs but leads
to inhibition of mitochondrial protein synthesis only
for carbomycin. In our opinion an alternative expla-

nation for the different experimental findings is much more likely. Protein synthesis by isolated rat-liver mitochondria is inhibited by chloramphenicol and not by erythromycin. The chloramphenicol inhibition is not altered if both inhibitors are added together. So, erythromycin and chloramphenicol do not compete in this case (4,5); they do on isolated ribosomes (Table II). Carbomycin, on the contrary, clearly competes with chloramphenicol (5). It can thus be inferred that both carbomycin and chloramphenicol reach the mito-chondrial ribosomes; obviously the mitochondrial mem-branes do not raise a barrier for either of the two drugs. Finally, a strong competition exists between erythromycin and carbomycin (4,5). Since erythro-mycin does not compete with chloramphenicol in isola-ted mitochondria whereas carbomycin does, we are for-ced to assume that the site for competition between carbomycin and erythromycin must be distinct from the ribosomal site. In our opinion this second site is li-kely to be located in the mitochondrial membranes. All experimental data can be reconciled with this idea: the sensitivity of protein synthesis to eryth-romycin in mitochondria of altered permeability (2); the sensitivity of mitochondrial ribosomes to eryth-romycin in poly(U) directed polyphenylalanine synthe-sis (9) or peptide bond formation (5); the lack of competition between erythromycin and chloramphenicol with isolated mitochondria; the competition between erythromycin and carbomycin. The latter two phenomena are most easily explained by assuming the presence in the mitochondrial membrane of a translocator for car-bomycin, to which also erythromycin can be bound, how-ever, without being subsequently transported. The translocator concept is also in good agreement with the probability that mitochondria are able to concen-trate carbomycin.

In conclusion, it is clear that the mitochondrial ribosomes as such are targets for the inhibition by a variety of macrolides. By the macrolides with the smal-ler lactone rings these targets can not be reached be-cause of the impermeability of the mitochondrial membrane. This holds for erythromycin and oleandomycin and also for lankamycin and methymycin (10). The macrolides with the larger lactone ring are transported through the mitochondrial membrane; for some representatives of the latter group (spiramycin, tylosin) another barrier appears to be present at the level of the cell membrane. The nature of the transport mechanism is presently being investigated in more detail.

ACKNOWLEDGEMENTS.

The authors wish to thank Prof. F.J. Loomeijer for his interest and encouragement and Mrs. Rya van der Koogh-Schuuring for her technical assistance. These studies were aided in part by the Netherlands Foundation for Chemical Research (S.O.N.) with financial aid from the Netherlands Organization for the Advancement of Pure Research (Z.W.O.).

REFERENCES.

1. A.M. Kroon, E. Agsteribbe and H. de Vries, in:"The Mechanism of Protein Synthesis and its Regulation", L. Bosch, Ed., North Holland Publ., Amsterdam, 539 (1972).
2. A.M. Kroon and H. de Vries, in:"Control of Organelle Development", P.L. Miller, Ed., Cambridge University Press, Cambridge, 181 (1970).
3. N.R. Towers, H. Dixon, G.M. Kellerman and A.W. Linnane, Arch.Biochem.Biophys. 151, 361 (1972).
4. N.R. Towers, G.M. Kellerman and A.W. Linnane, Arch.Biochem.Biophys. 155, 159 (1973).
5. H. de Vries, A.J. Arendzen and A.M. Kroon, Biochim.Biophys.Acta 331, 264 (1973).
6. S. Pestka, Ann.Rev.Microbiol. 25, 487 (1971).
7. H. Teraoka, Biochim.Biophys.Acta 213, 535 (1970).
8. L.A. Grivell, L. Reijnders and H. de Vries, FEBS Letters 16, 159 (1971).
9. M. Greco, G. Pepe and C. Saccone, this volume.
10. A.M. Kroon and A.J. Arendzen, unpublished observations.

Part III

Synthesis of Mitochondrial Proteins

MITOCHONDRIAL PRODUCTS OF YEAST ATPASE AND CYTOCHROME OXIDASE

Alexander Tzagoloff, Anna Akai and Meryl S. Rubin

The Public Health Research Institute of the City of New York, Inc., 455 First Avenue, New York, N.Y. 10016, U.S.A.

There is currently evidence that at least two enzymes of the mitochondrial inner membrane contain subunit proteins that are synthesized in the mitochondrion (1-4). Four proteins of the rutamycin-sensitive ATPase complex are translated on mitochondrial ribosomes (1). Similarly, studies from several laboratories (2-4) indicate that cytochrome *c* oxidase may contain as many as three proteins which are derived from mitochondrial protein synthesis. To date, however, it is not known whether these proteins play a role in catalytic function of whether they are structural proteins which are only important in the assembly of the enzymes.

There are essentially two experimental ways in which this problem can be examined. In a genetic approach the loss of function in a mutant may be correlated with a missing or defective component. The alternative is to purify the protein and hope that in the isolated state partial function of recognizable prosthetic groups will be retained. We have chosen to take the latter route simply because for technological reasons it seemed the more simple of the two. Our progress on the purification of the mitochondrially synthesized proteins of yeast ATPase and cytochrome *c* oxidase, their characterization and possible function will be the subject of this paper.

RESULTS AND DISCUSSION.

Mitochondrially synthesized component of yeast ATPase and cytochrome c oxidase. Basically, the same

Subunits of ATPase and cyt.oxidase

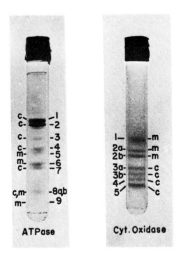

Fig. 1.   Polyacrylamide gels of yeast rutamcyn-sensitive ATPase and cytochrome *c* oxidase. The purified enzymes were separated on 7.5 % acrylamide gels in the presence of 0.1 % SDS. The subunits are numbered according to the previously adopted conventions (1,4). Subunit 9 of the ATPase is not stained by amido black. The synthetic origins of the various subunits are designated by the letter C for cytoribosomal synthesis and M for mitochondrial synthesis.

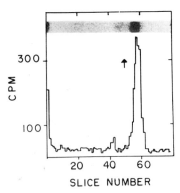

Fig. 2.   Mitochondrial product of the ATPase. The protein (25 µg) purified from submitochondrial particles was mixed with a small amount of purified [³H]labeled mitochondrial product. The mixture was separated on a 7.5 % acrylamide gel in the presence of 0.1 % SDS. The gel was stained amido black to localize the protein, sliced into 1 mm sections and counted. The top of the gel is on the left. The arrow indicates the position of the bromophenol tracking dye. From Sierra and Tzagoloff (12).

protocol has been used in several laboratories to identify mitochondrial products of the ATPase and cytochrome c oxidase.

1. Yeast or *Neurospora crassa* are labeled *in vivo* with a radioactive amino acid in the presence of the cyto-ribosomal inhibitor cycloheximide.

2. The enzymes are then isolated either by standard purification procedures (1,2) or by immunochemical precipitation with antiserum to the purified enzyme (1,3,4).

3. The labeled proteins are identified by electrophoretic separation on polyacrylamide gels in the presence of sodium dodecyl sulfate.

Protein components synthesized on cytoplasmic ribosomes have been identified by the same procedure, except that the labeling is performed in the presence of chloramphenicol or some other inhibitor of mitochondrial protein synthesis (3,5).

Gels of purified rutamycin-sensitive ATPase and of cytochrome c oxidase showing the subunit polypeptides and their synthetic origin are presented in Fig. 1. Of the ten components of the ATPase complex, four are synthesized in the presence of cycloheximide and have been designated as mitochondrial products. Subunit 9 of the complex is not visible in the gel shown because it comigrates with phospholipid which interferes with the amido black staining. The remaining six subunits of the ATPase are translated on cytoplasmic ribosomes and their synthesis is not blocked by chloramphenicol (5). Analogous experiments with cytochrome c oxidase have shown that only the three largest molecular weight subunits of this enzyme are mitochondrially produced (4). Mason and Schatz also found that the three largest proteins of yeast cytochrome c oxidase are labeled *in vivo* in the presence of cycloheximide (3).

Resolution of the ATPase complex. The rutamycin-sensitive ATPase of yeast has been resolved into three functional components. The basic catalytic unit of the complex is a water soluble ATPase, F_1, which was first isolated by Pullman *et al.* (6). OSCP, another compo-

nent with a recognizable function, is a protein involved in the binding of F_1 to the membrane (7,8). The
third component is a membrane lipoprotein factor which
consists of four proteins. The membrane factor is required to confer rutamycin-sensitivity on F_1 (9,10).
Both F_1 (subunits 1,2,3,4 and 8a) and OSCP (subunit 7)
are synthesized on cytoplasmic ribosomes (5,8,11). In
contrast, the four protein subunits of the membrane
factor (subunit 5,6,8b and 9) are synthesized in the
mitochondrion.

One of the proteins of the membrane factor (subunit
9) has recently been obtained in homogeneous form (12).
The protein can be purified from the ATPase complex
or from submitochondrial particles by extraction with
a 2:1 mixture of chloroform:methanol followed by chromatography on thin layer. From either source, the isolated protein appears as a single band on SDS gel
electrophoresis (Fig. 2) and has the same migration as
subunit 9 of the native ATPase. Amino acid analyses of
the purified protein have revealed it to be extremely
hydrophobic (*cf*. Table I). Non-polar amino acids, as
defined by Capaldi and Vanderkooi (13) account for more than 75 % of the total residues. This value exceeds
the values reported for other proteins which are soluble in organic solvents, such as brain proteolipid
(13,14) and the isoprenoid alcohol phosphokinase isolated by Sandermann and Strominger (15).

At present we cannot assign a function to the purified mitochondrial product. Several facts suggest,
however, that subunit 9 may contain the binding site
for the ATPase inhibitors, rutamycin and dicyclohexylcarbodiimide (DCCD). In previous studies we have found
that the synthesis of subunit 9 parallels the ability
of the membranes to confer rutamycin-sensitivity on
F_1 (1,10). More significant are the findings of Cattell *et al*. (16), that in beef-heart mitochondria DCCD
binds specifically to a low molecular weight protein
whose solubility properties are similar to those of
subunit 9. These observations have been extended by
Stekhoven *et al*. (17) who found a similar protein in
the rutamycin-sensitive ATPase of beef-heart mitochon

TABLE I.

AMINO ACID COMPOSITION OF THE MITOCHONDRIAL PRODUCT OF THE ATPase.
Purified subunit 9 obtained from the rutamycin-sensitive ATPase was hydrolyzed
for 24 h in 6N HCl in the presence of 0.08 M mercaptoethanol. The minimal composition was based on the assumption that the protein contains one residue of
arginine. The tyrosine peak was small and could only be approximated. (From
ref. 12).

Residue	Mole percent	Minimal composition
Lysine	3.4	2
Histidine	0	0
Arginine	1.5	1
Aspartic acid	4.7	3
Threonine	4.6	3
Serine	6.9	4
Glutamic acid	3.1	2
Proline	3.1	2
Glycine	14.2	9
Alanine	13.4	9
Valine	7.5	5
Methionine	3.1	2
Isoleucine	10.9	7
Leucine	15.5	10
Tyrosine	1.0	1
Phenylalanine	7.2	5

Non polar residues = 76 %

Minimum molecular weight = 8,900

dria.

Our efforts to reconstitute a rutamycin-sensitive
ATPase activity from F_1, OSCP and the purified subunit
9 have not been succesful. This, however, is not surprising since there is some evidence that in its native form the protein is combined with some other component which is lost during the purification. Thus, in
submitochondrial particles and in the native membrane
factor, subunit 9 has an apparent molecular weight of
45,000 (18). This form of the protein is destroyed by
exposure to organic solvents, alkali of acid (1,18),
and the molecular weight is reduced to 7,500. The nature of the material to which subunit 9 is complexed
in the native enzyme is not known.

Resolution of cytochrome c oxidase. Several preparations of yeast cytochrome *c* oxidase have been repor-

PURIFICATION OF CYT. OXIDASE SUBUNITS

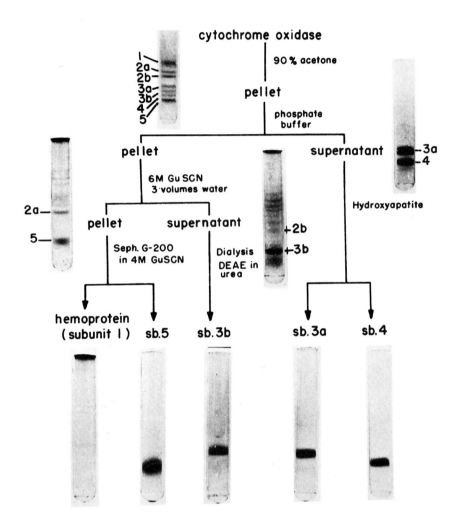

Fig. 3. Purification scheme used to purify subunits 1, 3a, 3b, 4, and 5 of yeast cytochrome c oxidase. The gels shown were run in 7.5 % polyacrylamide gels in the presence of 0.1 % SDS.

ted to contain six to seven non-identical subunit proteins (19,20). The active enzyme has a molecular
weight of approximately 230,000 (20). Unfortunately,
procedures which depolymerize the enzyme lead to irreversible loss of enzymatic activity.

We have recently made use of the powerful depolymerizing agent, guanidine thiocyanate (21) to isolate
some of the subunit proteins of yeast and beef-heart
cytochrome c oxidases. Since guanidine thiocyanate
does not appear to dissociate heme a from the protein,
it has been possible to purify a protein which we believe is the hemoprotein component of the oxidase.

In Fig. 3 is presented a scheme of the procedure we
have used to purify the four low-molecular weight subunits of yeast cytochrome c oxidase and a protein
fraction which is highly enriched in heme a. Acetone
washed enzyme is extracted with dilute phosphate buffer to remove subunits 3a and 4 which can then be further separated on hydroxyapatite. The insoluble material remaining after the phosphate extraction is solubilized by 6M guanidine thiocyanate. Dilution of the
guanidine thiocyanate solution with three volumes of
water precipitated virtually all of the hemoprotein,
subunits 2a and 5. The precipitate is redissolved in
6 M guanidine thiocyanate and chromatographed on Sephadex G-200 in the presence of the chaotrope. The Sephadex G-200 column separates the hemoprotein from
subunit 5. The supernatant after dilution of the initial 6M guanidine thiocyanate solution is used to obtain pure subunit 3a. The scheme yields homogeneous
preparations of subunits 3a, 3b, 4 and 5 as judged by
polyacrylamide gel electrophoresis. The identity of
the purified proteins has been established by co-electrophoresis with standard yeast cytochrome c oxidase.
Subunits 2a and 2b have not yet been obtained in pure
form.

Following precipitation with acetone, the largest
protein of the oxidase (subunit 1) is aggregated and
no longer subject to depolymerization in 1 % SDS. Un-
like the other components which undergo depolymerization, subunit 1 remains in a polymeric form at the top

EFFECT OF
ACETONE ON SUBUNIT I

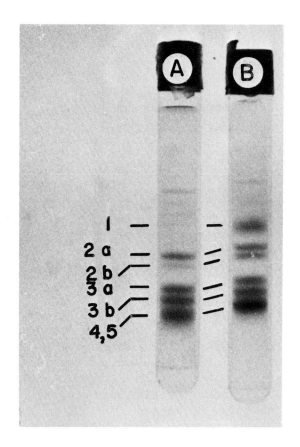

Fig. 4. Polyacrylamide gels of yeast cytochrome *c* oxidase with and without acetone treatment. A: the enzyme was washed with 9 volumes of acetone. The protein
was then dissolved in the standard 1 % SDS solvent and separated on a 7.5 % a-
crylamide gel in the presence of 0.1 % SDS. 50 µg of protein was applied to the
gel. B: same as A except that the oxidase was mixed with the SDS depolymerizing
solvent without prior acetone washing.

TABLE II.

RECOVERY OF HEME *A* AND PROTEIN.
Yeast cytochrome *c* oxidase was fractionated according to the scheme of Fig. 2.
Heme *a* was estimated by the pyridine hemochromogen method.

Fraction	Total protein (mg)	Total Heme *a* (nmoles)
Cytochrome *c* oxidase	46.5	425
Crude hemoprotein (precipitate after dilution of GuSCN)	18.5	370
Purified hemoprotein (from Sephadex G-200)	14.0	325

of the gel (Fig. 4). Although we have had some succes
in depolymerizing the protein at higher SDS concentrations, this has not been reproducible and most of
the time the protein barely penetrates the gel. The
hemoprotein which is purified by the guanidine thiocyanate procedure also is not depolymerized by SDS.
Gels of this protein do not show any bands within the
gel even when as much as 200 µg of protein are applied
to the gel. All of the amido black stain in gels of
this protein is seen in the uppermost few millimeters
of the gel. Since we do not recover subunit 1 in any
of the subfractions obtained by the scheme shown in
Fig. 2, we currently think that the purified hemoprotein is identical to subunit 1. As will be seen later
this is confirmed by our amino acid analyses.

Almost all of the heme *a* of yeast cytochrome *c* oxidase is recovered in the hemoprotein peak from Sephadex G-200 as is seen from the results summarized in
Table II. The small loss seen in the crude hemoprotein
fraction is due to incomplete precipitation of the
protein after dilution of the 6M guanidine thiocyanate solution with water. In Table III are presented the
specific heme *a* contents of the purified proteins.
None of the small molecular weight subunits (3a,3b,4 and
5) contain any detectable heme. It is also unlikely that
subunits 2a and 2b have any heme associated with them since fractions which are enriched in these components do
not contain spectrally detectable heme *a*. The specific
heme *a* content of the purified hemoprotein ranged from
23-25 mumoles of heme *a* per mg protein, suggesting a

413

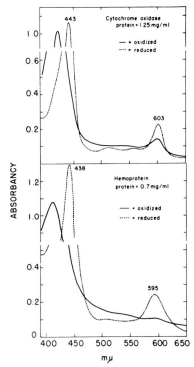

Fig. 5. Spectra of yeast cytochrome *c* oxidase and of the purified hemoprotein. Top panel: yeast cytochrome *c* oxidase (1.25 mg per ml). Bottom panel: purified hemoprotein (0.7 mg per ml). Spectra of the two proteins were recorded in the oxidized form (————) and then reduced with $Na_2S_2O_4$ (-------).

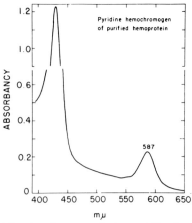

Fig. 6. Spectrum of the pyridine hemochromogen of the purified hemoprotein. The protein (0.35 mg) was dissolved in 1 ml of a 1:1 mixture of pyridine and 0.2 N KOH. The sample was reduced with $Na_2S_2O_4$ and the spectrum recorded.

414

TABLE III.

SPECIFIC HEME A CONTENTS OF PURIFIED SUBUNITS.
The subunits of yeast cytochrome c oxidase were purified according to the scheme of Fig. 2.

Subunit	Heme a (nmoles per mg protein)
Hemoprotein (subunit 1)	23-25
3a	0
3b	0
4	0
5	0

minimal molecular weight of 40,000 to 43,000. This value approximates the molecular weight of subunit 1 as determined on 10 % polyacrylamide gels (19,20).

The spectrum of the purified hemoprotein shows bands which absorb at wavelengths characteristic of certain types of denatured cytochrome c oxidase (Fig. 5). The peaks of the α- and Soret bands in the reduced form of the hemoprotein occur at 595 and 438 mμ respectively as opposed to active cytochrome c oxidase which has absorption maxima at 603 and 443 mμ. This shift towards the red has also been reported by Lemberg (22) when cytochrome c oxidase is denatured at pH 9.5 or in 8M urea. The pyridine hemochromogen of the hemoprotein (Fig. 6) is identical to that of heme a extracted from cytochrome c oxidase.

Although the heme a is tightly linked to the isolated protein, it is possible that the prosthetic group may have migrated during the depolymerization in guanidine thiocyanate. We have, therefore, carried out the purification with cytochrome c oxidase in which the heme a was covalently linked to the protein as the Schiff adduct (23). Cytochrome c oxidase was brought to pH 12 and reduced with borohydride. Less than 10 % of the total heme a remained extractable in acidic acetone, which quantitatively removes heme a from unreacted cytochrome c oxidase. When this preparation was used as the starting material in the purification, again all of the heme a was recovered in the fractions which we have identified as subunit 1. None

of the low-molecular weight subunits (2a–5) had heme
associated with them. This experiment tends to confirm the notion that subunit 1 is a hemoprotein but
does not completely rule out a displacement of heme a
from some other protein in the native enzyme and subsequent attachment to subunit 1 since such a migration
could occur during the alkaline treatment which is
necessary for formation of the Schiff base. Here the
biochemical approach seems to be faced with a variant
of the Heisenberg Principle. All the conditions which
have been found to dissociate cytochrome c oxidase also cause some change in the absorption properties of
the heme chromophore. Since dissociation is a necessary prelude to purification, a migration of the prosthetic group in a spectrally modified enzyme can not
be strictly excluded.

The previous finding that subunit 1 is a mitochondrial product (4) made it of interest to compare its
amino acid composition to that of subunit 9 of the
ATPase another mitochondrially produced protein. The
results of our preliminary analyses are presented in
Table IV. With the exception of tryptophan and cysteine, which were not determined, all the commonly occurring amino acids are represented in the protein. On
the basis of its gross amino acid composition, the hemoprotein of cytochrome c oxidase appears to be more
polar than subunit 9 of the ATPase. The polarity calculated according to Capaldi and Vanderkooi (13) is
0.38, i.e. 38 % of its residues may be classified as
polar amino acids. This value is somewhat lower than
that of most soluble proteins which are statistically
distributed around a polarity of 0.45 (13).

In an attempt to further substantiate the identity
of the purified hemoprotein with subunit 1 of yeast
cytochrome c oxidase, we have also done amino acid
analyses on subunit 1 purified on polyacrylamide gel.
Yeast cytochrome c oxidase was resolved into its sub-
component proteins on 7.5 % polyacrylamide gels in the
presence of SDS. The stained band corresponding to
subunit 1 was excised from the gel and hydrolyzed in
6N HCl according to the method of Houston (24). The

TABLE IV.

AMINO ACID COMPOSITION OF CYTOCHROME *C* OXIDASE HEMOPROTEIN.
The protein was hydrolyzed for 24 h in 6N HCl in the presence of 0.05 M mercaptoethanol. The minimal composition was calculated on the basis of 9 residues of arginine per mole of protein.

Residue	Mole percent	Minimal composition
Lysine	3.6	11
Histidine	3.2	10
Arginine	2.9	9
Aspartic acid	8.7	26
Threonine	6.5	19
Serine	6.8	20
Glutamic acid	6.6	20
Proline	5.1	15
Glycine	7.9	24
Alanine	8.4	25
Valine	7.3	21
Methionine	2.3	7
Isoleucine	7.2	22
Leucine	12.1	36
Tyrosine	4.8	14
Phenylalanine	6.6	20

Non polar residues = 62 %

Minimum molecular weight = 38,300

similarity of the amino acid compositions of the purified hemoprotein and of subunit 1 (Table V) isolated from polyacrylamide gels, further confirms the identity of the two.

CONCLUDING REMARKS.

The importance of mitochondrial protein synthesis in the development of a functional organelle has been appreciated for a long time. There is substantial evidence that mitochondrial products are few in number - recent estimates are in the range of five to ten polypeptides (18,25-27) - and that this synthetic activity is required only for the elaboration of a few enzymes localized in the inner membrane. Present evidence suggests that three major enzymes of the inner membrane contain mitochondrially synthesized proteins: ATPase, cytochrome *c* oxidase and coenzyme QH_2-cyto-

TABLE V.

COMPARISON OF THE AMINO ACID COMPOSITION OF THE PURIFIED HEMOPROTEIN AND OF SUB-
UNIT 1 ISOLATED FROM GELS.
Yeast cytochrome c oxidase was depolymerized in 1 % SDS and separated on 7.5 %
polyacrylamide gels in the presence of 0.1 % SDS. A total of 400 µg of protein
was applied on four separate gels. The gels were stained with amido black to localize the proteins. The stained band corresponding to subunit 1 was removed
with a razor blade from each of the four gels. The gel slices were hydrolyzed
in 6N HCl in the presence of 0.08 M mercaptoethanol according to the procedure
of Houston (24). Four blank gels were treated in an analogous fashion. To these
slices was added a standard mixture of amino acids. Both sets of gels were hy-
drolized for 24 h at 110°. The analyses were carried out on duplicate samples.
The constants were calculated from the standard mixture of amino acids which
was exposed to the hydrolysis conditions in the presence of the blank acrylamide slices. This method completely destroys tyrosine and methionine and partially histidine. The amido black stain produces excessive ammonia which overlaps with the arginine peak making its estimation difficult. The mole percent
values reported for the hemoprotein were recalculated by omitting the three
amino acids which could not be analyzed by this procedure.

Residue	Mole percent	
	Subunit 1 from gels	Hemoprotein
Lysine	4.8	4.1
Histidine	3.2	3.6
Arginine	–	–
Aspartic acid	8.6	9.8
Threonine	6.2	7.2
Serine	9.2	7.3
Glutamic acid	6.2	7.3
Proline	5.6	5.6
Glycine	11.1	8.7
Alanine	9.1	9.3
Valine	7.7	7.9
Methionine	–	–
Isoleucine	7.7	8.1
Leucine	12.6	13.5
Tyrosine	–	–
Phenylalanine	6.8	7.1

chrome c reductase (28,29). In retrospect, it is therefore surprising that the identity and role of the
mitochondrial products has taken a rather long time
to be established. In large measure this has been due
to the intractable nature of the proteins involved.

In this communication we have attempted to present
the results of our efforts at purifying and characte-
rizing some of the mitochondrially produced proteins
of two well defined enzyme complexes of the inner mem-

brane. It is anticipated that eventually such studies will reveal their role in the function and assembly of the enzymes. Our results to date suggest that some of the proteins made in the mitochondrion may in fact be catalytic subunits. If our identification of subunit 1 as the hemoprotein of cytochrome c oxidase is correct, this would clearly indicate that mitochondria synthesize the major catalytic component of cytochrome c oxidase. In the case of the ATPase, this is still questionable. Although the fundamental catalytic unit (F_1) of the ATPase is made on cytoplasmic ribosomes, knowledge of the machanism of ATP hydrolysis by the rutamycin-sensitive complex is insufficient to exclude a catalytic role for subunit 9. The presumptive evidence that this protein may have the rutamycin and DCCD binding sites favors the notion that it has some catalytic function since both reagents are potent inhibitors of the ATPase.

Finally, our preliminary analyses of the amino acid composition of the two purified proteins indicate that mitochondrial products as a class may be characterized by their high content of nonpolar amino acids. This could explain the numerous reports on the hydrophobic properties of mitochondrially synthesized proteins.

ACKNOWLEDGEMENTS.

This research was supported in part by Grants No. 1R01-HE 13003 and GM-18868 from the National Institutes of Health, U.S. Public Health Service. One of us (A.T.) is a recipient of a Research Career Award AM 42365 from the National Institution of Health, U.S. Public Health Service. M.S.R. is a recipient of a fellowship from the Damon Runyon Memorial Fund for Cancer Research (DRF-736).

REFERENCES.

1. A. Tzagoloff and P. Meagher, J.Biol.Chem.247, 594 (1972).
2. H. Weiss, W. Sebald and Th. Bücher, Eur.J.Biochem. 22, 19 (1971).
3. T.L. Mason and G. Schatz, J.Biol.Chem. 248, 1355

(1973).

4. M.S. Rubin and A. Tzagoloff, J.Biol.Chem. <u>248</u>, (1973) in press.

5. A. Tzagoloff, A. Akai and M.F. Sierra, J.Biol. Chem. <u>247</u>, 6511 (1972).

6. M.E. Pullman, H.S. Penefsky, A. Datta and E. Racker, J.Biol.Chem. <u>235</u>, 3322 (1960).

7. D.H. MacLennan and J. Asai, Biochem.Biophys.Res. Commun. <u>33</u>, 441 (1968).

8. A. Tzagoloff, J.Biol.Chem. <u>245</u>, 1545 (1970).

9. B. Bulos and E. Racker, J.Biol.Chem. <u>243</u>, 3891 (1968).

10. A. Tzagoloff, J.Biol.Chem. <u>246</u>, 3050 (1971).

11. G. Schatz, J.Biol.Chem. <u>243</u>, 2192 (1968).

12. M.F. Sierra and A. Tzagoloff, Proc.Nat.Acad.Sci. USA, (1973) submitted for publication.

13. R.A. Capaldi and G. Vanderkooi, Proc.Nat.Acad.Sci. USA <u>69</u>, 930 (1972).

14. L.P. Eng, F.C. Chao, B. Gerstl, D. Pratt and M.G. Tavaststjema, Biochemistry <u>7</u>, 4455 (1968).

15. H. Sandermann and J.L. Strominger, Proc.Nat.Acad. Sci. USA <u>68</u>, 2441 (1971).

16. K.J. Cattell, C.R. Lindop, I.G. Knight and R.B. Beechy, Biochem.J. <u>125</u>, 169 (1971).

17. F.S. Stekhoven, R.F. Waitkus and Th.B. van Moerkerk, Biochemistry <u>11</u>, 1144 (1972).

18. A. Tzagoloff and A. Akai, J.Biol.Chem. <u>247</u>, 6517 (1972).

19. T.L. Mason and G. Schatz, J.Biol.Chem. <u>248</u>, 1346 (1973).

20. M.S. Rubin and A. Tzagoloff, J.Biol.Chem. <u>248</u>, (1973) in press.

21. C. Moldow, J. Robertson and L. Rothfield, J. Membr.Biol. <u>10</u>, 137 (1972).

22. R. Lemberg and T.B.G. Pilger, Proc.Roy.Soc.B. <u>159</u>, 436 (1964).

23. S. Takemori and T.E. King, J.Biol.Chem. <u>240</u>, 504 (1965).

24. L.L. Houston, Anal.Biochem. <u>44</u>, 81 (1971).

25. G.S.P. Groot, W. Rouslin and G. Schatz, J.Biol. Chem. <u>247</u>, 1735 (1972).

26. P.O. Weislogel and R.A. Butow, J.Biol.Chem. <u>246</u>, 5113 (1971).
27. D.Y. Thomas and D.H. Williamson, Nature New Biol. <u>233</u>, 196 (1971).
28. H.R. Mahler, in:"Reviews in Biochemistry", Chemical and Rubber Co. Press, in press.
29. A. Tzagoloff, M.S. Rubin and A. Akai, J.Biol.Chem. submitted for publication.

INITIATION, IDENTIFICATION AND INTEGRATION OF MITO-CHONDRIAL PROTEINS

H.R. Mahler[*], F. Feldman, S.H. Phan, P. Hamill and K. Dawidowicz[†]

Department of Chemistry, Indiana University, Bloomington, Indiana 47401, U.S.A.

Among other features of their system for gene expression mitochondria of all species examined have conserved the mode of chain initiation in protein synthesis characteristic of their prokaryotic ancestors (1-5): fMet-tRNA$_F^{Met}$ is the entity responsible, and f-Met is retained at the N-terminal position of the nascent polypeptide until it is completed (6-11). These observations suggested to us the possibility of using radioactive [^{14}C] or [^{3}H] formate to probe some of the outstanding questions concerning the function and autonomy of mitochondria in the biosynthesis of their membrane proteins (4,12-14). The potential power and limitations of the method are indicated in Scheme 1. Incorporation of formate can be used to study chain initiation, its rate, extent and regulation, either directly (reaction 1) or - in the presence of puromycin (puro) - by following the synthesis of fMetpuro (reaction 1a) (15). It can be employed to determine the nature and to characterise the properties of the ribonucleoprotein particles responsible for mitochondrial protein synthesis (mitoribosomes) and to distinguish them from the analoguous system in the cyto-

---

[*] Recipient of Research Cancer Award K06 05060 from the Institute of General Medical Sciences. The studies described have been supported by a Research Grant GM 12228 from the same source.

[†] Present address: Av. Rio Orinoco, Ed. Residencia Lido. Apt. 32, Cumbres de Curume, Caracas, Venezuela.

SCHEME 1

FLOW OF FORMATE DURING PROTEIN SYNTHESIS IN MITOCHONDRIA

formate
↓
$[f\text{-}FH_4]$ ⇐⟹ "C_1" pool ⟶ X ⟶ PROTEINS

$\text{Met-tRNA}_F^{\text{Met}}$
↓

(1) $f\text{Met-tRNA}_F$ $\cdots$AUG-mRNA, $\dfrac{30S \quad 50S \quad \text{ribosomes}}{\text{GTP, IFs}}$ ⟶ $[f\text{Met-tRNA}_F \cdot \text{Rb}]$ (initiation)

(1a)   PS ⊢— puro
↓
$f\text{Met-puro}+\ldots$

(2) $[f\text{Met-tRNA}_F \cdot \text{Rb}]$ $\dfrac{\text{aa-tRNAs}}{\text{GTP, EFs}}$ ⟶ $[f\text{Met}\!\!-\!\!aa_i\text{-tRNA}_i]$ $+\ldots$(monosome)
nascent chains

(3)   repeat (1) + (2) $\dfrac{\text{aa-tRNAs}}{\text{GTP, IFs, EFs}}$ ⟶ $[f\text{Met}\!\!-\!\!aa_j\text{-tRNA}_j]_m + \ldots$(polysome)

(4) $[f\text{Met}\!\!-\!\!aa_{n-1}]$ $\dfrac{\text{terminator}}{\text{TFs}}$ ⟶ $f\text{Met}\!\!-\!\!aa_{n-1}$ $+\ldots$(termination and release)

(5) $f\text{Met-HN-}aa_1 aa_2\!\!-\!\!aa_n(\text{CO}_2^-)$ $\xrightarrow{\textit{deformylase}}$ $\text{Met-}aa_1 aa_2\!\!-\!\!aa_n$
formate   processing by aminopeptidases

Other components
(6)   INNER MEMBRANE ⟵———— aa——$aa_n + \text{Met} + aa_1 + \text{etc}.$

IFs, EFs and TFs = initiation, elongation and termination factors

PS   = peptide synthetase (peptidyl transferase)

⟶

Fig. 1. Comparison of mitochondrial (Fig. 1A, left) and cell sap (Fig. 1B, right) polysomes. Spheroplasts of strain A364A, obtained from cells growing exponentially in YM-1 3 % lactate, were labeled and processed as described previously (12,13). Labeling was at 23° in YM-5, 3 % lactate by exposure to $[^3H]$leucine (5 µCi/ml) and $[^{14}C]$formate (2.5 µCi/ml) for 10 min when the reaction was stopped by the simultaneous addition of cycloheximide (100 µg/ml) and chloramphenicol (4 mg/ml). Spheroplasts were lysed by blending in 0.5 M sorbitol-NMT, mitochondria and cell sap fractions isolated, and mitochondria washed with EDTA and lysed in 2 % Triton X-100, NMT. The lysates were analyzed in linear 0.3 M to 1.4 M sucrose-NMT gradients by centrifugation for 4 h at 26,000 rpm in the SW27 rotor of the L2 Beckman centrifuge. Samples were collected from the top and A_{260} and radioactivity, insoluble in hot TCA, determined.

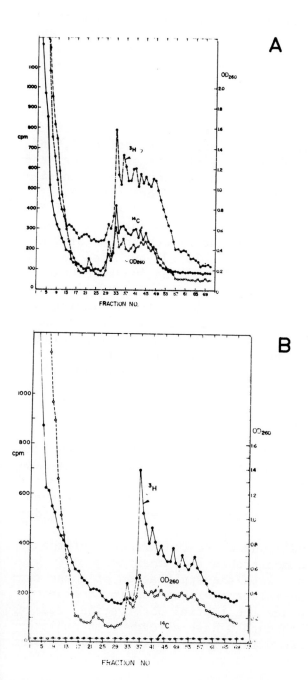

sol (reactions 2 and 3). The absolute requirement of all these reactions for mRNA suggests as a corollary that they may in turn be used as an operational assay for functional mRNA. If formyl groups are retained even after completed chains are released from their ribosomal site of synthesis (i.e. if reaction 5 catalysed by deformylase (16,17) is absent) then the way is open to explore the steps responsible for their integration into the innern membrane (reaction 6), and to investigate the dependence of this process on, and its possible regulation by, other entities synthesized either in the mitochondria or elsewhere (18-21). It should also prove feasible to use the presence of N-terminal formyl residues (radioactive or not) as an unambiguous diagnostic for the mitochondrial origin of polypeptide subunits in isolated, purified enzymes or enzyme complexes derived from the inner membrane (20-23). The most important assumptions and limitations inherent in this approach are the following: a) the absolute specificity of localization inside the mitochondria of the sequence of reactions depending on formate; b) the absence of any other process anywhere that might give rise to the incorporation of formate into mitochondrial proteins; or c) should such a process be identified, a ready means to distinguish its product from N-terminal fMet on polypeptide chains and d) the absence of a deformylase (reaction 5) in mitochondria is a prerequisite for the study of reactions subsequent to the release of the completed polypeptide. Since the unambiguous interpretation of all the experiments to be described Linge on this one point it has been checked explicitly. We detect no deformylase activity in our standard yeast strains under conditions where the presence of the enzyme is readily demonstrable in *Escherichia coli.*

This paper deals with the application of these concepts to mitochondrial gene expression in *Saccharomyces cerevisiae.* Since the initial observations dealing with reactions 1 through 3 have already been published (4,12,13), these will only be dealt with in passing. Our main emphasis will be on the subsequent steps dealing with N-terminal formate in membrane proteins and its potential utility in their identification and measurement.

RESULTS AND DISCUSSION.

Formate is a specific tag for nascent chains on mitochondrial polysomes. A recent example is shown in Fig. 1. We used spheroplasts derived from exponentially growing derepressed cells of strain A364A, exposed them to a simultaneous 10 min pulse of $[^{14}C]$formate and $[^3H]$leucine, isolated ribonucleoprotein particles from both the cell sap and from purified mitochondria, and analyzed them on sucrose gradients. While leucine incorporation coincides with the absorbancy profile in both instances, incorporation of formate is restricted to the mitochondrial system (Fig. 1A - Left). Formate (and leucine) incorporation is prevented by exposure of the spheroplasts to chloramphenicol or ethidium bromide prior to the pulse; formate-bearing nascent chains can be discharged by treatment with RNAse or transferred to puro producing N-formyl-peptidylpuros.

Formation of fMet-puro is a measure of initiation. If puromycin is added prior to the addition of labeled formate to suspensions of cells or spheroplasts, the antibiotic functions as an alternate acceptor and fMet-puro becomes the sole product (reactions 1 plus 1a). This fact has been demonstrated by means of high voltage electrophoresis of the compound formed from $[^{14}C]$formate plus $[^3H]$methionine which in addition showed no transfer of $[^{14}C]$label to methionine (Fig. 2). Since this abortive reaction continually regenerates all the components required for chain initiation it can be used as an alternate assay for this process, and indirectly for the maximal rate attainable in polypeptide chain synthesis. As shown in Table I the reaction is specific, linear over long periods of time and can be used to measure mitochondrial protein syntesis in different states of catabolite repression. An essential negative control is shown also, the complate absence of the reaction in a DNA^0 petite (24-27).

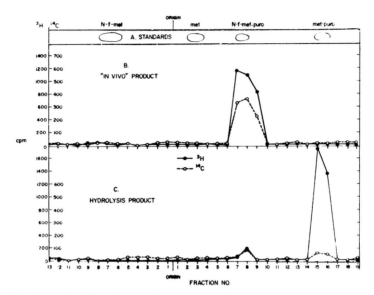

Fig. 2. Identification of mitochondrial product formed by exposing spheroplasts to formate in the presence of puro. Incubation of spheroplasts with [^{14}C]formate and [^{3}H]methionine plus puromycin (0.5 mM) for 30 min was followed by isolation of mitochondria and extraction with ethylacetate (12). The extract was divided in two: one part was evaporated to dryness, the residue taken up in pyridine and subjected to electrophoresis (Panel B, center). The second (Panel C) was taken to dryness and subjected to methanolysis in 0.5 M HCl for 48 h at 22°, prior to electrophoresis. The latter used Whatman 3 MM paper in 0.05 M pyridine acetate pH 3.5 and was for 1 h at 2000 V.

TABLE I.

MITOCHONDRIAL CHAIN INITIATION MEASURED BY FORMATION OF f*Met.

Experiment	Strain	C Source	Fraction	time (min)	cpm in fraction
I	D310-4D(ρ^+)	3 % Lactate	Mt	o	60
				5	2860
				10	5600
				30	24000
				60	49000
			Sup	10	650
				60	1890
II	IL8-8C($\rho^+ C^R_E E^R$)	3 % Lactate	Mt	10	5095
		2 % Glucose			705
	IL8-8C/R5($\rho^- C^R_E E^R$)				90
	IL8-8C/R2(ρ^0)				96

428

Additional experiments have shown that none of the ρ-
mutants tested are competent to carry out chain initiation, including ones which resemble their wild type parent by the size and base composition of their
mitochondrial DNA, as well as the retention of two
antibiotic resistance markers for chloramphenicol and
erythromycin. Thus this more restrictive parameter
complements the previously documented inability of
such mutants to perform protein synthesis properly
(28,29,30).

*Chain initiation and polysomes operationally defi-
ne mRNA function.* We have employed the formation and
persistance of mitochondrial polysomes, their ability
to form nascent chains labeled with formate, and formation of fMet-puro as alternate and mutually suppor-
tive assays for functional mRNA. By applying these
tests to ts-136, a temperature sensitive mutant with
a lesion in the supply of all nuclear transcripts at
the nonpermissive temperature (31), in conjunction
with ethidium bromide as a means for inhibiting the
synthesis of mtRNA (32-34), we have tried to determine the origin of mt mRNA (13). This problem has become acute, mainly because of some very cogent questions
posed by Dawid (35,36). Some of the pertinent experiments are summarized in Table II. We see first that we
can indeed interfere selectively with the appearance
of RNA either in the cell sap or the mitochondria.
Such interference leads to the decay of the respective
polysomes and polysomal functions within 30 min at 36°
for cytosol and - provided that ethidium bromide is
present - either at 36° or 23° for mitochondria. After
a shift down from the non-permissive to the permissive
temperature, RNA, polysomes and polysomal functions
reappear in the cytosol whether ethidium bromide is
present or not. However, in the presence of this inhibitor there is no trace of activity in the mitochondria. We have concluded from this series of experiments that mitochondria and their DNA are fully competent to transcribe the mRNA to be used by their
translational apparatus and need not depend on the import of nuclear mRNA for this purpose. Similar conclu-

TABLE II

RESPONSE OF FUNCTIONAL mRNA TO SELECTIVE BLOCKS

t (°C)	Etd Br	RNA synthesis[a]		Polysomes[b]		Protein Synthesis[c]		Initiation (Mt)[c]	
		Mt	Cell Sap	Mt	Cell Sap	Mt	Cell Sap	Formate	fMet–Puro
36°	–	2500	100[d]	+	–	92	3.5	100	75
	+	0[d,e]		–	–	2	f	29	22
36°→23°	–	1000	10,000[d]	+	+	100	17	100	100
	+	<20	8,000[d]	–	+	f	f	f	

Mt = mitochondria. a: incorporation of uracil into RNA (cpm x 20 min^{-1} in cell fraction); b: presence of active polysomes (*cf.* Fig. 1 for properties); c: as per cent of wild type activity at that temperature; incorporation of [^{3}H]leucine and [^{14}C]formate; d: after a lag of 20-30 min; e: decay of counts previously incorporated, with a $t_{\frac{1}{2}} \approx 20$ min; f: not done.

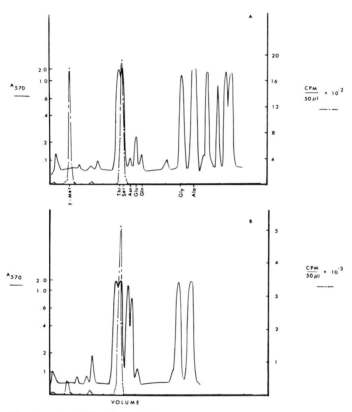

Fig. 3. Identification of mitochondrial amino acid residues labeled by formate. Cells were grown and labeled with 2.5 mc [^{3}H]formate, in Figure B (bottom) they were pre-incubated with chloramphenicol (4 mg/ml, 15 min), prior to labeling. Incorporation was terminated by chasing with 0.5 mmoles sodium formate for 15 min at 30°. Mitochondria were isolated, frozen and thawed, sonicated, and proteins precipitated, washed and digested enzymatically. For amino acid analysis, 40 % of the material in Fig. 3A and 100 % of the material in Fig. 3B was utilized. Total labeled formylmethionine obtained, after correction for volumes analyzed was 303,000 cpm in Fig. 3A and 20,300 cpm in Fig. 3B.

430

sions have also been reached independently by Penman *et al.* (37) for HeLa cells. The discovery of an ever increasing number of mitochondrial mutations to antibiotic resistance (*e.g.* 24,25,38), together with the demonstration that at least some of them do indeed affect mitochondrial membrane protein (25,29,39,40) also argue in support of the hypothesis.

fMet remains attached to some polypeptides after their transfer from mt ribosomes to mt membranes. Our earlier studies suggested to us that at least some of N-terminal fMet of nascent chains remained attached when these were released from the ribosomes, and even later after they became integrated into the inner membrane. The latter observation was in agreement with ones reported earlier in a note by Polz and Kreil concerning the proteins of mitochondria from honey bee thorax (41). However, as we extended the exposure to labeled precursor from short pulses to longer periods we also beaame aware that under these conditions fMet was not the only labeled product incorporated into proteins. Although the second process of formate incorporation originated in the cytosol, it did produce mitochondrial labeling since the bulk of the mitochondrial proteins come from the former source. However, a more detailed analysis suggested a way out of the apparent dilemma. We subjected the mixture of total inner membrane proteins, obtained from cells exposed to continuous labeling by formate for 60 min to complete enzymatic digestion, using a procedure currently being developed by Prof. F.R.N. Gurd and his students,and then examined the products with the amino acid analyzer. The results of one such experiment are shown in Fig. 3: only two radioactive peaks are present in the hydrolizate. One can be assigned to serine, the other is eluted with authentic fMet and also co-migrates with this compound after high voltage electrophoresis. Since the formate can be removed from fMet-peptides by mild hydrolysis under conditions to which the peptide bond is completely resistant (42,43), identification of polypeptides carrying this marker becomes possible.

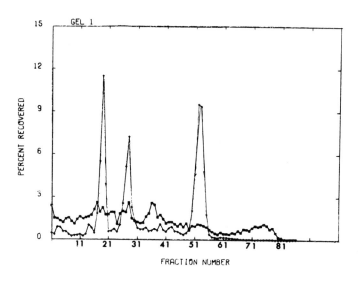

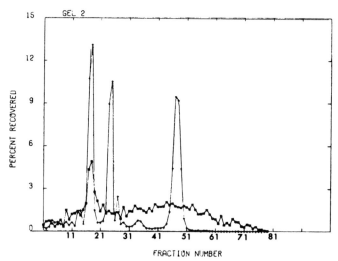

Fig. 4. The differential method for the identification of f*Met on mitochondrial polypeptides. *Saccharomyces cerevisiae*, strain 4D, was grown on SSM-3 % lactate supplemented with adenine and labeled with either [^{14}C]formate (5 µC/ml) or [^{3}H]formate (5 µC/ml) for 2 h at 30°. Mitochondrial membrane proteins were isolated, mixed with [^{3}H] or [^{14}C]labeled protein standards (bovine serum albumin, ovalbumin, and cytochrome *c*) and dissociated by boiling at 100° for 3 min in the presence of 3 % SDS and 0.4 M mercaptoethanol. Dissociated protein was then applied to electrophoresis gels as described in Methods. Gel 1 (top) shows the pattern from [^{14}C]labeled membranes (*———*) and [^{3}H]standards (+———+). Gel 2 shows the result of 24 h of methanolysis of such a preparation in 0.5 N methanolic HCl.

432

We have just begun to exploit the potentialities of this system for the identification of membrane proteins synthesized by mitochondria. This can be done by comparing the pattern of dissociated polypeptides on SDS polyacrylamide gels, before and after mild acid hydrolysis, either against internal standards or by the use of double labeling techniques (one isotope without, the other with hydrolysis). An example of the first type of experiment os presented in Fig. 4. These and similar experiments have led us to an important working hypothesis: that proteins carrying their label in formate do not contain any labeled serine and the converse. If this hypothesis is correct the conclusion would be that the mitochondrial serine pool is not readily accessible to formate in the medium. The advantages of this approach are obvious. We need no longer rely on the use of so called site specific inhibitors, which, in addition to possible undesirable side effects must, of necessity, disrupt any regulatory interplay between the two systems of gene expression (18-21, 44-46).

The formation of polypeptides by the mitochondrial system is regulated by products synthesized in the cytosol. In order to study the possible existance of such regulatory events we again made use of a temperature sensitive mutant. This strain, ts-187 (47), is blocked in polypeptide chain initiation on the ribosomes of the cytosol at the restrictive temperature. Under these conditions (Fig. 5A) protein synthesis, as measured by the incorporation of histidine, ceases not only in the cytosol, but also - after a possible lag - in the mitochondria. In this fraction we follow the incorporation of formate as well. Incidentally, the complete cessation of all formate labeling of the proteins in the post-mitochondrial supernatant after the shift-up, together with its continued incorporation into the mitochondria, permits the inference that under these conditions the incorporation of formate is restricted to N-terminal fMet. When instead of measuring protein synthesis we assay for initiation by means of fMetpuro, the shift-up results in no measu-

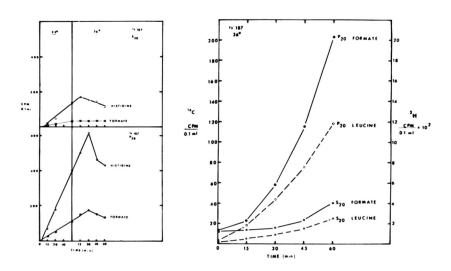

Fig. 5A. *Saccharomyces cerevisiae*, strain ts-187 was grown at 23° on YM-1 3 % lactate to midexponential phase, harvested, washed, and spheroplasts prepared but with all treatments at 23° rather than 30°. Spheroplasts were resuspended at 10 times the original concentration in YM-5 3 % lactate supplemented with 1 M sorbitol, preincubated with adenine sulfate, L-methionine, and L-serine and labeled with 5 μC/ml [³H]histidine and 0.5 μC/ml [¹⁴C]formate at 23°. 10 ml aliquots were removed as shown, chloramphenicol (4 mg/ml), cycloheximide (100 μg/ml) and formate (1 mmole) added, the cells iced, and 10 ml unlabeled carrier cells treated in the same way added. Spheroplasts were transferred to the non-permissive temperature after 60 min and further sampled as shown. Fractions were isolated, mitochondria, resuspended in 1 ml buffer A, 0.5 M sorbitol, and postmitochondrial supernatant (15 ml) were sampled as shown and analyzed.

Fig. 5B. Spheroplasts were prepared as in Fig. 5A. Chloramphenicol (2 mg/ml) was then added and spheroplasts incubated for 18 h with shaking at 23°. Chloramphenicol was removed by centrifuging at 23°, the spheroplasts were washed and resuspended into fresh YM-5 lactate supplemented with 1 M sorbitol and preincubated 10 min with adenine sulfate, L-methionine, and L-serine. The spheroplasts were then transferred to 36°, preincubated 30 min at 36°, labeled and sampled as in Fig. 5A but with [³H]leucine in place of [³H]lysine.

rable inhibition for periods as long as 60 min. Our
interpretation is that the continued availability of
polypeptides synthesized in the cytosol is required for
the maintenance of polypeptide formation (*e.g.* elongation, release or integration) by the mitochondrial
system. In other words mitochondrial protein synthesis is subject to direct translational control by an
entity dependent on protein synthesis in the cytosol.
If this hypothesis is correct we would expect to be
able to circumvent this control by a prior accumulation of these hypothetical polypeptides. In order to
do this we have interposed a period of exposure to
chloramphenicol prior to the shift-up. The results of
this treatment are shown in Fig. 5B, right-hand panel,
and appear to bear out the prediction. An analogous
requirement for cytoplasmic entities for the continued
synthesis or integration of mitochondrial proteins during derepression has been shown previously by Tzagoloff (22,48).

*N-terminal fMet is present on subunits of isolated
enzymes.* In order to determine whether we can use retention of formate in N-terminal position for the
identification of those subunits that are synthesized
within the mitochondria we have had recourse to cytochrome *c* oxidase. This appeared an obvious choice since a variety of studies, including our own, had suggested first of all the requirement for a close collaboration of the two systems of protein synthesis in
the formation of active enzyme (20, 49-51). More recently, site-specific inhibitors have been used for
the identification of certain of the dissociated subunits as being of mitochondrial origin (21,22,52-54).
For our studies we have used two types of highly purified enzyme (Table III). Both represent further modification and applications of standard techniques to a
method of purification previously described from our
laboratory (55). It differs from the other preparations recently reported (56,57) - and also discussed
in this volume - by the use of Tween 20 throughout
the fractionation. Preparation A shows a subunit composition very similar to those reported for yeast from

TABLE III

PROPERTIES OF CYTOCHROME C OXIDASE PREPARATIONS

| Preparation | Specific Activity or content (per mg protein) | | | | | Spectra | | | | |
	k (min^{-1})	k^a (min^{-1})	heme a (nmoles)	Cu (ng atoms)	µg P	ox A_{280}/A_{420}	red λ_{max}	A_{280}/A_λ	red + Co λ_{max}	A_{280}/A_λ
A	(600)	1500	> 8							
B	534		10.3	16.1	2.22	3.40	603 443	15.6 2.69	596 431	16.9 2.32
B_x^b	1300		13.7		0.80	2.67	603 443	11.5 1.93	596 431	9.09 2.17

a: in the presence of crude asolectin; b: after isooctane extraction.

TABLE IV.

SUBUNIT COMPOSITION OF CYTOCHROME OXIDASE.

Preparation Conditions	A	B	B_x
SPAGE[a], 15 %	29,20.5,14.0 11.3,9.6,7.1	28[c],15.5, 11.8,10.1	15.5,11.8,10.1
dto 7.5 %	27,21.5,14.0 12.4,11.7,10.5		
dto, extrapolated[b]		26.5[c],14.2, 11.8,10.0	14.2,11.8,10.0
isoelectric focusing[d]		5.8,6.2,6.7, 8.1,8.7	5.8,6.2,8.1

a SDS polyacrylamide (% indicated) gel electrophoresis (values in kdaltons);

b by the method of Ferguson;

c present in less than stoichiometric amounts;

d pI's shown.

TABLE V

FORMATE LABELING PATTERN OF PURIFIED CYTOCHROME C OXIDASE

| | Subunits of cytochrome c oxidase labeled with [^{3}H]formate | | | | | | |
| | Preparation A   MW(kdaltons) | | | | Preparation B   MW(kdaltons) | | |
Purified enzyme unlabeled	Nonhydrolyzed immunoprecipitate	Nonhydrolyzed Normal serum	Hydrolyzed immunoprecipitate	Purified enzyme unlabeled		Nonhydrolyzed	Hydrolyzed
29.0 (I)	25.0			26.5*	(I)	25.0	
20.5 (Ia)	21.3	20.5					
14.0 (II)	14.5		14.2	14.2	(II)	13.8	13.7
11.3 (III)	12.5		12.3	11.8	(III)	11.8	11.6
9.6 (IV)	9.4		9.8	10.0	(IV)	9.9	10.0
7.1 (IVa)	8.5	7.1	7.5				

*Extractable with iso-octane.

Schatz' and Tzagoloff's laboratories and for *Neurospora* by Weiss *et al.* (56-58). Preparation B appears quite different, especially after isooctane extraction, the last step in its purification. Note that by all criteria it qualifies as a cytochrome *c* oxidase of high purity. Its subunit composition and stoichiometry as determined by SDS electrophoresis and isoelectric focusing on polyacrylamide gels is given in Table IV.

We have labeled cells by exposure to formate and used them in two parallel experiments. In the first we prepared submitochondrial particles, solubilized the enzyme and purified it by immunoprecipitin. For this step we employed antibody against enzyme of Type A. For the second we have mixed the labeled with a large amount of carrier cells and taken them through the complete purification procedure. The enzymes so obtained were dissociated and their subunits separated by electrophoresis either directly of after prior hydrolysis for the removal of N-terminal formyl groups. The results are shown in Table V.

The molecular weights of the purified subunits (from Table IV) are given for each preparation as well as these of the labeled subunits found. It can be seen that 6 labeled subunits are obtained from preparation A and 4 from preparation B, with molecular weights in reasonable agreement with the unlabeled purified proteins. Table V also indicates that 2 polypeptide chains (Ia and IVa) precipitated in the control serum also. These polypeptides with molecular weights of 20,500 and 7100 also are found in the purified proteins and may represent contaminants which easily precipitate and are co-purified with authentic cytochrome *c* oxidase. The low molecular weight subunits II, III, and IV are not affected by deformylation and, therefore, are presumably not synthesized on mitochondrial ribosomes. The small subunit (IVa) seen in both the non hydrolyzed immunoprecipitate of cytochrome *c* oxidase as well as the control serum also is not formylated. Subunit I, with MW about 26,000, however, completely loses all label after deformylation and is, the-

refore, presumably an authentic mitochondrially ini-
tiated and synthesized product. Subunit Ia is also de-
formylated and must, therefore also be considered a
mitochondrial product. The data, however, do not yet
permit us to say if it is an authentic subunit of cytochrome c oxidase or was merely co-purified with it.

The data from the two preparations strongly indicate that peptides I and Ia are authentic mitochondrially synthesized products while peptides II, III,
and IV are authentic (cytoplasmically synthesized)
subunits of cytochrome c oxidase. Peptides I and Ia are
associated with purified cytochrome c oxidase but are
apparently not required for its enzymatic activity.
Instead they may well be concerned with the integration of the enzyme into the inner membrane.

REFERENCES.

1. J. Lucas-Lenard and F. Lipmann, Ann.Rev.Biochem.
   40, 409 (1971).
2. A.M. Kroon, E. Agsteribbe and H. de Vries, in:
   "The Mechanism of Protein Synthesis and its Regulation", L. Bosch, Ed., Nort Holland Publ.Co.,
   539 (1972).
3. H. Küntzel, Current Topics in Microbiology and
   Immunology 54, 92 (1971).
4. K. Dawidowicz and H.R. Mahler, in:"Gene Expression
   and Its Regulation", F.T. Kenney, B.A. Hamkalo,
   G. Favelukes and J.T. August, Eds., Plenum Press,
   New York, 503 (1973).
5. G.S. Getz, in:"Membranes Molecular Biology", C.F.
   Fox and A. Keith, Eds., Sinauer Assoc., Inc.,
   Stamford, Conn., 386 (1972).
6. A.E. Smith and K.A. Marcker, J.Mol.Biol. 38, 241
   (1968).
7. K.A. Marcker and A.E. Smith, Bull.Soc.Clin.Biol.
   (Paris) 51, 1453 (1969).
8. F. Sala and H. Küntzel, Eur.J.Biochem. 15, 280
   (1970).
9. J.C. Galper and J.E. Darnell, J.Mol.Biol. 57, 363
   (1971).
10. R. Bianchetti, C. Lucchini and M.L. Santirana,

Biochem.Biophys.Res.Commun. 42, 97 (1971).
11. A. Halbreich and M. Rabinowitz, Proc.Nat.Acad. Sci. USA 68, 294 (1971).
12. H.R. Mahler, K. Dawidowicz and F. Feldman, J.Biol. Chem. 247, 7439 (1972).
13. H.R. Mahler and K. Dawidowicz, Proc.Nat.Acad.Sci. USA 70, 111 (1973).
14. F. Feldman and H.R. Mahler, in preparation.
15. P. Leder and H. Bursztyn, Biochem.Biophys.Res. Commun. 25, 223 (1966).
16. M. Capecchi, Proc.Nat.Acad.Sci. USA 55, 1517 (1966).
17. D.M. Livingston and P. Leder, Biochemistry 8, 435 (1969).
18. H.R. Mahler, P. Perlman, C. Henson and C. Weber, Biochem.Biophys.Res.Commun. 31, 474 (1968).
19. C.P. Henson, P. Perlman, C.N. Weber and H.R. Mahler, Biochemistry 7, 4445 (1968).
20. H.R. Mahler, P.S. Perlman and B.D. Mehrotra, in: "Autonomy and Biogenesis of Mitochondria and Chloroplast", N.K. Boardman, A.W. Linnane and R.M. Smillie, Eds., NorthHolland Publ.Co., 492 (1971).
21. T. Mason, E. Ebner, R.O. Poyton, J. Saltzgaber, D.C. Wharton, L. Mennucci and G. Schatz, in:"Mitochondria/Biomembranes and Bioenergetics", North Holland, Amsterdam, 53 (1972).
22. A. Tzagoloff, M.S. Rubin and M.R. Sierra, Biochim.Biophys.Acta 301, 71 (1973).
23. H.R. Mahler, Critical Review in Biochemistry, in press.
24. M. Bolotin, D. Coen, J. Deutsch, B. Dujon, P. Netter, E. Petrochilo and P.P. Slonimski, Bull. de Inst.Past. 69, 215 (1971).
25. A.W. Linnane, J.M. Haslam, H.B. Lukins and P. Nagley, Ann.Rev.Microbiol. 26, 163 (1972); cf. also P. Nagley and A.W. Linnane, Biochem.Biophys.Res. Commun. 39, 989 (1970).
26. E.S. Goldring, L.I. Crossman and J. Marmur, J. Bacteriol. 107, 377 (1971).
27. H.R. Mahler, P.S. Perlman, P.P. Slonimski, M.J. Deutsch, H. Fukuhara and C. Faye, Fed.Proc. 30,

1149 (1971).

28. S. Kuzela and E. Grecna, Experientia 25, 776 (1968).
29. A.W. Linnane and J.M. Haslam, Current Topics in Cellular Regulation 2, 101 (1970).
30. G. Schatz and J. Saltzgaber, Biochem.Biophys.Res. Commun. 37, 996 (1969).
31. H.T. Hutchison, L.H. Hartwell and C.S. McLaughlin, J.Bacterol. 99, 807 (1969).
32. D.J. South and H.R. Mahler, Nature 216, 1226 (1968).
33. E. Zylber, C. Vesco and S. Penman, J.Mol.Biol. 44, 195 (1969).
34. H. Fukuhara and C. Kujawa, Biochem.Biophys.Res. Commun. 41, 1002 (1970).
35. I.B. Dawid, J.Mol.Biol. 63, 201 (1972).
36. I.B. Dawid, Devel.Biol. 29, 139 (1972).
37. S. Perlman, H.T. Abelson and S. Penman, Proc.Nat. Acad.Sci. USA 70, 350 (1973).
38. D.E. Griffiths, in:"Mitochondria/Biomembranes", G.S. van den Bergh, P.Borst and E.C. Slater, Eds., North Holland, Amsterdam, 95 (1972).
39. P.R. Avner and D.E. Griffiths, Eur.J.Biochem. 32, 301; 312 (1973).
40. C. Shannon, R. Enns, L. Short, K. Burchiel and R.S. Criddle, J.Biol.Chem., in press (1973); P.L. Molloy, N. Howell, D.T. Plummer, A.W. Linnane and H.B. Lukins, Biochem.Biophys.Res.Commun. 52, 9 (1973).
41. G. Polz and G. Kreil, Biochem.Biophys.Res.Commun. 39, 516 (1970).
42. J.M. Adams and M.R. Capecchi, J.Biol.Chem. 236, 1955 (1961).
43. J.C. Sheehan and D.H. Yang, J.Amer.Chem.Soc. 80, 1154 (1958).
44. M. Ashwell and T.S. Work, Ann.Rev.Biochem. 39, 251 (1970).
45. G.S.P. Groot, W. Rouslin and G. Schatz, J.Biol. Chem. 247, 1735 (1972).
46. Z. Barath and H. Küntzel, Nature New Biol. 240, 195 (1972).

47. L.H. Hartwell, J. Bact. 93, 1662 (1967).
48. A. Tzagoloff and P. Meagher, J.Biol.Chem. 247, 594 (1972); A. Tzagoloff, A. Akai and M.F. Sierra, ibid 247, 6511 (1972).
49. W.L. Chen and F.C. Charalampous, J.Biol.Chem. 244, 2767 (1969).
50. J. Kraml and H.R. Mahler, Immunochem. 4, 213 (1967); see also P. Schakespeare and H.R. Mahler, Arch.Biochem.Biophys. 151, 496 (1972).
51. H.R. Mahler and P.S. Perlman, Biochemistry 10, 2979 (1971).
52. T.L. Mason and G. Schatz, J.Biol.Chem. 248, 1355 (1973).
53. M.S. Rubin and A. Tzagoloff, submitted.
54. W. Sebald, H. Weiss and G. Jackl, Eur.J.Biochem. 30, 413 (1972); see also H. Weiss, W. Sebald and Th. Bücher, ibid 22, 19 (1971).
55. P. Shakespeare and H.R. Mahler, J.Biol.Chem. 246, 7649 (1971).
56. T.L. Mason, R.O. Poyton, D.C. Wharton and G. Schatz, J.Biol.Chem. 248, 1346 (1973).
57. M.S. Rubin and A. Tzagoloff, submitted.

THE BIOSYNTHESIS OF MITOCHONDRIAL RIBOSOMES IN *SACCHAROMYCES CEREVISIAE**

G.S.P. Groot

Section for Medical Enzymology, Laboratory of Biochemistry, University of Amsterdam, Eerste Constantijn Huygensstraat 20, Amsterdam, The Netherlands.

Mitochondria contain their own genetic system, which makes a small but very essential contribution to the biosynthesis of the mitochondrial inner membrane (1). We have studied the possibility that one of the constituents of this system - the mitochondrial ribosome (2) - contains mitochondrial translation products. This study was prompted by the fact that recently a good technique for the isolation and purification of mitochondrial ribosomes has been devised (3). Secondly, yeast mutants resistant to antibiotics are available, which possess altered mitochondrial ribosomes (4). By analogy with bacterial mutants, where it has been shown that such mutations are caused by alteration in certain ribosomal proteins, it was expected that alteration of certain mitochondrial ribosomal proteins might be the cause of the resistance to the antibiotic. Since the antibiotic-resistant mutants show cytoplasmic inheritance, it could be expected that some mitochondrial ribosomal proteins are coded for by mtDNA and are possibly synthesized on mitochondrial ribosomes.

Several authors (5-7) have shown that most mitochondrial ribosomal proteins are synthesized on cell-sap ribosomes. Therefore, the biosynthesis of the mitochondrial ribosome might be a very complicated task involving ribosomal RNA (rRNA), synthesized in the mitochondria, the bulk of the ribosomal proteins synthesized on cell-sap ribosomes and possibly some pro-

*Presented by M.B. Katan.

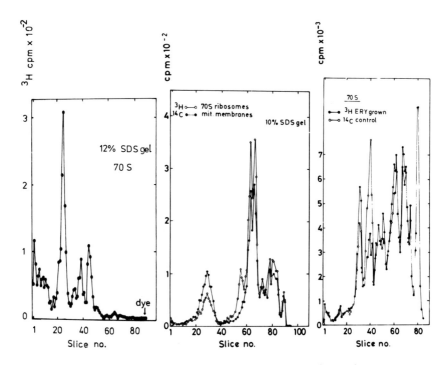

Fig. 1. Yeast cells (strain D273-10B) were grown and 20 g (wet weight) were labelled with 2 mCi [³H]leucine in the presence of 100 µg/ml cycloheximide as has been described before (8). After 2 h non-radioactive leucine was added at 10 mM final concentration. 15 min later the cycloheximide was removed by washing the cells 3 times with cold 40 mM P$_i$ buffer-5 mM leucine. The cells were then transferred to 3 L of growth medium, pre-warmed at 28° C, and subsequently incubated for 2 h. After conversion of the cells to protoplasts, the mitochondria were isolated (9), washed 3 times and incubated under conditions optimal for protein synthesis (3). After addition of puromycin (50 µg/ml final concentration) the mitochondria were lysed and ribosomes were prepared according to Grivell *et al.* (3). The ribosomal proteins were analysed on 12 % acrylamide gels containing SDS as has been described before (8).

Fig. 2. Mitochondrial ribosomes were labelled with [³H]leucine as described in Fig. 1. In a separate experiment 20 g cells of strain D273-10B were labelled with 100 µCi [¹⁴C]leucine in the presence of 100 µg/ml cycloheximide. The mitochondria were isolated and sonicated as described before (8). Aliquots of the two preparations were mixed, lysed in 2 % SDS and electrophoresed on 10 % acrylamide gels containing SDS. ●—●: [¹⁴C]labelled membrane fragments; O—O: [³H]labelled mitochondrial ribosomes.

Fig. 3. Yeast cells (D273-10B) were grown a) in 0.6 % yeast nitrogen base *plus* 1 % glucose in the presence of 50 µCi [¹⁴C]leucine and b) in 0.67 % yeast nitrogen base *plus* 1 % glucose in the presence of 2 mCi [³H]leucine and 4 mg/ml erythromycin. Both batches of cells were converted to protoplasts. The protoplasts were mixed and mitochondrial ribosomes were prepared as described in Fig. 1. The ribosomal proteins were further analysed on 12 % acrylamide gels containing SDS as described in Fig. 1. O—O: [¹⁴C]labelled mitochondrial ribosomal proteins from the control culture; ●—●: [³H]labelled mitochondrial ribosomal proteins from the culture grown in the presence of erythromycin.

teins synthesized on mitochondrial ribosomes. This complexity might cause difficulties in the interpretation of pulse-label experiments in the presence of inhibitors, when complete mitochondrial ribosomes are analysed (7). We have, therefore, used the following experimental approach which circumvents these difficulties. Cells were grown and labelled with radioactive leucine in the presence of cycloheximide (CHI). After termination of the incorporation of radioactive leucine by the addition of excess non-radioactive leucine, the cells are washed free of CHI and allowed to grow for another generation. Subsequently, mitochondrial ribosomes are prepared and analysed.

RESULTS AND DISCUSSION

In Fig. 1 the pattern of radioactivity recovered in mitochondrial ribosomes is shown. Besides a main peak with a molecular weight of 30-40,000 two other peaks are observed. The pattern resembles very much the label pattern of total mitochondrial protein synthesis. We have, therefore, compared the label recovered in the 73S mitochondrial ribosomes with that from complete mitochondrial membranes labelled in the presence of CHI (Fig. 2). It is clear that both patperns look very much the same and that the 73S ribosomes are probably contaminated despite the A_{260}/A_{280} = 1.85 and A_{260}/A_{230} = 2.0.

As an alternative approach we have compared the mitochondrial ribosomal proteins isolated from control cells grown in the presence of [^{14}C]leucine and from cells grown in the presence of 4 mg/ml erythromycin (ERY) with [^{3}H]leucine. The sodium dodecylsulphate (SDS) gel analysis of 73S ribosomal proteins is shown in Fig. 3. In this case it can be seen that polypeptides with a molecular weight of 30-40,000 are missing in the mitochondrial ribosomes isolated from the ERY-grown cells. This experiment thus complements the previous one, showing that normally 73S mitochondrial ribosomes do contain polypeptide chains synthesized by themselves. The question whether these are true ribosomal proteins or contaminants can be answer-

SDS GELS OF mt RIBOSOMAL SUBUNITS.

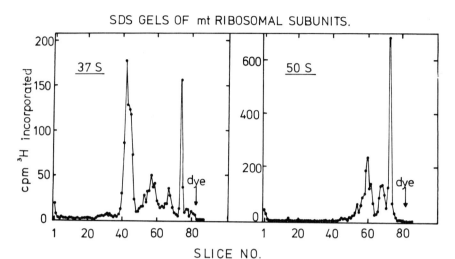

Fig. 4. Mitochondrial ribosomes were labelled with [^{3}H]leucine as described in Fig. 1. The ribosomes were separated into subunits on an isokinetic sucrose gradient, the fractions containing the large and small subunit were pooled and the subunits were precipitated with Carbowax. The ribosomal proteins were analysed on 10 % gels containing SDS.

UREA GELS OF mt RIBOSOMAL SUBUNITS.

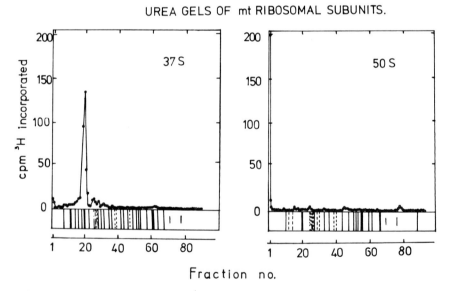

Fig. 5. Comparison of polypeptides of mitochondrial and of cytoplasmic site synthesis. The polypeptide fraction with an apparent molecular weight of about 40,000 was separated from whole mitochondrial membrane proteins by preparative SDS-gel electrophoresis. The experiments are described in the text.

ed by further purification of the ribosomes. We have, therefore, separated the 73S ribosomes into subunits and analysed the ribosomal proteins on SDS gels. Fig. 4 shows the SDS gels of the ribosomal proteins of the 50S and 38S subunit. In both cases the labelling was carried out in the presence of CHI using the experimental setup mentioned in the introduction. The figure shows that one peak of molecular weight 35,000 is recovered on the 38S subunit, there appears a small amount of labelled material of a lower molecular weight. The same pattern of components with a low molecular weight is present on the 50S subunit. It is assumed that material not specifically present in one of the subunits is due to contamination.

Fig. 5 shows the acid–urea gel patterns of the same material. Also here the label in the 38S subunit is present in one peak indicating that we are probably dealing with one protein. The contaminating material present in the 50S subunit does not penetrate the gel. As a control, in Fig. 6 the label patterns of the ribosomal proteins of 38S and 50S subunits are given. Here ribosomal proteins isolated from control cells grown in the presence of [^{14}C]leucine are compared with those isolated from cells grown in ERY in the presence of [^{3}H]leucine. It is clear that no differences are found in the 50S ribosomal proteins, in agreement with the results shown before. Also labelled material with a molecular weight of 35,000 is partially absent from the 38S ribosomal proteins isolated from the cells grown in the presence of ERY.

The crucial question as to the nature and identity of this protein is whether it is a true ribosomal protein, essential for ribosomal function or a variable non–ribosomal contaminant. We have tried to answer this question by studying mitochondrial protein synthesis in mitochondria isolated from cells grown in the presence of ERY. It must be emphasized that in this strain the concentration of ERY used (4 mg/ml) blocks mitochondrial protein synthesis completely. Table I summarizes the results of such experiments. Two sets of conditions were tested, namely with and

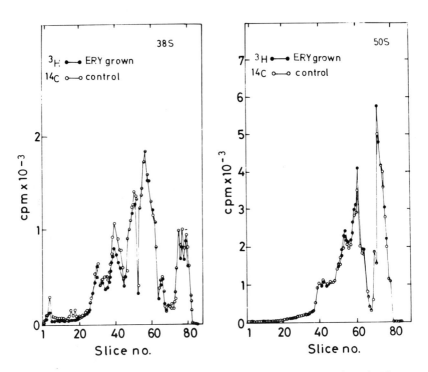

Fig. 6. An aliquot of the mixture of ribosomes, obtained as described in Fig. 3, was separated into subunits and concentrated as described in Fig. 5. The ribosomal proteins were analysed on 10 % gels containing SDS.

TABLE I.

PROTEIN SYNTHESIS IN ISOLATED MITOCHONDRIA FROM CONTROL AND ERYTHROMYCIN-GROWN CELLS.
Protein synthesis in mitochondria isolated from yeast cells (strain DT-XII) was measured as described by Grivell (11). [^{3}H]leucine was added at 2 ᵤCi per incubation; [^{14}C]leucine at 0.1 ᵤCi per incubation. The ATP-regenerating system consisted of 5 mM phosphoenolpyruvate, 2 mM ADP and 10 ᵤg pyruvate kinase. ERY= erythromycin.

Additions	− ATP-regenerating system		+ ATP-regenerating system	
	Control	ERY-grown	Control	ERY-grown
	cpm [^{3}H]leu inc./mg/30 min		cpm [^{14}C]leu inc./mg/30 min	
None	34 200	4500	3100	2940
CAP (50 ᵤg/ml)	570	750	480	460
CHI (10 ᵤg/ml)	33 700	3500	2840	2680

without an ATP-regenerating system, since it can be expected that mitochondria from ERY-grown cells can not perform oxidative phosphorylation due to the lack of a respiratory chain and part of the phosphorylation enzymes. In the presence of an ATP-regenerating system mitochondria from ERY-grown cells incorporate radioactive amino acids at approximately the same rate as control cells. The incorporation is insensitive to CHI and sensitive to chloramphenicol (CAP) and, therefore, can be regarded as true mitochondrial protein synthesis.

Furthermore, mitochondrial ribosomes isolated from ERY-grown cells behave exactly the same as ribosomes from control cells when their sedimentation is tested on sucrose gradients under dissociating conditions (Fig. 7).

There is, however, still the remote possibility that the protein found in the 38S subunit is a true ribosomal protein, but that synthesis of this protein on mitochondrial ribosomes is relatively resistant to CAP or ERY. We have tested this possibility by comparing the products of mitochondrial protein synthesis, measured *in vivo* in the presence of CHI, with the products made in the mitochondria under conditions where mitochondrial protein synthesis is inhibited 75 %. We expect then a preferential synthesis of material banding in SDS gels at the position correspon-

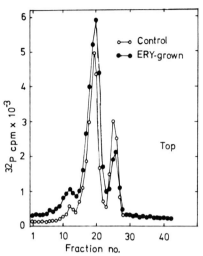

Fig. 7. Cells of strain D273-10B were grown in a P_i-deficient medium containing 1 mCi [^{32}P]P_i in the presence or absence of 4 mg/ml erythromycin. Mitochondrial ribosomes were prepared and analysed on an isokinetic sucrose gradient containing 15-25 % sucrose (w/w), 500 mM NH_4Cl, 10 mM Tris-Cl (pH 7.5)
and 10 mM $MgCl_2$.

Is CAP inhibition uniform ?

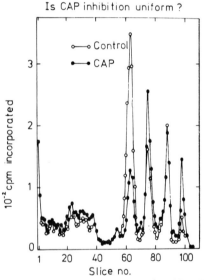

Fig. 8. Yeast cells of strain D273-10B were grown as described before. One aliquot (4 g wet weight) was labelled with 25 μCi [^{14}C]leucine in the presence of
100 μg/ml cycloheximide. Another aliquot (4 g wet weight) was labelled with 0.5
mCi [^{3}H]leucine in the presence of 100 μg/ml cycloheximide and 2 mg/ml chloramphenicol. This amount of chloramphenicol inhibits cycloheximide-resistant protein synthesis by 75 % in this strain. After the incorporation of labelled leucine had been stopped, the cells were mixed, mitochondria were prepared and membranes were analysed on 10 % acrylamide gels containing SDS.

ding to a molecular weight of 35,000.

Fig. 8 shows that this is not the case. On the contrary, the synthesis of material with a molecular weight of 35,000 is relatively depressed. As a conclusion I would like to say that it is most likely that all ribosomal proteins necessary for a functional mitochondrial ribosome are synthesized on cell-sap ribosomes. They have then to be imported into the mitochondria, where they can form functional ribosomes in combination with the rRNA transcribed from mtDNA. The nature of the protein synthesized on mitochondrial ribosomes and sticking to 38S subunits, is at the moment not clear. It is clearly not necessary for ribosomal function as far as we can measure. We could consider the possibility that it is a fractional protein or an initiation factor, which must then be present in vast excess to the amount of ribosomes. From the data known about fractional proteins in bacterial ribosomes this seems unlikely. An alternative possibility is that this protein forms a link between the ribosome and the mitochondrial membrane. The mitochondrial ribosome would then clearly be different from other membrane-bound ribosomes which are linked to the membrane via the large subunit. Finally, our results imply that all mitochondrial mutants that are resistant against antibiotics acting on the 50S subunit, are mutants where the rRNA is changed (excluding the unlikely possibility that messenger RNA transcribed from mtDNA is exported and translated on cell-sap ribosomes). Grivell and co-workers have analysed some of these mutants and indeed found no differences in the proteins from normal and mutant ribosomes (12).

ACKNOWLEDGEMENTS

I am indebted to Dr. L.Grivell for his help and advice in the isolation of ribosomes and in the study of protein synthesis by isolated mitochondria. This work was supported in part by a grant from The Netherlands Foundation for Chemical Research (S.O.N.) with financial aid from The Netherlands Organization for the Advancement of Pure Research (Z.W.O.) to P.Borst.

451

REFERENCES

1. P. Borst, Ann.Rev.Biochem. 41, 333 (1972).
2. P. Borst and L.A. Grivell, FEBS Letters 13, 73 (1971).
3. L.A. Grivell, L. Reijnders and P. Borst, Biochim. Biophys.Acta 247, 91 (1971).
4. L.A. Grivell, P. Netter, P. Borst and P.P. Slonimski, Biochim.Biophys.Acta 312, 358 (1973).
5. H. Küntzel, Nature 222, 142 (1969).
6. H. Schmitt, FEBS Letters 26, 215 (1972).
7. P.M. Lizardi and D.J.L. Luck, J.Cell Biol. 54, 56 (1972).
8. G.S.P. Groot, W. Rouslin and G. Schatz, J.Biol. Chem. 247, 1735 (1972).
9. L. Kovác, G.S.P. Groot and E. Racker, Biochim. Biophys.Acta 256, 55 (1972).
10. P. Traub, S. Mizushima, C.V. Lowry and M. Nomura, in: Methods in Enzymology (S.P. Colowick and N.O. Kaplan, Eds.), Vol.XX, part C, Academic Press, New York, pp. 403 (1971).
11. L.A. Grivell, Biochem.J. 105, 44c (1967).
12. L.A. Grivell, this volume.

COOPERATION OF MITOCHONDRIAL AND CYTOPLASMIC PROTEIN SYNTHESIS IN THE FORMATION OF CYTOCHROME *C* OXIDASE

Walter Sebald, Werner Machleidt & Joachim Otto

Institut für Physiologische Chemie und Physikalische Biochemie der Universität, 8000 München 2, Goethestrasse 33, Federal Republic of Germany.

ASSEMBLY OF CYTOCHROME *C* OXIDASE IN THE PRESENCE OF CYCLOHEXIMIDE AND CHLORAMPHENICOL.

Cytochrome *c* oxidase from *Neurospora crassa* is resolved by SDS-gel electrophoresis into 7 protein components. The site of synthesis of these components was investigated by *in vivo* incorporation of radioactive amino acids under the action of inhibitors. In the presence of chloramphenicol (Fig. 1, ref. 1) the labelling of components 4, 5, 6 and 7 proceeds, whereas the labelling of components 1, 2 and 3 is strongly inhibited, as compared to the control label. In the presence of cycloheximide (Fig. 2, ref. 4) a labelling of only components 1, 2 and 3 is observed. This suggests a mitochondrial site of synthesis of the three large components, and a cytoplasmic site of synthesis of components 4 to 7. Similar results have been reported by Mason *et al.*(2) and recently by Tzagoloff *et al.* (3). It must be mentioned, however, that in *Neurospora* the greater part of the labelled polypeptides, synthesized in the presence of either cycloheximide or chloramphenicol, appeared in the isolated enzyme only after washing out the inhibitors and further growth of the cells under normal conditions. The labelling patterns presented in Fig's 1 and 2 were obtained in this way. Cytochrome *c* oxidase isolated immediately from the poisoned cells exhibits an incomplete labelling pattern. In the presence of chloramphenicol none of the components becomes significantly labelled (1), and in the presence of cycloheximide a major label is

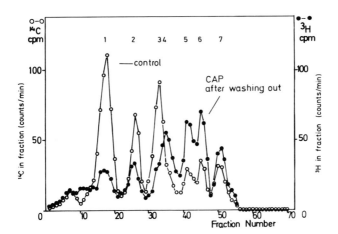

Fig. 1. Chloramphenicol-resistant labelling of cytochrome *c* oxidase (wash-out experiment). Hyphae grown for 17 h were labelled for 1 h with [^{14}C]leucine, [^{14}C]isoleucine and [^{14}C]phenylalanine each 0.025 mCi per ltr. Thereafter 4 mg chloramphenicol per ml was added and 5 min later [^{3}H]leucine, [^{3}H]isoleucine and [^{3}H]phenylalanine each 0.5 mCi per ltr. 30 min after adding the [^{3}H]labelled amino acids the cells were washed free of the chloramphenicol and incubated for another hour before isolation of cytochrome *c* oxidase. Gel electrophoresis was performed in the presence of SDS. (o) [^{14}C]control label; (●) [^{3}H]-radioactivity incorporated in the presence of chloramphenicol.

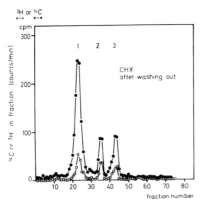

Fig. 2. Cycloheximide-resistant labelling of cytochrome *c* oxidase (wash-out experiment). Hyphae grown for 18 h were incubated for 2 min with 0.1 mg cycloheximide per ml. Then [^{14}C]leucine (0.05 mCi per ltr) was added and 60 min later [^{3}H]leucine (0.5 mCi per ltr). 5 min after application of the [^{3}H]amino acid the cells were washed with fresh culture medium containing 2 mM unlabelled leucine. The cells were aerated in the chase medium for another 4 h before isolation of cytochrome *c* oxidase. Gel electrophoresis was performed in the presence of SDS. Unspecific labelling after washing-out of the inhibitor can be calculated as described in ref. 4. (●) [^{14}C]radioactivity; (o) [^{3}H]radioactivity.

observed only in component 3 (Fig. 3A; ref 5). These incomplete labelling patterns are to expect, if both types of polypeptides - those of mitochondrial and those of cytoplasmic origin - are essential building stones of the complex enzyme. In this case, all protein components are necessary for the assembly. In the presence of the inhibitors the assembly ceases, if the precursor polypeptides, whose synthesis is inhibited, are exhausted. The pool sizes of these precursor polypeptides have been determined by Schwab *et al.* (6). And indeed, the labelling pattern observed in the presence of chloramphenicol is explained by the assumption, that the pool of component 3 is limiting, which has an half life of only 3.5 min. Whereas in the presence of cycloheximide the pool of component 7 may be limiting, which has an half life of 7 min.

Remarkably, cycloheximide-resistant labelling of whole cytochrome *c* oxidase is enlarged, if the cells have been previously incubated with chloramphenicol, and thereafter were washed free again from the inhibitor. After this treatment, which was first described with yeast by Tzagoloff (7), besides component 3 also components 1 and 2 become highly labelled in the presence of cycloheximide (Fig. 3B). During the chloramphenicol incubation the precursor polypeptides of cytoplasmic origin accumulate (1). Possibly, this allows a longer assembly of cytochrome *c* oxidase in the presence of cycloheximide, and thus, also components 1 and 2 become labelled, which have larger pools than component 3.

ISOLATION OF THE MITOCHONDRIAL TRANSLATION PRODUCTS OF CYTOCHROME *C* OXIDASE.

Components 1, 2 and 3 can be isolated by preparative SDS-gel electrophoresis (4). Preparations of cytochrome *c* oxidase were used containing 14-15 nmoles heme *a* per mg protein. Such a preparation is shown in Fig. 4, trace a, after electrophoretic separation and staining with coomassie blue. As can be seen, minor bands are still present. These are most probably contaminations, because they are more prominent in prepa-

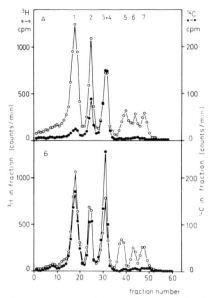

Fig. 3. Change of uncomplete cycloheximide-resistant labelling pattern of cyto-chrome c oxidase after stimulation of mitochondrial protein synthesis. Hyphae grown for 17 h in a 2 ltr culture were labelled for 1 h with [3H]leucine, [3H]-isoleucine and [3H]phenylalanine each 0.25 mCi per ltr. Thereafter the culture was divided into equal portions. One half was incubated with 4 mg chloramphenicol per ml for 30 min. Then both cultures were washed separately with fresh culture medium, and aerated in 1 ltr medium for another 15 min. To both cultures 0.1 mg cycloheximide per ml was added, and 2 min later [14C]leucine, [14C]isoleucine and [14C]phenylalanine each 0.025 mCi per ltr. After 60 min the cells were mixed with ice and cytochrome c oxidase was isolated and separated by SDS-gel electro-phoresis. (o) [3H]control label; A: (●) [14C]radioactivity incorporated in the presence of cycloheximide; B: (●) [14C]radioactivity incorporated in the presence of cycloheximide after a transitory incubation with chloramphenicol.

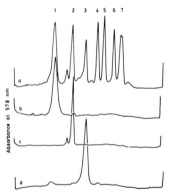

Fig. 4. Densitometric tracings of (a) cytochrome c oxidase and isolated compo-nents 1 (b), 2 (c) and 3 (d) after SDS-gel electrophoresis and coomassie brilliant blue staining.

456

rations of lower purity. The isolated components (Fig. 4, traces b, c and d) show only a minor contamination in the case of component 2. A small amount of protein of higher molecular weight is observed with component 3. This band seems to be an aggregate, because it vanishes nearly completely after preincubation at alkaline pH.

TYPICAL PROPERTIES OF THE MITOCHONDRIAL TRANSLATION PRODUCTS IN *NEUROSPORA CRASSA*.

Now one may ask, how pure are the isolated components ? It can not yet be decided, if each component represents only one polypeptide chain of mitochondrial origin. But it can be checked, if these components are contaminated with polypeptides of cytoplasmic origin. Fortunately, the mitochondrial translation products exhibit typical properties, by which they can be clearly distinguished from the mitochondrial membrane proteins with a cytoplasmic site of synthesis. This is demonstrated in the experiments compiled in Fig. 5 A-F. Mitochondrial membrane proteins were labelled by *in vivo* incorporation of [^{3}H]leucine in the presence of chloramphenicol. Thereafter the inhibitor was washed out, and [^{14}C]leucine was incorporated in the presence of cycloheximide. The cycloheximide-resistant label appeared mainly in three fractions with apparent molecular weights of about 40,000, 30,000 and 20,000 (4). These fractions were separated by preparative SDS-gel electrophoresis. The experiments shown in Fig. 5 are concerned only with the fraction of the molecular weight of about 40,000. But similar results were obtained with the other fractions. After electrophoresis on a 10 % gel (Fig. 5A) both labels are found in nearly the same positions. However, when the electrophoresis was performed on a 15 % SDS-gel (Fig. 5B) the two labels appear to be separated. The electrophoretic mobility of the mitochondrial translation product is reduced. The apparent molecular weight seems to be enlarged by about 10 %. A still better separation of the two types of polypeptides is observed, when gel electrophoresis is performed in a formic acid medium (4).

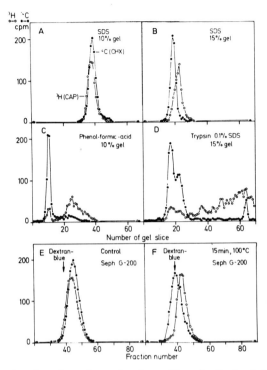

Fig. 5. Comparison of polypeptides of mitochondrial and of cytoplasmic site of synthesis. The polypeptide fraction with an apparent molecular weight of about 40,000 was separated from whole mitochondrial membrane proteins by preparative SDS-gel electrophoresis. The experiments are described in the text.

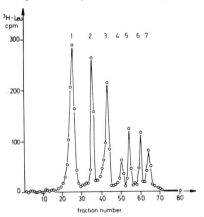

Fig. 6. Distribution of homogeneous [³H]leucine label among the protein components of cytochrome *c* oxidase. Cytochrome *c* oxidase was labelled with [³H]leucine by growing cells of a 15 h culture for 3 h in the presence of the radioactive amino acid. The enzyme protein was separated by SDS-gel electrophoresis.

In this acidic system, the mitochondrial translation products migrate only very slowly (Fig. 5C). In another experiment (Fig. 5D) the fraction was digested with trypsin (0.7 mg per ml) in the presence of 0.1 % SDS. As seen after SDS-gel electrophoresis, the polypeptides of cytoplasmic origin have been completely degraded. The mitochondrial translation product, however, is digested only slowly. A large amount of the original protein is still present, besides two quite defined peptide fractions of high and low molecular weight. Possibly, this indicates, the the lysine and arginine residues are clustered at the end(s) of the polypeptide chains. When the fraction is submitted to chromatography on Sephadex G-200 in the presence of 0.5 % SDS, the mitochondrial translation product is eluted from the column slightly behind the other proteins (Fig. 5E). However, when the fraction has been heated in the presence of 2 % SDS, the mitochondrial translation product emerges with the void volume (Fig. 5F). It has been aggregated by the heat treatment, while the other proteins are no influenced. Similar experiments, as shown in Fig. 5 A-F, have been performed with isolated components 1, 2 and 3 of cytochrome *c* oxidase (4,8). The components completely exhibited the properties of a mitochondrial translation product. No material could be detected, that behaved as the polypeptides of cytoplasmic origin. Hence, it is concluded that these components are not seriously contaminated.

AMINO ACID COMPOSITION OF COMPONENTS 1, 2 AND 3 AND OF WHOLE CYTOCHROME *C* OXIDASE PROTEIN.

The amino acid composition of cytochrome *c* oxidase from *Neurospora* (Table I) is very similar to that of the beef heart enzyme (9). Only in the case of threonine and isoleucine major differences are observed. The whole protein of the *Neurospora* enzyme has a polarity of 39 %. According to Capaldi and Vanderkooi (10), a polarity below 40 % is typical for an intrinsic membrane protein, which can be solubilized only by detergents. The amino acid composition of components

TABLE I.

AMINO ACID COMPOSITION OF CYTOCHROME OXIDASE AND COMPONENTS 1, 2 AND 3.

Amino acid	Amount in					
	cytochrome c oxidase from		components (*Neurospora*)			
	beef[a] heart	*Neurospora*	1	2	3	4-7[b]
Aspartic acid	7.36	7.67	6.45	8.55	7.21	8.32
Threonine	7.92	5.10	4.80	4.15	6.51	5.24
Serine	7.78	7.39	10.09	9.36	7.55	4.47
Glutamic acid	7.56	8.45	4.28	9.80	6.13	11.71
Proline	6.97	6.63	6.84	7.33	4.69	6.86
Glycine	7.85	8.56	10.20	6.81	9.56	7.80
Alanine	8.22	8.20	7.53	4.90	7.46	10.45
Cysteine	1.03	n.d.	n.d.	n.d.	n.d.	n.d.
Valine	6.75	6.28	7.14	7.71	6.38	4.97
Methionine	2.05	2.84	1.27	2.05	1.28	4.91
Isoleucine	5.65	7.29	8.72	9.16	9.41	4.63
Leucine	11.37	10.64	13.37	11.68	12.59	7.47
Tyrosine	4.26	4.45	4.18	4.89	3.65	4.74
Phenylalanine	6.24	6.59	8.25	5.38	9.15	5.00
Lysine	4.04	3.26	1.51	2.50	1.24	5.61
Histidine	2.79	2.97	2.56	2.64	4.62	2.82
Arginine	2.94	3.67	2.81	3.07	2.56	4.97
Tryptophan	4.04	n.d.	n.d.	n.d.	n.d.	n.d.
NH_3	9.17	n.d.	n.d.	n.d.	n.d.	n.d.
Total	105.1	100.02	100.0	99.98	99.99	99.97
Polarity[c]	40.39	38.51	32.5	40.07	35.82	43.14

a) From reference (9). The mean values of two preparations were calculated.
b) Calculated from the amino acid composition of components 1, 2 and 3 and of cytochrome c oxidase on the basis of the contributions of components 1 (28.6 %), 2 (17.9 %) and 3 (14.2 %) to the total amino acids of cytochrome c oxidase (see table II).
c) Calculated as in reference (10). Tryptophan and cysteine, which had not been determined, were not included in the calculation of these values.

1, 2 and 3 (Table I) exhibits large differences in the contents of single amino acids, *e.g.* in the case of glutamic acid, phenylalanine and histidine. Common to

the three mitochondrial translation products is the
high content of hycrophobic amino acids and the low
content of basic amino acids, especially of lysine.
The polarity of component 2 (40 %) is similar to that
of the whole enzyme protein, however, the polarities
of components 1 (33 %) and 3 (36 %) are much lower.
In the last column of Table I, a calculated mean amino acid composition of components 4-7 is shown (4).
These proteins have in average a polarity of 43 %.
This value corresponds to that of soluble proteins,
and it is higher than that of whole cytochrome *c* oxidase. From the amino acid compositions it may be concluded, that cytochrome *c* oxidase consists of two types of proteins: of the hydrophobic components 1, 2
and 3, which are provided by mitochondrial protein
synthesis, and of the polar components 4-7, which are
contributed by cytoplasmic protein synthesis. One may
speculate, that the hydrophobic components of the enzyme are embedded into the inner mitochondrial membrane, while the polar components are protruding into the
matrix or interchristae-space.

QUANTITATIVE SUBUNIT COMPOSITION OF CYTOCHROME *C* OXI-DASE.

By means of SDS-gel electrophoresis, it became possible for the first time to clear up qualitatively the
subunit composition of cytochrome *c* oxidase. And one
may ask, if the electrophoretic patterns obtained,
could be evaluated also quantitatively. In the literature, the amounts of protein present in the individual components have been tentatively estimated from the
distribution of coomassie blue stain (11) or uniform
[^{3}H]leucine label (12). Both approaches would lead to
erroneous results with the *Neurospora* enzyme, because
the leucine contents of components 1, 2 and 3 are
nearly twice as high, as the average content of components 4-7 (Table I), whereas the coomassie blue stains
components 4-7 in average twice as strongly as the
components 1, 2 and 3 (4). Within each of these groups
leucine contents and binding capacity for the dye seem
to be more uniform. An SDS-gel of a cytochrome *c* oxi-

TABLE II.

CONTRIBUTION OF COMPONENTS 1, 2 AND 3 TO THE TOTAL PROTEIN OF CYTOCHROME *C* OXIDASE.
The individual contributions of components 1-7 to the total [^{3}H]leucine of cytochrome *c* oxidase were evaluated from electrophoretic separations as shown in Fig. 6. The specific radioactivity of leucine (counts per min per nMol leucine) was nearly the same in whole cytochrome *c* oxidase and components 1, 2 and 3 (4). The leucine contents (μMol per mg protein) of cytochrome *c* oxidase and components 1, 2 and 3 were 0.97, 1.25, 1.05 and 1.16 respectively (from Table I). Tryptophan and cysteine, which had not been determined, were not included in the calculation of these values.

Component	[^{3}H]leucine % of total	Protein % of total	Protein per mol. weight of 150 000	Mol. weights determined on 15 % SDS-gels (4)
1	35.9	27.9	42 000	41 000
2	19.6	19.0	27 000	28 500
3	16.9	14.2	21 500	21 000
4	4.9			
5	8.0			
6	7.1	39.9	60 000	51 500[a]
7	7.5			

a) sum of components 4-7.

dase, labelled homogenously with [^{3}H]leucine, is shown in Fig. 6. The components are completely separated. Hence, the amounts of [^{3}H]leucine present in each component can be precisely determined. The relative amounts obtained are presented in Table II. The protein contents were calculated from these values taking into account the individual leucine contents determined by amino acid analysis. Components 1, 2 and 3 contain together 60 % of the whole enzyme protein. Remarkably, the relative amounts of the individual components occur in nearly the same proportions as the molecular weights determined on the SDS-gels. This strongly suggests, that components 1, 2 and 3 are present in equimolar amounts.

On the basis of the molecular weights of components 1, 2 and 3 and on the individual protein contents, it is concluded, that the minimal molecular weight of whole cytochrome *c* oxidase is 150,000 daltons. The cytochrome *c* oxidase preparations used in the present experiments contained 14-15 nmoles heme *a* per mg protein. This corresponds to a molecular weight of about 70,000 per heme group. This supports the view, that the smallest structural unit of cytochrome *c* oxidase contains two heme groups.

REFERENCES.

1. W. Sebald, H. Weiss and G. Jackl, Eur.J.Biochem. 30, 413 (1972).
2. T.L. Mason and G. Schatz, J.Biol.Chem. 248, 1455, (1973).
3. A. Tzagoloff, M.S. Rubin and M.F. Sierra, Biochim. Biophys.Acta 301, 71 (1973).
4. W. Sebald, W. Machleidt and J. Otto, Eur.J.Biochem. in press (1973).
5. H. Weiss, W. Sebald and Th. Bücher, Eur.J.Biochem. 22, 19 (1971).
6. A.J. Schwab, A.J., W. Sebald and H. Weiss, Eur.J. Biochem. 30, 511 (1972).
7. A. Tzagoloff, J.Biol.Chem. 246, 3050 (1971).
8. W. Sebald and G. Jackl, in "Abstr.Commun.Meet.Fed. Eur.Biochem.Soc. 8", abstr. 659 (1972).
9. M. Kuboyama, F.C. Young and T.E. King, J.Biol. Chem. 247, 6375 (1972).
10. R.A. Capaldi and G. Vanderkooi, Proc.Nat.Acad.Sci. U.S.A. 69, 930 (1972).
11. J.J. Keirns, C.S. Yang and M.V. Gilmour, Biochem. Biophys.Res.Commun. 45, 835 (1971).
12. T.L. Mason, R.O. Poyton, D.C. Wharton and G. Schatz, J.Biol.Chem. 248, 1346 (1973).

STUDIES ON THE CONTROL OF MITOCHONDRIAL PROTEIN SYNTHESIS IN YEAST

Diana S. Beattie, Leu-Fen H. Lin and Robert N. Stuchell

Department of Biochemistry, Mount Sinai School of Medicine of the City University of New York, New York, N.Y. 10029, U.S.A.

The process of mitochondrial assembly requires that proteins synthesized on cytoplasmic ribosomes be integrated into the mitochondrial membrane in conjection with proteins synthesized on the unique mitochondrial ribosomes. Under normal growth conditions, membrane formation proceeds in an orderly manner such that a constant level of different enzymes is maintained indicating that a mechanism may exist to control the synthesis of mitochondrial proteins at the two different intracellular sites. Such controls have been proposed for protein synthesis in both the cytoplasm and the mitochondria. For example, addition of inhibitors of mitochondrial transcription and translation to cultures of *Neurospora* stimulated the synthesis of certain mitochondrial enzymes in the cytoplasm suggesting that mitochondrial gene products might act as repressors (1). Likewise, the synthesis of certain mitochondrial proteins was stimulated by products of cytoplasmic protein synthesis (2,3), while the absence of these proteins resulted in a decrease in the rate of mitochondrial protein synthesis (3,4).

Previously, the kinetics of appearance of enzymatic activity of different segments of the respiratory chain was studied in yeast cells undergoing glucose derepression (5). The increase in activity of both NADH-cytochrome c reductase and cytochrome c oxidase in repressed cells was inhibited completely by chloramphenicol, but was not affected for several hours af-

ter addition of cycloheximide. These results suggest
that the synthesis of mitochondrial proteins on cytoplasmic ribosomes may precede that on mitochondrial
ribosomes and that these proteins may accumulate in
the mitochondria. This observation lead to the present study in which the possible control of mitochondrial protein synthesis by the accumulation of mitochondrial proteins previously synthesized in the cytoplasm has been investigated in yeast. The results obtained indicate that the rate of mitochondrial protein
synthesis, measured both *in vitro* and *in vivo* was stimulated under these conditions. In addition, analysis
by sodium dodecyl sulfate (SDS) polyacrylamide gel
electrophoresis revealed significant differences in
the proteins labelled by mitochondria *in vivo* when the
cells were grown in inhibitors of protein synthesis.
Activity of different mitochondrial enzyme complexes
was also examined to determine whether the observed
stimulation of mitochondrial protein synthesis was reflected in increased activity of any segment of the
respiratory chain.

METHODS.

Yeast cells were grown for either 9 or 13 h in 5 %
glucose (5) at which time the culture was divided into
three equal parts. To one part was added chloramphenicol (4 mg/ml), to the second was added cycloheximide
(20 µg/ml), while the third was used as the control.
After another three hours of growth, all three cultures were harvested, washed two times and transferred
to an equal volume of fresh medium containing 0.8 %
glucose and allowed to grow for another one to three
hours. In some experiments, the cells grown in chloramphenicol were transferred to fresh medium which contained cycloheximide and allowed to grow for one to
three hours. Cycloheximide was also present in the medium during the harvesting and washes of the cells for
these experiments. Yeast mitochondria were prepared
in 0.25 M mannitol (5).

Amino acid incorporation by isolated yeast mitochondria was measured in the presence of either ATP

TABLE I.

EFFECT OF GROWTH IN INHIBITORS ON AMINO ACID INCORPORATION.
Cells were grown for 3 h (I), then washed and grown for 1 h in fresh medium (II).
Mitochondria were isolated and incubated with [^{14}C]leucine (3). CAP: chloramphenicol; CHI: cycloheximide.

Addition to medium		Protein Synthesis			
I	II	ATP-generating		ATP-succinate	
		cpm/mg	% change	cpm/mg	% change
None	None	837	–	3510	–
CAP	None	1200	+143	5150	+146
CHI	None	583	– 30	2510	– 29
CAP	CHI	1770	+211	7040	+200

and a regenerating system, consisting of phosphoenol-
pyruvate and pyruvate kinase, or ATP and succinate
(3). Yeast cells were labelled *in vivo* with [^{3}H]leuci-
ne in the presence of cycloheximide (100 µg/ml). The
mitochondrial membranes were then isolated and prepared for SDS-polyacrylamide gel electrophoresis as previously described (3). Enzyme assays in isolated mitochondria were performed as described by Kim and
Beattie (5).

RESULTS AND DISCUSSION.

In a previous study (3), we reported that the rate
of amino acid incorporation *in vitro* was greatly decreased in mitochondria obtained from yeast cells
grown in either chloramphenicol or cycloheximide. One
explanation for the 50-60 % decrease observed after
growth in chloramphenicol is that some of the drug is
still present in the mitochondria bound to the ribosome despite the extensive washings of the cells and mitochondrial pellet. After growth for one hour in fresh
medium, the rate of incorporation *in vitro* measured in
either the ATP-regenerating or ATP-succinate systems
was increased nearly 50 % in mitochondria isolated
from cells which had been preincubated in chloramphenicol (Table I). A similar stimulation was observed
when mitochondria were labelled *in vivo* in the presence of cycloheximide. These results suggest that the
products of cytoplasmic protein synthesis which had
accumulated during growth in chloramphenicol may act
to stimulate the actual rate of mitochondrial protein
synthesis. In contrast, the rate of amino acid incor-

poration was decreased approximately 30 % in mitochondria obtained from cells which had been preincubated in cycloheximide and then allowed to grow for one hour in fresh medium. Under these conditions, the pool of proteins synthesized on cytoplasmic ribosomes would have been severely depleted by growth in cycloheximide. An even greater stimulation of amino acid incorporation *in vitro* was observed in mitochondria isolated from yeast cells which were grown for three hours in chloramphenicol and then allowed to grow for one hours in medium containing cycloheximide. When the cells continued to grow for three hour in cycloheximide, however, the isolated mitochondria had greatly lowered rates of amino acid incorporation. Thus, the proteins which are synthesized on cytoplasmic ribosomes and are, perhaps, destined to be integrated into the mitochondrial membrane may control mitochondrial protein synthesis in several significant ways. The rate of mitochondrial protein synthesis is not only optimized by their presence but also stimulated by their accummulation. Furthermore, some protein synthesized in the cytoplasm may place additional restraints on mitochondrial protein synthesis since blocking the further synthesis of cytoplasmic proteins resulted in even greater rates of amino acid incorporation.

The products of mitochondrial protein synthesis in yeast grown in the presence of inhibitor as described above were analyzed by SDS-polyacrylamide gel electrophoresis. As seen in Fig. 1, the stimulation of mitochondrial protein synthesis observed both *in vitro* and *in vivo* after growth in chloramphenicol was not reflected in an increased synthesis of all the labelled proteins observed after gel electrophoresis. The labelling in the four peaks of molecular weights 56000 to 31000 were significantly lower than that of the control, while the counts in the peak of molecular weight 18000 were stimulated nearly 2-fold in mitochondrial membranes obtained from cells grown first in chloramphenicol and then allowed to grow for one hour in fresh medium. A completely different pattern was observed in mitochondrial membranes obtained from

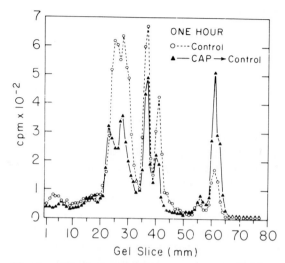

Fig. 1. SDS-gel electrophoresis profiles of mitochondrial membranes obtained from yeast cells grown in chloramphenicol for 3 h, washed and transferred to fresh medium containing 0.1 % glucose and allowed to grow for 1 h (▲—▲). Control cells incubated for the same time without chloramphenicol (0----0).

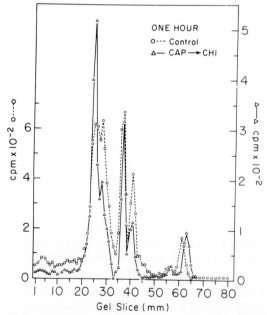

Fig. 2. SDS-gel electrophoresis profiles of mitochondrial membranes obtained from yeast cells grown in chloramphenicol for 3 h, washed and transferred to fresh medium containing cycloheximide (Δ—Δ). Control cells incubated for the same times without drugs (0----0).

cells grown first in chloramphenicol and then in cycloheximide for one hour (Fig. 2). No change in the
amount of radioactivity in the two lowest molecular
weight peaks of that of molecular weight 37000 was observed; however, labelling of the peak of molecular
weight 56000 was increased 2-fold and those of 48000
and 31000 were lowered 50 %.

Activity of the various enzymes and enzyme complexes of the respiratory chain was measured in yeast
cells previously grown in inhibitors to determine
whether the stimulation of mitochondrial protein synthesis observed both quantitatively and qualitatively was reflected in a change of activity in any part
of the respiratory chain. Previous studies (5-8) had
indicated that both cytochrome c oxidase and the cytochrome b-c_1 region of the respiratory chain, measured as either NADH-cytochrome c or coenzyme QH_2-cyto-
chrome c reductases (9,10) contain products of both
the mitochondrial and cytoplasmic systems for protein
synthesis. The inhibitor experiments described in this
study were performed at a time during growth in 5 %
glucose when the activity of both these enzyme complexes is rapidly increasing as the cells undergo glucose derepression. As seen in Fig. 3, growth for three
hours in chloramphenicol resulted in a decrease in the
antimycin-sensitive mitochondrial NADH-cytochrome c
reductase activity compared to the control activity
which had nearly doubled during that time. These results suggest that formation of some part of the respiratory chain between NADH and cytochrome c requires
products of mitochondrial protein synthesis, perhaps
the membrane fraction containing cytochrome b described by Weiss (7). After the cells were removed from the
medium containing chloramphenicol, this enzymic activity increased after an initial lag at a similar rate
as the control. During growth in the fresh medium,
however, the increase in enzymic activity of the
chloramphenicol-pretreated cells was at no time comparable to the nontreated control cells despite the
stimulation of mitochondrial protein synthesis under
the former growth conditions. In contrast, when the

cells were grown sequentially in chloramphenicol followed for one hour in cycloheximide, NADH-cytochrome c reductase activity was nearly doubled compared to the activity of cells grown first in chloramphenicol and then allowed to grow in fresh medium without any inhibitor. The NADH-cytochrome c reductase activity did not increase further when the cells were allowed to grow in cycloheximide longer than one hour. As mentioned previously the rate of mitochondrial protein synthesis had decreased to a low rate when cells were grown in cycloheximide for 3 h.

NADH-cytochrome c reductase activity involves both complexes I and III of the respiratory chain, i.e. NADH-coenzyme Q and coenzyme QH_2-cytochrome c reductases. Both of these enzymes were assayed in the cells grown in inhibitors as described above. The addition of chloramphenicol to the medium had no effect on the increase in NADH-coenzyme Q reductase activity (Fig. 4). In addition, after the cells were transferred into cycloheximide the activity of this enzyme complex did not increase any further suggesting that Complex I contains products of cytoplasmic protein synthesis but does not contain proteins synthesized on the chloramphenicol-sensitive mitochondrial ribosomes. Formation of complex III of the respiratory chain, however, does require products of mitochondrial protein synthesis, since growth for three hours in chloramphenicol resulted in a decrease in coenzyme QH_2-cytochrome c activity as compared to the control cells. Furthermore, transfer of the cells into medium containing cycloheximide for one hour after the previous growth in chloramphenicol resulted in a doubling of activity compared to that observed in cells grown in chloramphenicol and then allowed to grow in fresh medium without inhibitors. The activity of this enzyme complex also did not increase when the cells were in medium containing cycloheximide for more than one hour. Thus, it is apparent that the response of NADH-cytochrome c reductase activity to growth in inhibitors is a result of changes of the control of synthesis in the cytochrome b-c_1 region in the respiratory

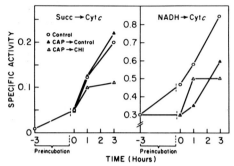

Fig. 3. Specific activities in μmoles cytochrome *c* reduced per min per mg of antimycin-sensitive NADH and succinate-cytochrome *c* reductases in mitochondria obtained from yeast cells grown for 3 h in chloramphenicol, washed and transferred to fresh medium containing 0.1 % glucose (▲—▲) or glucose plus cycloheximide (△—△). Control cells were grown for the same times without drugs (O----O).

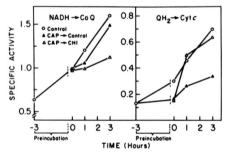

Fig. 4. Specific activities in μmoles substrate reduced or oxidized per min per mg of NADH-coenzyme Q and coenzyme QH₂-cytochrome *c* reductases in mitochondria obtained from yeast cells grown as described in legend to Fig. 3.

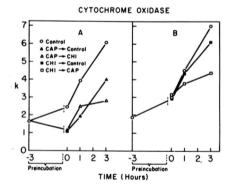

Fig. 5. Specific activity of cytochrome *c* oxidase expressed as the initial first order rate constant (*k*) per min mg in the mitochondrial suspension obtained from yeast cells grown as described in legend to Fig. 3. Cells were also grown for 3 h in cycloheximide, then washed and transferred to medium with (◻—◻) or without (◼—◼) chloramphenicol.

chain.

A surprising observation was the completely different response to inhibitors of succinate-cytochrome c reductase, which transports electrons through cytochromes b and c_1 in common with NADH-cytochrome c reductase. Succinate-cytochrome c reductase activity was not decreased in cells treated with chloramphenicol nor did subsequent growth in cycloheximide result in any increase in enzymic activity compared to either control (Fig. 3). One explanation for this different response is the significantly lower activity of succinate-cytochrome c reductase compared to that of NADH-cytochrome c reductase. The activity of succinate dehydrogenase in yeast mitochondria is also much lower than that of NADH dehydrogenase (5). Hence, the absolute amount of cytochromes b and c_1, and the consequent rate of electron transport through this segment of the respiratory chain may not be rate-limiting in the assay of succinate-cytochrome c reductase activity. It should be noted in this context that the increase in coenzyme QH_2-cytochrome c reductase activity observed during glucose derepression is absolutely dependent on proteins synthesized both in the mitochondria and in the cytoplasm (9,10).

The response of cytochrome c oxidase to growth in inhibitors was also studied (Fig. 5). As anticipated from the results of earlier studies (6,8,11), there was no increase in this activity when the cells were grown in the presence of chloramphenicol. The enzymic activity also increased at a rate similar to that of the control when the cells were transferred to fresh medium without chloramphenicol. Furthermore, after the cells were transferred to cycloheximide-containing medium, the enzymic activity increased at an identical rate as the control for one hour. Subsequently, no further increases were observed suggesting that sufficient proteins were synthesized in the cytoplasm during growth in chloramphenicol to result in an increase in cytochrome c oxidase activity during this time interval. No stimulation of cytochrome c oxidase compared to the control was observed. Fig. 5 also shows the effects of preincubation in cycloheximide followed by transfer to fresh medium with or without chloramphenicol. Very little accumulation of protein synthesized in the mitochondria was observed.

CONCLUSIONS.

Mitochondrial proteins synthesized in the cytoplasm may accumulate when yeast cells undergoing glucose de-repression are grown in medium containing chloramphenicol. A significant increase in the rate of mitochondrial protein synthesis, measured both *in vitro* and *in vivo*, was observed when the cells were allowed to grow in fresh medium, with or without cycloheximide, after the preincubation in chloramphenicol. SDS-gel electrophoresis revealed significant differences in the labelling patterns of mitochondrial membrane proteins obtained either from control cells, from cells preincubated in chloramphenicol or from cells transferred to medium containing cycloheximide after the preincubation in chloramphenicol. Under the latter growth conditions, complex III of the respiratory chain, measured as either coenzyme QH_2 of NADH-cytochrome *c* reductase was the only enzymatic activity to increase. No change in cytochrome *c* oxidase or NADH dehydrogenase activities was observed suggesting that the mitochondrial synthesis of proteins in the cytochrome $b-c_1$ region of the respiratory chain is subject to control by proteins synthesized in the cytoplasm.

REFERENCES.

1. Z. Barath and H. Küntzel, Proc.Nat.Acad.Sci.U.S.A. 69, 1371 (1972).
2. A. Tzagoloff, J.Biol.Chem. 246, 3050 (1971).
3. N.G. Ibrahim, R.N. Stuchell and D.S. Beattie, Eur. J.Biochem. in press (1973).
4. A.S.T. Millis and Y. Suyama, J.Biol.Chem. 247, 4063 (1972).
5. I.C. Kim and D.S. Beattie, Eur.J.Biochem. in press (1973).
6. H.R. Mahler and P.S. Perlman, Arch.Biochem.Biophys. 148, 115 (1972).

7. H. Weiss, Eur.J.Biochem. <u>30</u>, 469 (1972).
8. T.L. Mason and G. Schatz, J.Biol.Chem. <u>248</u>, 1355 (1973).
9. L.H. Lin, I.C. Kim and D.S. Beattie, manuscript in preparation.
10. M.S. Rubin and A. Tzagoloff, Fed.Proc. <u>32</u>, 641 Abs. (1973).
11. M.J. Vary, P.R. Stewart and A.W. Linnane, Arch. Biochem.Biophys. <u>141</u>, 430 (1970).

THE BIOSYNTHESIS OF MITOCHONDRIAL CYTOCHROMES

E.Ross, E.Ebner, R.O.Poyton, T.L.Mason, B.Ono and G. Schatz

Section of Biochemistry, Molecular and Cell Biology Cornell University, Ithaca, New York 14850, USA.

The biogenesis of mitochondria involves a concerted action of two distinct protein synthesizing systems: the cytoplasmic system and the mitochondrial system. In order to understand the process of mitochondrial assembly, one therefore has to answer the following questions:
1. Where is each mitochondrial protein synthesized ?
2. How are the two protein-synthesizing systems coordinated ?

We are attempting to answer these questions by studying the biosynthesis of mitochondrial cytochromes. These hemoproteins are important components of the mitochondrial inner membrane and exhibit well-defined functions. Since they possess typical absorption bands and, in some cases, easily measurable catalytic properties, they can be purified. Finally, the formation of cytochromes aa_3, b and c_1 requires both cytoplasmic and mitochondrial protein synthesis and thus resembles the biogenesis of mitochondria as a whole (1). This raises the possibility that the interplay between mitochondrial and cytoplasmic protein synthesis can be explored at the level of individual membrane enzymes.

This paper deals with the biosynthesis of cytochrome c oxidase (cytochrome aa_3) and cytochrome c_1 in *Saccharomyces cerevisiae*. In order to identify the site of synthesis, we have measured the incorporation of radioactive amino acids into these proteins in the presence of specific inhibitors which block either cytoplasmic or mitochondrial protein synthesis. In

order to investigate the <u>assembly</u> of these proteins
into the mitochondrial membranes, we have studied
yeast mutants in which the biogenesis of cytochrome
c oxidase was blocked by nuclear or extrachromosomal
mutations.

THE BIOSYNTHESIS OF CYTOCHROME *C* OXIDASE

Recent studies in three different laboratories have shown that cytochrome *c* oxidase from *Neurospora
crassa* (3,4) or *S.cerevisiae* (5,6,7) can be separated
into seven polypeptides by SDS-polyacrylamide gel
electrophoresis. Previous discrepancies concerning
the number of polypeptides (3,8) appear to be resolved; the relatively small disagreements about the molecular weights of the individual polypeptides are
almost certainly the result of different electrophoretic procedures (*cf*. ref. 6).

In our laboratory, cytochrome *c* oxidase was purified from submitochondrial particles of commercial
baker's yeast by solubilization with cholate, fractionation with ammonium sulfate, and chromatography
on DEAE-cellulose in the presence of Triton X-100 (6).

TABLE I
POLYPEPTIDE COMPOSITION OF CYTOCHROME *C* OXIDASE FROM *SACCHAROMYCES CEREVISIAE*

"Subunit" No.	Apparent molecular weight (*cf*. 6,10)	Site of synthesis (*cf*. 5,9)	Remarks (*cf*. 7,9,10,11)
I	42,000	mitochondria	Hydrophobic. Synthesis or integration prevented by lack of oxygen or by nuclear mutation
II	34,500	mitochondria	pet E11 and pet 1030.
III	23,000	mitochondria	Hydrophobic. Synthesis or integration prevented by nuclear mutations pet E11, pet 1030 and pet 494.
IV	14,000	cytoplasm	Hydrophilic
V	12,500	cytoplasm	Hydrophilic
VI	12,500	cytoplasm	Probably hydrophilic. Non-ideal behavior during SDS-acrylamide gel electrophoresis.
VII	4,500	cytoplasm	Probably hydrophilic. Not detectable in mitochondria from cytoplasmic petite mutants.

The purified enzyme contained approximately 10 nmoles of heme a per mg protein; its polypeptide composition is listed in Table I (first two vertical columns).All seven polypeptides are present in roughly equimolar amounts except for polypeptide I which seems to be present in excess. However, as pointed out earlier (6), this excess may reflect an artifact of our measurements.

At present we do not know whether all seven polypeptides are necessary for the catalytic activity of the solubilized enzyme. However, the following observations suggest that these polypeptides form a specific complex that may function in the oxidation of ferrocytochrome c *in vivo*.

1. Cytochrome c oxidase preparations prepared from different organisms by widely different procedures exhibit essentially identical polypeptide compositions (4-7, 12).
2. All seven polypeptides are specifically precipitated from a crude mitochondrial lysate by antisera against the holo-enzyme or against polypeptides VI + VII (6,10).
3. One or more of the three large polypeptides (I-III) are missing in nuclear yeast mutants which have specifically lost functional cytochrome c oxidase (*cf*. below).
4. Antisera against polypeptides VI + VII inhibit the enzymic activity of purified cytochrome c oxidase (10).
5. Preliminary experiments suggest that, in the native enzyme, heme a is bound to one or more of the four small polypeptides (IV - VII) (13).

Although by no means conclusive, 4. and 5. also suggest that the catalytic function of the enzyme is associated with the small subunits. The role of the large ones is still uncertain. As discussed below, the large polypeptides may bind the small ones to the mitochondrial inner membrane.

In order to study the biosynthesis of those polypeptides, yeast cells were labeled with [3H]leucine in the presence of cycloheximide (an inhibitor of cy-

toplasmic protein synthesis) or erythromycin (an inhibitor of mitochondrial protein synthesis). Labeled
cytochrome *c* oxidase components were then isolated
from crude mitochondrial extracts by immunoprecipita-
tion and analyzed by SDS-polyacrylamide gel electrophoresis. Labeling of polypeptides I, II and III was
insensitive to cycloheximide and sensitive to erythromycin. The four small components IV – VII were not
labeled in the presence of cycloheximide but became
labeled in the presence of erythromycin (5,9,13).

Our cytochrome *c* oxidase preparation thus contains
three polypeptides which are translated on mitochondrial ribosomes and four polypeptides which are translated on cytoplasmic ribosomes. Additional experiments
have shown that labeling of polypeptides I and II is
absolutely dependent on oxygen (5,9).

More recently, the individual polypeptides were
purified by dissociating the holoenzyme with guanidine hydrochloride or SDS and separating the polypeptides by gel filtration, fractional precipitation, or
ion-exchange chromatography in the presence of urea
(Fig. 1). These experiments revealed that at least
two (and probably all) of the cytoplasmically-synthesized polypeptides are quite hydrophilic whereas the
mitochondrially-synthesized polypeptides are quite
hydrophobic. Antibodies against the individual polypeptides have been prepared and are now used to probe the location of these subunits in the mitochondrial inner membrane as well as the assembly of individual subunits during the biogenesis of the holoenzyme.

THE BIOSYNTHESIS OF CYTOCHROME C_1

Cytochrome c_1 was isolated from yeast submitochondrial particles by a modification of the procedure
described by Yu *et al.* (14) for beef-heart. The yeast
particles were extracted with cholate and the solubilized succinate-cytochrome *c* reductase portion of the
respiratory chain was purified by repeated fractionation with ammonium sulfate. Succinate dehydrogenase
was then denatured by exposure to alkali and cytochrome c_1 was cleaved from cytochrome *b* by high concentra-

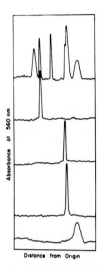

Fig. 1. Electrophoretic analysis of yeast cytochrome c oxidase and of isolated cytochrome c oxidase "subunits". The isolation of cytochrome c oxidase and the analysis by SDS-polyacrylamide gel electrophoresis were as described earlier (6). The isolation of individual subunits will be reported in detail elsewhere (10).

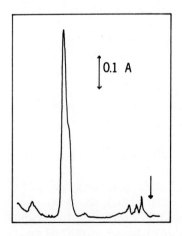

Fig. 2. Analysis of purified cytochrome c_1 from *Saccharomyces cerevisiae* by SDS-polyacrylamide gel electrophoresis. The isolation of cytochrome c_1 will be described in detail elsewhere (20). The arrow denotes the position of the tracking dye bromphenol blue. The gels (12 % acrylamide) were stained with Coomassie Brilliant Blue to visualize the protein band and scanned at 600 nm.

tions of mercaptoethanol in the presence of bile salt
and ammonium sulfate. Finally, cytochrome c_1 was purified by chromatography on DEAE-cellulose in the
presence of cholate. Preparations obtained by this
procedure were spectrally pure and contained up to
32 nmoles of heme c_1 per mg protein. The reduced cytochrome did not react with CO or molecular oxygen
and was, therefore, not grossly denatured. Since the
heme of cytochrome c_1 is covalently bound to the protein, the heme-carrying polypeptide could be unambiguously identified. Upon SDS-acrylamide gel electrophoresis, cytochrome c_1 yielded a single major protein band which coincided with the position of the
heme color. However, the band exhibited a shoulder
at the low molecular weight edge, regardless of whether protein or heme color was assayed. The extent of
this shoulder varied between different preparations
and could be diminished by raising the SDS concentration during the initial denaturation from 2.5 to
5 %. We suspect, therefore, that the partial splitting of this band reflects an artifact but further
evidence is needed to support this view.

The molecular weight of cytochrome c_1 as determined by analytical ultracentrifugation in the presence
of 0.5 % cholate was 27,000. SDS-polyacrylamide gel
electrophoresis in gels of varying porosity yielded
a value of 30,000 daltons. Assuming a mean value of
28,500 daltons and one heme group per polypeptide
chain one should expect a heme c_1-to-protein ratio of
35. This value is quite close to the experimentally
determined value of 32 (cf. above).

Initially, the DEAE-cellulose chromatography step
was carried out in the presence of Triton X-100 rather than in the presence of cholate. The enzyme prepared in the presence of Triton exhibited at most 27
nmoles of heme c_1 per mg protein and contained varying amounts of a heme-free polypeptide of molecular
weight 18,500. Since this smaller polypeptide is almost completely removed by DEAE-cellulose chromatography in the presence of cholate, it is probably a
contaminant.

In order to study the site of synthesis of cytochrome c_1, yeast cells were labeled with [^{3}H]leucine
in the absence and in the presence of specific inhibitors of cytoplasmic protein synthesis (Table II).

TABLE II

CYTOCHROME C_1 IS SYNTHESIZED ON CYTOPLASMIC RIBOSOMES.
Wild-type yeast cells (D273-10B; ATCC 24657) were labeled with [^{3}H]leucine in the
absence of antibiotics and in the presence of cycloheximide (100 μg/ml) or acri-
flavin (12.5 μg/ml) as described earlier (9) and the incorporation of radioactivity into submitochondrial particles (6) and purified cytochrome c_1 was determined
according to published procedures (15). Protein was determined by the method of
Lowry et al. (16).

Inhibitor present during labeling	Specific radioactivity (cpm/mg protein)	
	Submitochondrial particles	cytochrome c_1
None	3.7×10^5	1.3×10^4
Acriflavin	2.7×10^5	1.5×10^4
Cycloheximide	3.3×10^4	$< 10^2$ *

*below limit of detection.

Pulse-labeling of cytochrome c_1 was unaffected by a-
criflavin (a specific inhibitor of mitochondrial protein synthesis in yeast (15)) and completely blocked
by cycloheximide. The apoprotein of cytochrome c_1 is
therefore synthesized on cytoplasmic ribosomes.

Even though cytochrome c_1 is synthesized in the
cytoplasm, formation of the holo-cytochrome in growing cells requires mitochondrial protein synthesis
(1). This phenomenon could be explained by several
schemes, one of which is outlined in Fig. 3. According to this scheme, the apoprotein of cytochrome c_1
must first be combined with at least one mitochondrially-synthesized polypeptide ("polypeptide X") before it can accept its heme group. It is tempting to
speculate that "polypeptide X" is a component of cytochrome b; cytochrome b is not only tightly associated with cytochrome c_1 in vivo but also contains a
polypeptide made in mitochondria (17).

The simple scheme depicted in Fig. 3 invokes different "assembly lines" and a certain sequence of
assembly. This is reminiscent of the more intricate
"assembly lines" which operate during the morphogenesis of bacteriophages (18). Experiments with synchro-

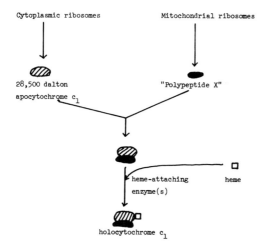

Cytoplasmic ribosomes   Mitochondrial ribosomes

28,500 dalton
apocytochrome c_1

"Polypeptide X"

heme-attaching   heme
enzyme(s)

holocytochrome c_1

Fig. 3. Hypothetical scheme outlining the biosynthesis of cytochrome c_1.

nized yeast cultures could perhaps help to shed more light on this point.

ANALYSIS OF CYTOCHROME C OXIDASE-LESS YEAST MUTANTS

In our attempts to study the regulation of mito-chondrial assembly, we decided to concentrate on in-dividual membrane polypeptides whose function is rea-sonably well established. We also attempted to avoid inhibitors as much as possible. As a first step, we investigated the effect of nuclear and extranuclear mutations on the assembly of the seven cytochrome c oxidase polypeptides. Eighteen respiratory-deficient yeast mutants were isolated from our wild-type strain D 273-10B by mutagenesis with ethylmethane sulfonate (19). Segregation analysis indicated that each mutant phenotype resulted from the mutation of a single nu-clear gene. All mutants except one efficiently re-tained mitochondrial DNA as measured by ability to complement mitochondrial DNA-less ρ^0 tester strains.

When the mutants were crossed pairwise and the re-sulting zygotes checked for functional respiration, the eighteen strains could be classified into seven complementation groups (19). Four of these exhibited a variety of pleiotropic lesions and will not be dis-cussed here. The remaining three groups were charac-

terized by a specific loss of cytochrome aa_3; all o-
ther mitochondrial functions tested (including oxidative phosphorylation in the span from succinate to
cytochrome c) were essentially normal (11,26).

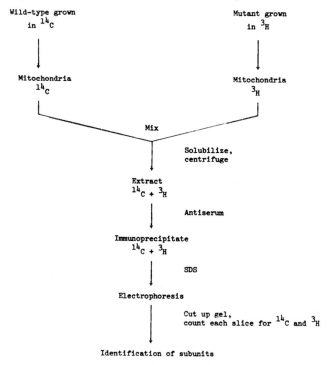

Fig. 4. Flow diagram of the radioimmunochemical method for identifying subunits
of cytochrome c oxidase in yeast mutants lacking the functional enzyme (11).

One member of each of those three complementation
groups (pet E11, 1030 and 494) was analyzed for residual cytochrome c oxidase polypeptides as outlined
in Fig. 4. It can be seen (Table III) that each of
the three nuclear mutants lacked at least one cytochrome c oxidase polypeptide which, in the wild-type,
is synthesized on mitochondrial ribosomes. All nuclear mutants possessed near-normal amounts of the
cytoplasmically-synthesized cytochrome c oxidase polypeptides.

The simplest explanation for these findings would
be that the mRNAs for the missing subunits are transcribed from a nuclear gene, but translated on mito-

TABLE III

ANALYSIS OF CYTOCHROME C OXIDASE POLYPEPTIDES IN CYTOCHROME C OXIDASE-LESS YEAST
MUTANTS

+ denotes that the polypeptide is present; − denotes that the polypeptide is undetectable by the procedure outlined in Fig. 4.

Yeast strain	Cytochrome c oxidase polypeptide					
	I	II	III	IV	V+VI*	VII
Wild-type	+	+	+	+	+	+
pet 494	+	+	−	+	+	+
pet E11	−	−	−	+	+	+
pet 1030	−	−	−	+	+	+
cytoplasmic petite mutant	−	−	−	+	+	−

*These two polypeptides are not adequately separated from each other by our rou-
tine electrophoretic procedure.

chondrial ribosomes. However, we favor the alternate
possibility that mutants pet E11, pet 1030 and pet
494 are "regulatory" or "assembly" mutants in which
the loss of cytochrome c oxidase polypeptides is the
secondary result of the loss or the alteration of
nuclearly-coded "organizer" proteins (26). This spe-
culation receives some support from the preliminary
observation that the pet 494 mutation can be suppres-
sed by a typical amber suppressor. Since it is highly unlikely that mitochondrial and cytoplasmic protein synthesis share a common suppressor activity,
these suppression studies suggest that the primary
effect of the pet 494 mutation involves an (as yet
unknown) protein which is made in the cytoplasm.

To complete these experiments, we also studied an
extrachromosomal petite mutant (D273-10B-1) which
lacks not only cytochrome aa_3 but cytochromes c_1 and
b as well (21). Since cytoplasmic petite mutants have
lost mitochondrial protein synthesis (22,23), it was
not surprising that the mutant lacked the three mitochondrially-synthesized cytochrome c oxidase polypeptides (Table III). However, the smallest of the cytoplasmically-synthesized polypeptides was also missing.
The remaining cytochrome c oxidase polypeptides were
only loosely bound to the mitochondria, and could be
easily detached by mild sonication. In this respect,
they differed sharply from the corresponding subunits

in wild-type mitochondria.

These data lead us to the following conclusions.

1. Polypeptide III appears to be an essential component of cytochrome c oxidase *in vivo*, since its absence in mutant pet 494 is paralleled by a specific loss of the functional enzyme.

2. One or more of the mitochondrially synthesized cytochrome c oxidase subunits are necessary for the tight binding of the cytoplasmically synthesized cytochrome c oxidase subunits to the mitochondrial inner membrane.

3. The synthesis (or the integration) of mitochondrially synthesized cytochrome c oxidase subunits can be prevented by nuclear mutations. Conversely, extrachromosomal mutations can impair or completely prevent the integration of cytoplasmically synthesized cytochrome c oxidase subunits.

On the basis of indirect evidence it was suggested earlier that mitochondrially-synthesized cytochrome c oxidase subunits regulate the assembly of cytoplasmically-synthesized subunits (2,4). The present results with the cytoplasmic petite mutant support this view.

CONCLUSION

Although these experiments are only a beginning, it is perhaps justified to draw some preliminary conclusions about the role of mitochondrial protein synthesis in the assembly of the mitochondrial inner membrane. Experiments on the biogenesis of cytochrome c oxidase (3-7,9,10) and the oligomycin-sensitive ATPase complex (7,24) indicate that mitochondria synthesize hydrophobic subunits of these membrane-bound oligomeric enzymes. The hydrophobic subunits are required for the correct assembly of the cytoplasmically-made "partner" proteins. Present evidence is consistent with the view that the mitochondrially-synthesized polypeptides are "integral" membrane proteins (25) which are synthesized close to their site of deposition. This may perhaps help to explain why mitochondria have retained their own protein synthesizing system.

ACKNOWLEDGEMENTS

This study was supported by Grants GM 16320 and GM 18197 from the U.S. Public Health Service, by a Helen Hay Whitney Fellowship to T.L. Mason, a NIH fellowship to R.O. Poyton and a NSF predoctoral fellowship to E. Ross.

REFERENCES

1. G.D. Clark-Walker and A.W. Linnane, J.Cell Biol. 34, 1 (1967).
2. W.L. Chen and F.C. Charalampous, J.Biol.Chem. 244, 2767 (1969).
3. H. Weiss, W. Sebald and T. Bücher, Europ.J.Biochem. 22, 19 (1971).
4. W. Sebald, H. Weiss and G. Jackl, Europ.J.Biochem. 30, 413 (1972).
5. T. Mason, E. Ebner, R.O. Poyton, J. Saltzgaber, D.C. Wharton, L. Mennucci and G. Schatz in: Mitochondria/Biomembranes (S.G. van den Bergh, P. Borst, L.L.M. van Deenen, J.C. Riemersma, E.C. Slater and J.M. Tager, Editors), North Holland, Amsterdam, p. 53, 1972.
6. T. Mason, R.O. Poyton, D.C. Wharton and G. Schatz, J.Biol.Chem. 248, 1346 (1973).
7. A. Tzagoloff, M.S. Rubin and M.F. Sierra, Biochim. Biophys.Acta 301, 71 (1973).
8. M.R. Rubin, Fed.Proc. 31, 3896 (1972).
9. T.L. Mason and G. Schatz, J.Biol.Chem. 248, 1355 (1973).
10. R.O. Poyton and G.Schatz, in preparation.
11. E. Ebner and G. Schatz, J.Biol.Chem. 248, September (1973).
12. H.Weiss, B. Lorenz and W. Kleinow, FEBS Letters 25, 49 (1972).
13. G. Schatz, G.S.P. Groot, T.L. Mason, W. Rouslin, D.C. Wharton and J. Saltzgaber, Fed. Proc. 31, 21 (1972).
14. C.A. Yu and T.E. King, J.Biol.Chem. 247, 1012 (1972).

15. G.S.P. Groot, W. Rouslin and G. Schatz, J.Biol. Chem. 247, 1735 (1972).
16. O.H. Lowry, N.J. Rosebrough, A.L. Farr and R.J. Randall, J.Biol.Chem. 193, 265 (1951).
17. H. Weiss, Europ.J.Biochem. 30, 469 (1972).
18. W.B. Wood, R.S. Edgar, J. King, I. Lielausis and M. Henninger, Fed. Proc. 27, 1160 (1968).
19. E. Ebner, L. Menucci and G. Schatz, J.Biol.Chem. 248, September (1973).
20. E. Ross and G. Schatz, in preparation.
21. P.P. Slonimski, La formation des enzymes respiratoires chez la levure, Masson, Paris 1953.
22. G. Schatz and J. Saltzgaber, Biochem.Biophys.Res. Commun. 37, 996 (1969).
23. S. Kuzela and E. Grecna, Experientia 25, 776 (1963)
24. G. Schatz, J.Biol.Chem. 243, 2192 (1968).
25. S.J. Singer in: Structure and Function of Biological Membranes (L.I. Rothfield, Editor), Academic Press, New York, 1971, p. 146.
26. E. Ebner, T.L. Mason and G. Schatz, J.Biol.Chem. 248, September (1973).

BIOGENESIS OF CYTOCHROME B IN *NEUROSPORA CRASSA*

Hanns Weiss and Barbara Ziganke

Institut für Physiologische Chemie und Physikalische Biochemie der Universität, 8000 München 2, Goethestrasse 33, Federal Republic of Germany.

Three aspects of cytochrome b in *Neurospora crassa* mitochondria will be discussed in this paper: 1) the molecular structure, 2) the site of biosynthesis, and 3) the process of assembly of protein and heme.

MOLECULAR STRUCTURE.

Cytochrome b was isolated from mitochondrial membranes by means of chromatography on oleyl polymethacrylic acid resin (1), DEAE-cellulose and Sephadex G-75, all steps being performed in a medium containing cholate and deoxycholate as detergents (2). The preparation shows the characteristic absorption spectra with maxima at 560 nm, 528 nm, 417 nm and 278 nm in the air oxidized form and at 561 nm, 530 nm, and 429 nm in the dithionite reduced form (Fig. 1). The heme is easily extracted from the protein by acid-acetone. It proves to be a protoheme.

The protein moiety of the cytochrome b preparation is resolved by SDS polyacrylamide gel electrophoresis into two main bands (Fig. 2). These bands can be attributed to polypeptides with the apparent molecular weights of approximately 10,000 and 30,000. However, after comparison of a great number of cytochrome b preparations the relative amount of these two bands proved to be inconstant, Thus, our original conception of both bands representing the subunits of one single protein (2) has to be corrected.

In order to decide, which of both bands can be attributed to cytochrome b the following experiment was performed. For homogeneous labelling of cell protein a

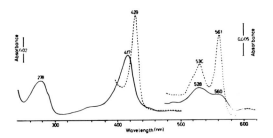

Fig. 1. Absorption spectra of the cytochrome *b* preparation. ———: air oxidized form, ----:dithionite reduced form.

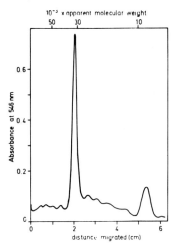

Fig. 2. Electrophoretic pattern of the cytochrome *b* preparation. 15 % polyacrylamide gels equilibrated with 0.5 % SDS and 0.1 M Tris-acetate, pH 8 were used. The protein bands were stained with Coomassie Brillant Blue. The molecular weight scale used in the abscissa was calibrated with bovine serum albumin, chicken egg albumin, chymotrypsinogen, myoglobin and cytochrome *c*.

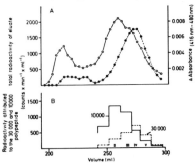

Fig. 3. Elution diagram of the cytochrome *b* preparation from the leucine-deficient mutant. o——o: [^{3}H]radioactivity; ●——●: cytochrome *b*; ———: 10,000 molecular weight band; ----: 30,000 molecular weight band.

492

leucine-deficient mutant (Leu 1) of *Neurospora crassa* was grown in Vogels minimal medium plus 2 % sucrose (3) supplemented with L-[^{3}H]leucine (5 mCi L-[^{3}H]leucine per 8 1, 0.01 mCi/mg leucine; inoculum: 3.3 x 10^9 conidia, growth period: 22 h, yield: approximately 4 g protein, 360 x 10^3 cpm/mg). The cytochrome *b* preparation from the labelled cells was submitted to recycling gelfiltration on a Sephadex G-75 column equilibrated with 1 M KCl, 0.05 M Tris-acetate pH 8.0, 0,25 % cholate and 0.25 % deoxycholate. Fig. 3 shows the elution diagram obtained after a total filtration length of about 4.5 m. Most of the [^{3}H]radioactivity which indicates protein is eluted as a broad and asymmetric peak (Fig. 3, 240th ml - 280th ml). However, only that part of the peak which appears at the higher elution volume overlaps with the cytochrome *b* peak.

In order to attribute the [^{3}H]radioactivity to the two bands of the cytochrome *b* preparation (Fig. 2) a series of consecutive fractions (Fig. 3B, II-V) eluted from the column was investigated by SDS gel electrophoresis. The distribution of radioactivity over the gels is represented in Fig. 4. As can be seen, the relative amount of the band in the 30,000 molecular weight region increases with increasing elution volume (Fig. 4, II-V), so that the appearance of the 30,000 molecular weight band coincides with the appearance of cytochrome *b* in the eluate (Fig. 3). Indeed, a linear dependency results between the specific content of the 30,000 molecular weights band and the specific content of spectrophotometrically measured cytochrome *b* (Fig. 5). If the straight line in the diagram of Fig. 5 is extrapolated to 100 %, the point corresponding to a pure preparation of the 30,000 molecular weight band, a specific heme content of 35-40 µmole/g protein results. This value indicates a minimum molecular weight of 25,000 - 30,000, a value which is in good agreement with the molecular weight of the band measured by means of SDS gelelectrophoresis (Fig. 2). Consequently this band is supposed to represent the heme binding polypeptide.

On the other hand, in a solution containing 1 M KCl,

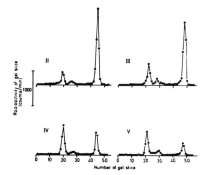

Fig. 4. Electrophoretic pattern of the sections of the [^{3}H]radioactivity peak overlapping the cytochrome b peak. The roman numbers indicate the sections of the [^{3}H]radioactivity peak eluted from the Sephadex column (cf. Fig. 3).

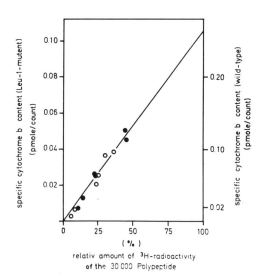

Fig. 5. Relation between the specific content of cytochrome b and the 30,000 molecular weight band. The results of two independent preparations are summarized. •——•: cytochrome b preparation from the labeled leucine-deficient mutant (cf. Fig. 3; specific radioactivity of the protein: 360 x 10^3 cpm/mg); o——o: cytochrome b from labeled wild-type cells (cf. Fig. 7A; specific radioactivity of the protein: 120 x 10^3 cpm/mg). The following absorbance coefficient was used for cytochrome b: $\Delta\varepsilon_{416-480} = 95$ mM^{-1} x cm^{-1}.

494

0.05 M Tris-acetate pH 8.0, 0.25 % cholate and 0.25 %
deoxycholate, a medium in which cytochrome *b* is not
desintegrated into polypeptide subunits and heme, cytochrome *b* passes a Sephadex G-75 column like a 55,000
molecular weight protein (Fig. 6). A similarly high
apparent molecular weight resulted from the sedimentation behaviour of cytochrome *b* in the KCl + bile acid
medium during ultracentrifugation (4). Consequently,
the siolated cytochrome *b* is assumed to be a dimeric
protein, containing two polypeptides with the apparent
molecular weight of approximately 30,000, each one
binding one protoheme.

SITE OF BIOSYNTHESIS.

The site of biosynthesis of the cytochrome *b* poly-
peptide(s) was investigated by the following labelling
experiment. An exponentially growing culture of *Neurospora crassa* (approximately 15 g protein) was divided into three equal portions. One portion remained untreated and served as a control. To the second portion
cycloheximide (0.1 mg/ml) was added, to the third portion chloramphenicol (4 mg/ml). Two min later an equal
amount of L-[^{3}H]leucine (2 mCi per 5 g protein) was
added to each portion and was allowed to be incorporated for two h. Then the three portions were harvested. Cytochrome *b* was isolated and submitted to recycling gel filtration on a Sephadex G-75 column equilibrated with the KCl + bile acid medium as described
above. Fig. 7 shows the elution diagram obtained with
the cytochrome *b* preparation from untreated cells
(Fig. 7 A), cycloheximide-treated cells (Fig. 7 B),
and chloramphenicol-treated cells (Fig. 7 C). With cytochrome *b* from untreated cells a broad and asymmetric
radioactivity peak is eluted (Fig. 7 A, 230th ml -
280th ml),which only partly coincides with the cytochrome *b* peak. This radioactivity peak results from
the overlapping of the 10,000 and 30,000 molecular
weight bands revealed by SDS gel electrophoresis (Fig.
8 A). With cytochrome *b* from cycloheximide-treated
cells, the radioactivity peak eluted from the Sephadex column exactly coincides with the cytochrome *b*

495

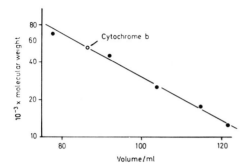

Fig. 6. Calibration curve of the molecular weight determination of cytochrome b. A Sephadex G-75 column (d=1.3 cm, h=150 cm) equilibrated with the KCl/bile acid medium (see text) was used. The elution volume was calibrated with bovine serum albumin, chicken egg albumin, chymotrypsinogen, myoglobin, and cytochrome c.

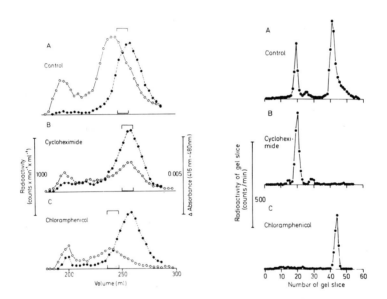

Fig. 7. Elution diagram of the cytochrome b preparations from cells labeled in the absence and in the presence of cycloheximide or chloramphenicol. o——o: [^{3}H]radioactivity; ●——●: cytochrome b.

Fig. 8. Electrophoretic pattern of cytochrome b preparations labeled in the absence and in the presence of cycloheximide or chloramphenicol. The sections marked with the horizontal brackets in Fig. 7 were submitted to gel electrophoresis.

496

peak (Fig. 7 B, 240th ml - 280th ml). SDS gel electrophoresis identifies this radioactivity peak as the 30,000 molecular weight band (Fig. 8 B). With cytochrome *b* from the chloramphenicol-treated cells the radioactivity peak is eluted before the cytochrome *b* peak. SDS gel electrophoresis identifies this peak as the 10,000 molecular weight band (Fig. 8 C).

This result indicates, that the 30,000 molecular weight band represents a mitochondrial translation product and it further supports the assumption that this band represents the heme binding polypeptide(s).

PROCESS OF ASSEMBLY.

In order to study the assembly of cytochrome *b* the labelling kinetics of the protein after a pulse with [^{3}H]leucine were compared with the labelling kinetics of the heme after a pulse with [^{59}Fe]. For labelling of the protein exponentially growing cells were first treated with a pulse of L-[^{14}C]leucine (0.05 mCi/g protein). Most of the [^{14}C]label was supposed to be at its definite position after 200 min. At this time L-[^{3}H]leucine was added (0.4 mCi/g protein). Thereafter, portions were taken from the culture at short time intervals, immediately deeply frozen and then disrupted (3) for the isolation of cytochrome *b*. In an approximation the [^{3}H] [^{14}C]ratio of the proteins corresponds to their relative specific [^{3}H]radioactivity (2,5).

In the average membrane protein the [^{3}H] [^{14}C]ratio shows a rapid increase within the first five min and reaches a constant level already after 10 min (Fig. 9 A). In contrast, in the 30,000 molecular weight band of cytochrome *b* the constant level is reached only after 120 min. Remarkably, the [^{3}H] [^{14}C]ratio follows a biphasic time course, showing an initial rapid increase followed by a slower increase. These two phases of the time course are also reflected by the semilogarythmic plot of Δ[^{3}H]/[^{14}C]; ([^{3}H]/[^{14}C] final level minus [^{3}H]/[^{14}C] time point) versus time. Two straight lines result (Fig. 9 B).

The interpretation of this biphasic time course of appearance of [^{3}H]radioactivity in the 30,000 molecu-

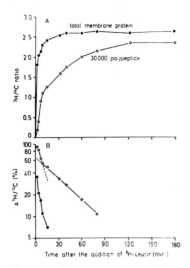

Fig. 9. Time course of labeling of the total mitochondrial membrane protein and of the 30,000 molecular weight band of the cytochrome b preparation. ●——●: total mitochondrial membrane protein; o——o: 30,000 molecular weight band.

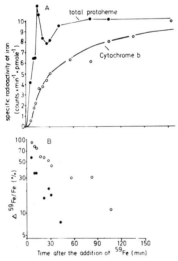

Fig. 10. Time course of labeling of total mitochondrial protoheme and cytochrome b. The protoheme was extracted from the mitochondrial membrane proteins by means of acetone + 0.1 M HCl and separated from heme a by means of chromatography on Sephadex LH-20 equilibrated with isopropanol + pyridine + 0.1 N NaOH (2:1:1) (4). ●——●: total mitochondrial protoheme; o——o: cytochrome b. In both time courses the decrease of the specific radioactivity due to the growth of the culture was accounted for, supposing a doubling time of 240 min.

lar weight band may be as follows: It is reasonable to assume an equally high translation rate of all proteins in exponentially growing cells. Thus the slower labelling rate of this band compared to the total membrane protein suggests the occurrence of precursor polypeptide(s). The larger the precursor pool to be passed, the more delayed will radioacticity appear in the final protein. The biphasic character of the time course indicates the occurrence of two precursor polypeptides of different pool size, which are connected in parallel. [^{3}H]leucine which passes the smaller precursor pool appears more rapidly in the end product, the final level of the [^{3}H] [^{14}C]ratio is reached in short time. [^{3}H]leucine which passes the larger pool appears more slowly in the end product, the final level of the [^{3}H] [^{14}C]ratio is reached less rapidly. The end product being, however, only one band at the gel, two interpretations are possible: 1) the band results from the overlapping of two polypeptides of equal size, each one being derived from its own precursor; 2) the band represents only one polypeptide, which itself is derived from two independent precursor polypeptides. We are not yet able to decide, which of both interpretations holds true. However, supported by the finding, that the apparent molecular weight of cytochrome *b* not disintegrated by SDS is about twice the molecular weight attributed to the band at the gel, we favour the first interpretation.

Now, at which step does the incorporation of the heme take place ? If protoheme is incorporated at the stage of the precursor polypeptide(s), then protoheme has to pass the same precursor pool(s). Thus, the delay of labelling between total mitochondrial protoheme and cytochrome *b* after a pulse with [^{59}Fe]should be equal to the delay of labelling between total membrane protein and the cytochrome *b* polypeptide(s) after a pulse with [^{3}H]leucine.

As can be seen in Fig. 10 A, after a [^{59}Fe]pulse (0.015 mCi/g protein, 1 mCi/ 65 µg Fe; cells were grown in minimal medium (3) prepared with substances of analytical grade, iron was omitted from the trace

499

element solution) of 5 minutes followed by a chase with the 100-fold amount of non labelled iron, radioactivity rapidly appears in the total mitochondrial protoheme. The partial decrease of specific radioactivity of protoheme followed immediately thereafter may be due to the well known fact, that part of the protoheme synthesized within the mitochondria is incorporated into cytoplasmic proteins. In contrast to total protoheme, in cytochrome b radioactive iron appears more slowly. The accuracy of this experiment does not allow the distinction of two phases in the time course. However, it is obvious, that the delay of appearance of [59Fe] label comparing total protoheme and cytochrome b is very similar to the delay of appearance of [3H] label comparing total membrane protein and the 30,000 molecular weight band of cytochrome b. Thus, the required but not fully conclusive condition is given for the idea that both pools are identical.

ACKNOWLEDGEMENTS.

We are greatly indebted to Theodor Bücher for advice and discussion. We thank Inge Gillmeier for excellent technical assistance. This work was supported by the Deutsche Forchungsgemeinschaft, Schwerpunktsprogramm "Biochemie der Morphogenese".

REFERENCES.

1. H. Weiss and Th. Bücher, Eur.J.Biochem. **17**, 561 (1970).
2. H. Weiss, Eur.J.Biochem. **30**, 469 (1972).
3. H. Weiss, G. von Jagow, M. Klingenberg and Th.Bücher, Eur.J.Biochem. **14**, 75 (1970).
4. B. Ziganke and H. Weiss, in preparation.
5. A.J. Schwab, W. Sebald and H. Weiss, Eur.J.Biochem. , 511 (1972).

PRECURSOR PROTEINS OF CYTOCHROME C OXIDASE IN CYTOCHROME C OXIDASE DEFICIENT COPPER-DEPLETED NEUROSPORA

Andreas J. Schwab

Institut für Physiologische Chemie und Physikalische Biochemie der Universität, D-8000 München 2, Goethestrasse 33, Germany.

It is already known from experiments with yeast and with mammals that copper deficiency leads to a lack of cytochrome c oxidase (1). In copper-depleted *Neurospora* , the cytochrome aa_3 content is below the detection limit (0.02 μmol/g mitochondrial protein), as can be seen from the spectrum shown in Fig. 1. In contrast, the b and c-type cytochrome contents are not substantially altered. Mitochondrial respiration is almost completely resistant (94-98 %) to cyanide and antimycin A; it is strongly inhibited by salicylhydroxamate, the residual activity (20 %) being almost entirely cyanide and antimycin A sensitive. Thus, copper-depleted *Neurospora* cells possess a branched respiratory chain with an additional oxidase similar to that demonstrated in the mi-1, mi-3 and cni-1 mutants and in chloramphenicol-inhibited wild type (2-5).

The recovery of copper-deficient *Neurospora* cells after addition of copper might be a good model for studying the assembly of cytochrome c oxidase. The enzyme appears only several hours after addition of copper; in yeast it has been shown that this process is inhibited by cycloheximide (6,7), so that protein synthesis is probably necessary.

In the following some evidence will be presented that copper-depleted cells synthesize precursor proteins of cytochrome c oxidase. In the experiment shown in Fig. 2, mitochondrial protein synthesis was investigated *in vivo*. Copper-depleted cells were sequentially treated with [^{14}C]leucine (uniform labelling of

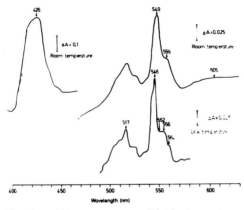

Fig. 1. Difference spectra (dithionite reduced *vs.* air oxidized) of mitochondria from copper depleted *Neurospora* cells at room temperature and at the temperature of liquid nitrogen.

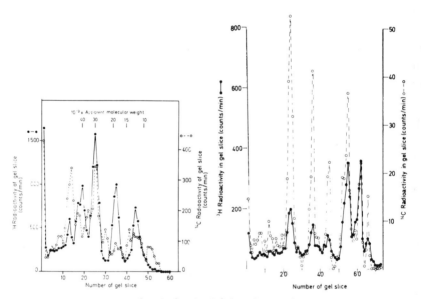

Fig. 2. Gel electrophoresis of mitochondrial membranes from copper depleted *Neurospora* cells labelled sequentially (see text) with [^{14}C]leucine (o---o) in the absence of cycloheximide (50 µCi per liter of culture medium) and with [^{3}H] leucine (o——o) in the presence of cycloheximide (1 mCi per liter).

Fig. 3. Appearance of radioactive leucine incorporated in the absence of copper in the cytochrome *c* oxidase formed after subsequent addition of copper. Copper-depleted *Neurospora* cells were labeled with [^{3}H]leucine (o——o) (2 mCi per liter of culture medium). After 30 min 1 µM copper was added. After 6 h of further growth the cells were mixed with an equal amount of cells labeled uniformly with [^{14}C]leu-cine (o---o) (50 µCi per liter) and harvested. Mitochondria were isolated and lysed with Triton X-100. Cytochrome *c* oxidase was precipitated with a specific antibody supplied by S. Werner. The precipitate was submitted to SDS gel electrophoresis. The [^{14}C]scale is expanded relative to the [^{3}H]scale by a factor of 15.8, equal to the [^{3}H]/[^{14}C]ratio in the mitochondria. The origin of electrophoresis is on the left.

all proteins), after 30 min with cycloheximide and after 2 min with [^{3}H]leucine, and harvested after a further 30 min of incubation. Under these conditions, the [^{3}H]leucine will label only proteins synthesized on the mitochondrial ribosomes (8). The specific radioactivities of the total mitochondrial membrane protein were similar to those obtained under analogous conditions with copper-sufficient cells. Sodium dodecylsulfate gel electrophoresis of the mitochondrial membrane protein led to the patterns shown. The [^{3}H]-radioactivity, representing protein of mitochondrial origin, shows four major peaks, which correspond to proteins with apparent molecular weights of approximately 12000, 20000, 30000 and 40000. These are similar to those found in copper-sufiicient cells, differing only in the relative peak areas (8). One may conclude that in copper-depleted cells, mitochondrial protein synthesis is not substantially affected. Therefore, the subunits of cytochrome *c* oxidase of mitochondrial origin must be already present in copper-depleted cells.

In the experiment shown in Fig. 3, copper-depleted cells were uniformly labelled by addition of [^{3}H]leucine of high specific radioactivity. After 30 min, when the radioactive leucine had been completely incorporated into proteins (9), copper was added, the cells were allowed to grow further for 6 h and were then harvested. The cytochrome *c* oxidase produced contained label in all subunits; this indicates that precursor proteins are synthesized in the absence of copper and are integrated into the enzyme protein after addition of copper. The presence of a precursor protein of cytochrome *c* oxidase in copper-depleted cells could be made probable also by an immunological test using the Ouchterloni technique. When an antibody to the isolated subunit of cytochrome *c* oxidase with the apparent molecular weight of 20000, was allowed to diffuse against Triton X-100 solubilized mitochondrial membranes from copper-depleted *Neurospora*, a single sharp precipitation line could be observed (S. Werner and A.J. Schwab, unpublished result).

The general conclusion may ne drawn that in the absence of copper the assembly of the enzyme stops at a stage where copper should be introduced. The heme group has not yet been introduced at this step. It should be noted that the fact that cytochrome c oxidase is a copper protein explains the lack of enzymic activity of cytochrome c oxidase in copper-depleted cells, but not the lack of the cytochrome aa_3 absorption spectrum.

REFERENCES.

1. H. Wohlrab and E.E. Jakobs, Biochem.Biophys.Res. Commun. 28, 991 (1967).
2. G. von Jagow and M. Klingenberg, FEBS Letters 24, 278 (1972).
3. G. von Jagow, H. Weiss and M. Klingenberg, Eur.J. Biochem. 33, 140 (1973).
4. A.M. Lambowitz, E.W. Smith and C.W. Slayman, J. Biol.Chem. 247, 4850 (1972).
5. D.L. Edwards and F. Kwiecinsky, J.Bacteriol. in press.
6. E. Keyhani and B. Chance, FEBS Letters 17, 127 (1971).
7. A. Light, FEBS Letters 19, 319 (1972).
8. W. Sebald, J. Otto and Machleidt, Eur.J.Biochem. in press.
9. A.J. Schwab, W. Sebald and H. Weiss, Eur.J.Biochem. 30, 511 (1972).

ANTIBODIES TO SUBUNITS OF CYTOCHROME *C* OXIDASE AND THEIR RELATION TO PRECURSOR PROTEINS OF THIS ENZYME

Sigurd Werner

Institut für Physiologische Chemie und Physikalische Biochemie der Universität, Goethestrasse 33, D-8000 München 2, Federal Republic of Germany.

In trying to understand the process of the assembly of cytochrome *c* oxidase, we need detailed information on the individual components to be assembled and their precursors. An immunological procedure was developed as a tool for the isolation and characterization of precursor proteins. The following approach was chosen: 1) Isolation of individual subunits of cytochrome *c* oxidase; 2) Preparation of specific antibodies to these polypeptides; 3) Precipitation of precursor proteins from solubilized mitochondria by antibodies.

Cytochrome *c* oxidase from *Neurospora crassa* was prepared by means of oleyl-polymethacrylic acid-resin (1). Individual subunits were isolated by large scale chromatographic methods using dextran and polyacrylamide resins (2). Fig. 1 shows a recycling chromatography of cytochrome *c* oxidase on Sephadex G-100. Proteins eluted with the first 3 peaks represent the mitochondrially synthesized subunits, peak 4 contains the cytoplasmic subunits. Using Sephadex G-100, peak 3 was clearly separated from the other peaks. Peak 3 is equivalent to subunit 3 with an apparent molecular weight of 21,000 as revealed by SDS gel electrophoresis. Re-chromatography of this fraction resulted in a pure (about 95 %) preparation of this subunit.

This polypeptide fraction was injected into rabbits to produce subunit specific antibodies. Immunoglobulin from the sera were isolated and tested. In the Ouchterlony analysis (globulin against subunit) one sharp precipitin line was obtained. The globulin did not cross-

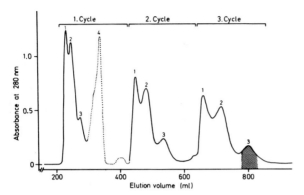

Fig. 1. Recycling gel chromatography on Sephadex G-100 of cytochrome c oxidase treated with SDS. 30 mg of oxidase were chromatographed on one column of gel (2.2 x 150 cm), equilibrated with 50 mM Tris-HCl pH 8.0, 0.5 % SDS. Polypeptides eluted with peak 4 (dotted line) were not recycled on the same column. The protein of peak 3, equivalent to subunit 3 of cytochrome c oxidase, was collected after 3 cycles (elution volume 800 ml).

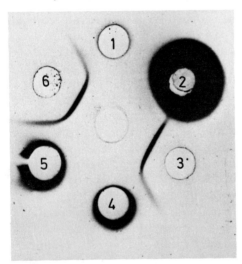

Fig. 2. Double diffusion in agar containing 1 % Triton X-100. Center well: subunit 3 (1 mg/ml); well 1: IgG to subunit 3 absorbed with subunit 3; well 2: subunit 3 with excess of subunit specific IgG; well 3 and 6: IgG to subunit 3 (2 mg/ml); well 4: IgG to subunit 3 absorbed with solubilized mitochondrial membranes; well 5: solubilized membranes (3 mg/ml).

506

react with other subunits of cytochrome c oxidase or with cytochrome b. Furthermore it did not precipitate a pure preparation of cytochrome c oxidase.

Fig. 2 shows a double diffusion test with solubilized mitochondrial membranes as antigen. A single line of precipitation was observed between membranes (well 5) and immunoglobulin (well 6). This precipitin line was identical to that obtained with a preparation of subunit 3 (center well). In another part of the gel (well 1) subunit specific immunoglobulin was absorbed with the 21,000 molecular weight polypeptide. The absorbed globulin did no longer react with subunit 3 (center well). An excess of subunit 3 was present as indicated by the weak precipitin line appearing between the absorbed globulin (well 1) and globulin in well 6. A similar experiment was carried out with mitochondrial membranes as absorbant (well 4). No precipitation could be detected against the center well. Finally in well 2 subunit 3 was succesfully absorbed by an excess of specific immunoglobulin.

These experiments show that a subunit specific antibody has been raised, which is able to recognize an antigen in solubilized mitochondrial membranes.

In order to identify this membrane antigen the following experiments were performed. An exponentially growing culture of Neurospora crassa was divided into two portions. To one portion cycloheximide was given, to the other chloramphenicol. 2 min later an equal amount of [^{3}H]leucine was added and incorporated for 25 min. The cultures were harvested and mitochondrial membranes were prepared. These were lysed with Triton X-100 and treated with subunit specific immunoglobulin. To increase the sensitivity of the test a second antibody – immunoglobulin from goat to rabbit γ-globulin – was added. In the case of mitochondrial membranes prepared from the cycloheximide poisoned cells, the radioactivity in the immuno-precipitate should reflect the mitochondrially synthesized precursor protein of subunit 3. The precipitate was dissolved in SDS and applied to polyacrylamide gel electrophoresis.

Two major bands (Fig. 3) with an apparent molecular

507

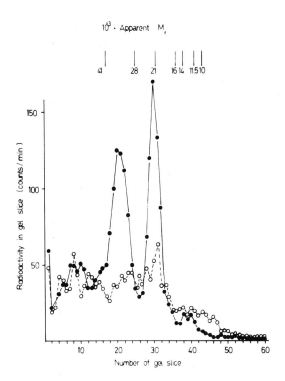

Fig. 3. Gel-electrophoretic analysis of immunoprecipitates from cells labeled with [³H]leucine in the presence of cycloheximide (●——●) or chloramphenicol (o——o). The gel was calibrated by co-electrophoresis of [¹⁴C]labeled subunits of cytochrome *c* oxidase. Apparent molecular weight of the different subunits according to (3).

weight of about 21,000 and 35,000 were observed. Electrophoresis of the immunoprecipitate from chloramphenicol treated cells showed a rather scattered radioactivity pattern, similar to that obtained in control experiments using non-specific immunoglobulin (not shown).

The reported data indicate that the subunit specific antibody recognizes at least two proteins of mitochondrial origin. In respect to their molecular weight the smaller one coincides with the isolated subunit. The larger protein however does not resemble any subunit isolated from cytochrome *c* oxidase. Obviously, these two proteins form no part of complete cytochrome *c* oxidase molecules in solubilized membranes. Therefore we suggest that they represent precursors of the enzyme.

ACKNOWLEDGEMENTS.

I am grateful to Prof. Th. Bücher for his encouraging support. Furthermore I thank Miss M. Wild for skilful technical assistance.

REFERENCES.

1. H. Weiss and Th. Bücher, Eur.J.Biochem. **17**, 561 (1970).
2. S. Werner, in preparation.
3. W. Sebald, W. Machleit and J. Otto, Eur.J.Biochem. in press (1973).

CYTOPLASMIC RIBOSOMES ASSOCIATED WITH YEAST MITOCHONDRIA

Rod E. Kellems, Venita F. Allison, and Ronald A. Butow,

Department of Biochemistry, The University of Texas Southwestern Medical School and Department of Biology, Southern Methodist University, Dallas, Texas, U.S.A.

Both nuclear and mitochondrial gene products are required for the formation of functional mitochondria (1,2) and this fact represents a major contribution towards our understanding of mitochondrial biogenesis. Presently, it is possible to account for about 10-15 % of the total mitochondrial protein as the contribution from the intrinsic mitochondrial protein synthesizing system (3). An important and, as yet, unresolved question naturally arises; how are the majority of mitochondrial proteins, coded for by nuclear genes, directed towards the developing mitochondrion and, in certain cases, how do these proteins traverse the seemingly impermeable double membrane barrier of the closed organelle ? A number of models have been suggested to account for these processes; among these include:
1. Diffusion
2. Translational and post-translational modification processes
3. Aggregation with lipid
4. Nuclear mRNA import and translation on mitochondrial ribosomes
5. Vectorial discharge of polypeptides into mitochondria from
   a) rough endoplasmic reticulum—mitochondrial continuum
   b) ribosomes directly associated with mitochondria
Diffusion, assuming selective product binding sites

with high affinity does not account for known mitochondrial impermeability to proteins of even modest
molecular weight (4,5). Conceivably, such permeability
barriers could be circumvented by modification of
translation products (for example by addition of hydrophobic residues) so as to facilitate penetration of
these products into the organelle. Similarly, the association of cytoplasmically synthesized proteins with
lipid and subsequent fusion of the protein-lipid aggregates with membrane could accomplish the same task.
The possibility that nuclear derived mRNA could be imported and translated on intrinsic mitochondrial ribosomes has been considered (6), but recent evidence indicates that such a process probably does not account
for a majority of total mitochondrial protein of nuclear origin (7). While some of the above models have
yet to be tested exhaustively, we have chosen to focus
our attention on the concept of vectorial discharge
of polypeptides into mitochondria. We decided to explore this possibility because in a variety of cell
types, particularly secretory cells, there is considerable experimental evidence for the transfer of specific translation products across a membrane barrier; in
particular, those proteins destined for secretion from
the cell are translated selectively on bound polysomes
and discharged vectorially across the endoplasmic reticulum (8). Since yeast do not have an extensively
developed ER-system we have considered specifically
the discharge of polypeptides from ribosomes that
might be associated directly with mitochondria. As an
overall effort to attempt clarification of the mechanism by which cytoplasmic ribosomes in yeast contribute to the synthesis of mitochondrial protein, we
hoped to find experimental support which would be consistant with, or alternatively rule out such a mechanism for import of protein into yeast mitochondria.

CYTOPLASMIC 80S RIBOSOMES ASSOCIATED WITH YEAST MITO-
CHONDRIA.

In a previous report from this laboratory (9) we
described a class of cytoplasmic-80S ribosomes in

yeast which co-purified with mitochondria and were distinguished from free cytoplasmic ribosomes of the post-mitochondrial supernatant fraction by, 1) resistance to dissociation in KCl and, 2) decreased capacity to catalyze poly(U)-directed phenylalanine incorporation. In addition, isolated mitochondria washed with EDTA were capable of accepting ribosomes at labile binding sites. In this communication, we shall describe recent experiments demonstrating: 1) a corralation between the amount of ribosomes associated with purified yeast mitochondrial fractions and the physiological state of the cell; 2) the capacity of these bound ribosomes to carry out vectorial translation analogous to the rough endoplasmic reticulum of secretory cells; 3) electron microscopic evidence suggesting the close association of cytoplasmic ribosomes with yeast mitochondria *in situ.*

80S RIBOSOMES ASSOCIATED WITH THE MITOCHONDRIAL FRACTION ARE 10-15 % OF THE CELLS' TOTAL,

If yeast cells grown on 2 % galactose are harvested in early log phase in the presence of cycloheximide to minimize ribosome run-off from polysomes (see below), mitochondria isolated after extensive washing and purification by isopycnic banding on sucrose gradients contain as much rRNA per mg mitochondrial protein as the post-mitochondrial supernatant fraction

TABLE I

RNA EXTRACTED FROM MITOCHONDRIA AND POST-MITOCHONDRIAL SUPERNATANT FRACTION
RNA was extracted with phenol-sodium dodecyl sulfate (SDS) in acetate-EDTA buffer pH 5.1 and precipitated with 2.5 volumes of ethanol.

Conditions	A_{260} Units extracted/mg protein	
	Mitochondria	Post-Mitochondrial Supernatant
1. Growing cells treated with cycloheximide (200 µg/ml)	6.7	6.5
2. Cells starved 1 hour	4.1	6.6
3. Mitochondria washed 2 times with 2 mM EDTA	2.0	
4. As in 1, except cells opened and fractionated in the presence of 350 mM KCl	8.3	5.5

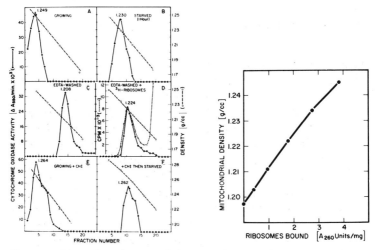

Fig. 1. Effect of various conditions on mitochondrial density. Mitochondria were sedimented to equilibrium in linear sucrose gradients. After fractionation, the gradients were analyzed for cytochrome *c* oxidase activity as previously described (9) and refractive index. A: mitochondria from growing cells; B: mitochondria from cells starved for one hour; C: mitochondria from growing cells washed with 2 mM EDTA; D: same as C except mitochondria (0.77 mg) were incubated before centrifugation, with 4 A_{260} units of [^{3}H]labeled ribosomes; E: mitochondria from growing cells, culture pretreated with 200 μg/ml cycloheximide prior to harvesting; F: mitochondria from growing cells treated with 200 μg/ml cycloheximide, then starved for 1 h.

Fig. 2. Effect of added ribosomes on mitochondrial density. EDTA-washed mitochondria (0.49 mg. protein) were incubated for 15 min at 30° with increasing amounts of [^{3}H]labeled cytoplasmic ribosomes. The incubation mixtures were then layered on 30-70 % sucrose gradients and the mitochondria sedimented to equilibrium.

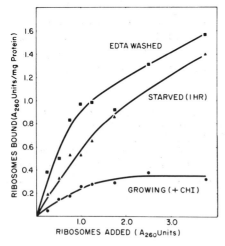

Fig. 3. Binding of [^{3}H]labeled cytoplasmic ribosomes to different mitochondrial preparations. Ribosome binding to a fixed amount of mitochondria from cycloheximide-treated growing cells (●—●), starved cells (△—△), and EDTA-washed mitochondria obtained from cycloheximide-treated growing cells (■—■) was carried out as previously described (9).

514

(Table I). Since under these conditions of derepressed growth mitochondria comprise 10-15 % of the total cell protein (10), it follows that the mitochondrial-associated class of cytoplasmic ribosomes represents a similar percentage of the cells' total ribosomes.

MITOCHONDRIAL DENSITY AS A FUNCTION OF BOUND RIBOSOMES

Mitochondria isolated from growing cells have a buoyant density in sucrose of 1.249 g per cc (Fig. 1A). However, when cells are starved for one hour and the mitochondria are then isolated the buoyant density has decreased to 1.230 g per cc (Fig. 1B). If mitochondria isolated from growing cells are washed with 2 mM EDTA, conditions known to remove ribosomes from membranes, the buoyant density shifts to 1.208 g per cc (Fig. 1C). Finally, when ribosomes are incubated with EDTA-washed mitochondria, the ribosomes will re-attach to the vacant ribosome binding sites and the mitochondria will move to a position of greater density (Fig. 1D). There is, in fact, a direct correlation between mitochondrial density and the amount of ribosomes bound. This was demonstrated by incubating a fixed amount of mitochondria with increasing amounts of [³H]labeled cytoplasmic ribosomes and measuring both the mitochondrial density and the amount of ribosomes bound (Fig. 2).

Since slow cooling of a culture probably allows some run-off of ribosomes from polysomes, we felt that the density of mitochondria isolated from growing cells should *increase* if cycloheximide was added to the culture prior to harvesting and isolation of mitochondria. Indeed, when mitochondria are isolated from growing cells treated with cycloheximide, the density shifts from 1.249 to 1.264 g per cc (Fig. 1E); moreover, the mitochondrial density shift observed upon starvation of cells could be prevented if cycloheximide was added prior to starvation (Fig. 1F). Recovery of polysomes in cells under these various conditions was as expected; that is, recovery of both bound and free polysomes increased when cycloheximide was added

to log phase cultures and polysomes disappeared following a one hour starvation which could be prevented
by prior addition of cycloheximide. We have shown previously that EDTA-stripped mitochondria are capable
of accepting ribosomes *in vitro* (9). In order to determine the extent to which ribosome binding sites
are occupied in growing cells, cycloheximide was added
to a log phase culture prior to harvesting and the mitochondria were then isolated, purified on sucrose
gradients and RNA extracted with phenol-SDS. As shown
in Table I, mitochondria isolated in this manner contain 6.7 A_{260} units of RNA per mg mitochondrial protein. That this value represents near saturation of
the ribosome binding sites is indicated by the data
presented in Fig. 3. Here it is shown that mitochondria isolated from growing cells that were treated
with cycloheximide will bind less than 0.4 A_{260} units
of ribosomes per mg mitochondrial protein. On the
other hand mitochondria isolated from starved cells
and EDTA-washed mitochondria contain less RNA (Table
I) and are capable of accepting significantly more
exogenously added ribosomes (Fig. 3).

Inasmuch as the consequence of ribosome binding is
to markedly shift the density of cytochrome c oxidase
in the gradient, our results suggest that ribosomes
are indeed bound to mitochondria or at least to structures intimately associated with mitochondria. It is
of considerable interest that, as is the case with
yeast described here, the association of ribosomes
with membrane of the endoplasmic reticulum of liver
cells (11), cells in culture (12), and with thylakoid
membranes of chloroplasts (13), decreases as the polysome content of the cell (or chloroplast) decreases.

EFFECT OF KC1 ON POLYSOME BINDING TO MITOCHONDRIA AND
ON THE RECOVERY OF MITOCHONDRIAL-ASSOCIATED 80S RIBO-
SOMES.

While a significant fraction of the total ribosomes
in the cell remains associated with mitochondria even
after extensive washing and purification on sucrose
gradients, the extent to which non-physiological as-

sociation of polysomes to mitochondrial membranes occurs is difficult to evaluate quantitatively. However, in an attempt to approach this problem we asked whether we could find conditions whereby polysome binding to mitochondria was effectively eliminated *in vitro* and what the effect of these conditions would be on the recovery of mitochondrial-associated 80S cytoplasmic ribosomes. As shown in Fig. 4, we have compared the effect of increasing concentrations of KCl on the binding of free cytoplasmic polysomes and the polysomes obtained from the mitochondrial fraction to a preparation of stripped (EDTA-washed) mitochondria. While the binding of bound and free polysome preparations differs slightly with respect to KCl sensitivity, binding of both preparations is completely inhibited at 350 mM KCl. Mitochondria were then isolated and purified from log phase cells to which cycloheximide was added prior to harvesting as before except that the isolation medium, including that in which the cells were broken, contained 350 mM KCl, As shown in Table I, the recovery of A_{260} units per mg mitochondrial protein was not dimished under these conditions when compared to the amount recovered following the usual method of preparation of mitochondria in the absence of KCl, In fact, the recovery was somewhat greater. This result could possibly be accounted for by some selective removal by high salt of more soluble mitochondrial proteins. While these results are consistant with the association of polysomes with mitochondria *in vivo* we cannot rule out the possibility that when the cells are broken in the presence of high salt the association between some polysomes and mitochondria is occurring more rapidly than the time required for establishment of a uniform ionic state.

VECTORIAL TRANSLATION.

One of the established and important functions of the rough endoplasmic reticulum in secretory cells of higher organisms is the capacity for vectorial discharge of nascent polypeptide chains from bound polysomes into the internal space of the endoplasmic reti-

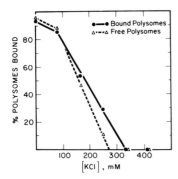

Fig. 4. Effect of KCl on polysome binding to EDTA-washed mitochondria. EDTA-washed mitochondria (1.9 mg) obtained from growing cells were incubated in the presence of increasing amounts of KCl for 15 min at 30° with 0.25 A_{260} units of either free cytoplasmic polysomes or cytoplasmic polysomes obtained from the mitochondrial fraction.

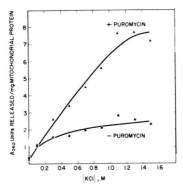

Fig. 5. Effect of KCl and puromycin on the release of ribosomes from mitochondria. Mitochondria were incubated for 15 min at 30° with increasing concentrations of KCl plus and minus 2.1 mM puromycin as indicated in the figure. The incubation mixtures were layered onto 15-30 % linear sucrose gradients and centrifuged for 2 h at 221,800xg max. Under these conditions, the mitochondria are pelleted and the amount of released ribosomes was quantitated in the gradient by flow scanning in an ISCO Model 224 UV Flow Analyzer.

Fig. 6. A representative selection of mitochondrial profiles seen in thin sections of growing yeast spheroplasts. Spheroplasts were prepared for electron microscopy by fixation in 3 % glutaraldehyde and post-fixed with 2 % osmium tetroxide. Sections were stained with uranyl acetate and lead citrate. Magnification:
a) 64,000 X; b) 86,400 X; c) 64,000 X; d) 64,000 X; e) 79,750; f) 97,500.

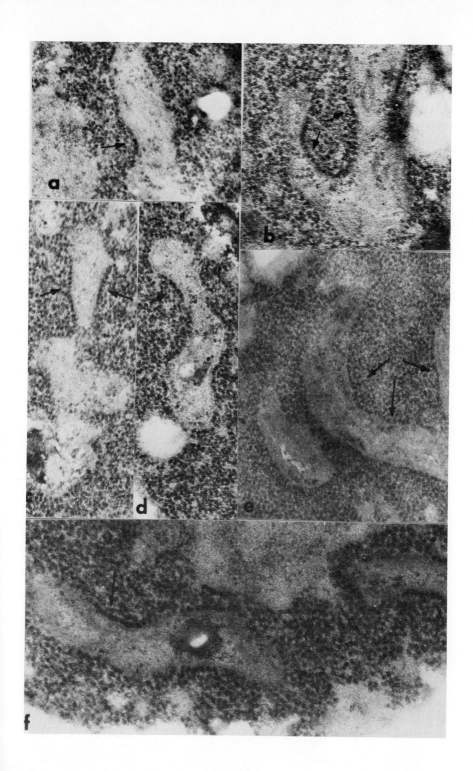

culum. Demonstration of this phenomenon, as first outlined by Redman and Sabatini (14), depends on assess-
ment of the distribution of nascent polypeptide chains
released from polysomes by puromycin, between the extra vesicular space and a detergent-releaseable compartment. We have carried out a similar series of experiments with mitochondria obtained from log phase
cells. With these preparations in an *in vitro* amino
acid incorporating system, about 80 % of the incorporation is inhibited by cycloheximide; the remaining incorporation is mostly inhibited by chloramphenicol and
undoubtedly represents the contribution by intrinsic
mitochondrial protein synthesis. Table II shows the
results of an average of three experiments designed to
test for vectorial translation in these mitochondrial
preparations. As shown, of the total puromycin dis-
chargeable counts, about half remain associated with
the mitochondrial and are releaseable by detergent and
half are released directly into the medium. These results are comparable to the data reported for the distribution of puromycin released nascent chains with
preparations of rough endoplasmic reticulum from rat
liver (15). Thus, by the criterion of this particular
experimental protocol, cytoplasmic polysomes associated with yeast mitochondria have some capacity to carry out vectorial translation. What remains to be established in the yeast system is the precise disposition
of the vectorially released nascent chains, that is,
whether they enter intra-membrane spaces or are interposed between the ribosome and membrane constituents.

Mitochondria (about 4 mg) from growing cells were
incubated in an *in vitro* amino acid incorporating sys-

TABLE II

VECTORIAL TRANSLATION

Total incorporation (DPM)		$183000xg_{max}$ Pellet % of total incorporation				% Puromycin release	% Vectorial
+PURO	-PURO	(A) -PURO -TRITON	(B) -PURO +TRITON	(C) +PURO -TRITON	(D) +PURO +TRITON	(B-D)	$\dfrac{(C-D)}{(A-C)+(C-D)}$
64,025- 154,207	61,073- 136,639	75.7 ($\pm$3.9)	77.2 ($\pm$5.3)	65.7 ($\pm$0.9)	54.4 ($\pm$3.9)	22.8 ($\pm$9.1)	53.1

tem containing [3H]leucine and 100 μg/ml chloramphenicol (Kellems and Butow, details to be described elsewhere). Puromycin (1.8 mM) was then added to half the reaction mixture and the incubation continued for an additional 12 min. Samples were removed, precipitated with 5 % TCA, heated at 90º for 10 min and counted to determine total incorporation. To the remaining material Triton X-100 was added as indicated in the Table to a final concentration of 2% and the material centrifuged for 90 min at 183,000xg$_{max}$. The pellets were precipitated with 5 % TCA, heated at 90º for 10 min and counted.

An additional point which deserves comment and further underscores the striking similarity between the yeast system described here and the rough endoplasmic reticulum is the role of nascent polypeptide in the association between the membrane and polysomes (16). In both systems nascent polypeptide is implicated in the attachment of polysomes to membranes. Fig. 5 shows that high salt alone removes only about one third of the attached ribosomes whereas virtually all ribosomes are removed by treatment with high salt *plus* puromycin. In other words, both systems display a nearly identical salt dependent association of ribosomes with membrane, probably involving ribosome association through the large subunit (17), and some salt independent association involving nascent polypeptide.

ELECTRON MICROSCOPIC EXAMINATION OF GROWING SPHEROPLASTS.

The data presented in Fig. 6 illustrate a number of representative views of mitochondrial profiles in preparations of growing spheroplasts. Although the resolution of membranes is limited probably because of some interaction between osmium and the sorbitol used for osmotic stabilization, ribosomes are clearly outlined; particularly, their alignment along the periphery of the mitochondria can be seen. Recently, Watson has reported the alignment of cytoplasmic ribosomes along the outer mitochondrial membrane in *S. cerevisiae* (18) and Keyhani has made similar observations in the red yeast, *Rodotorulla rubra* (19). It has also been shown that in *Dictostylium* under certain physiological conditions, cytoplasmic ribosomes are aligned along the outer mitochondrial membrane (20). So far, reports of cytoplasmic ribosomes associated with mitochondria *in situ* have been restricted to lower eukaryotes.

CONLUDING REMARKS.

The physiological advantages to the cell of topographical segregation of polysomes, particularly regarding mitochondrial biosynthesis, are obvious. The experiments we have presented here provide a basis for further tests for translational segregation in yeast as has now been well documented in secretory cells. We have shown that in a number of respects, the properties of polysome interaction with membranes of purified mitochondrial fractions of yeast are similar to those of the rough endoplasmic reticulum in secretory cells of higher eukaryotes. Conclusive evidence for vectorial translation as a mechanism for import into mitochondria of mitochondrial proteins encoded for by nuclear genes must await demonstration of the *selective* synthesis of known mitochondrial proteins by the class of mitochondrial - associated cytoplasmic polysomes.

ACKNOWLEDGEMENTS.

We thank Miss Martha J. Ferguson for expert technical assistance. This work was supported by Grants from the National Institutes of Health and the American Cancer Society.

REFERENCES.

1. A.W.Linnane, J.M.Halsam, H.B.Lukins and P.Nagley, Ann.Rev.Microbiology **26**, 163 (1972).
2. P.Borst, Ann.Rev.Biochemistry **41**, 333 (1972).
3. D.S.Beattie, Sub-Cell.Biochem. **1**, 1 (1971).
4. L.Wojtczak and H.Faluksa, Biochim.Biophys.Acta **193**, 64 (1969).
5. D.J.Luck and E.Reich, Proc.Nat.Acad.Sci.U.S.A. **52**,

931 (1964).

6. R.F.Swanson, Nature New Biol. 231, 31 (1971).
7. H.R.Mahler and K.Davidowicz, Proc.Nat.Acad.Sci. U.S.A. 70, 111 (1973).
8. P.N.Campbell, Fed.Eur.Biochem.Soc.Lett. 7, 1 (1970).
9. R.E.Kellems and B.A.Butow, J.Biol.Chem. 247, 8043 (1972).
10. R.S.Criddle and G.Schatz, Biochemistry 8, 322 (1969).
11. P.J.Goldblatt, Sub-Cell.Biochem. 1, 14 (1972).
12. S.Y.Lei, V.Krsmanovic and J.Brawerman, J.Cell. Biol. 49, 683 (1971).
13. N.Chua, G.Blobel, P.Siekevitz and G.Palade, Proc. Nat.Acad.Sci.U.S.A. 70, 1554 (1973).
14. C.M.Redman and D.D.Sabatini, Proc.Nat.Acad.Sci. U.S.A. 56, 608 (1966).
15. T.M.Andrews and J.R.Tata, Biochem.J. 121, 683 (1971).
16. M.R.Adelman, D.D.Sabatini and G.Blobel, J.Cell. Biol. 56, 206 (1973).
17. C.Baglioni, I.Bleiberg and M.Zauder, Nature New. Biol. 232, 8 (1971).
18. K.Watson, J.Cell Biol. 55, 721 (1972).
19. E.Keyhani, J.Cell.Biol. (in press).
20. D.A.Cotter, L.Y.Miura-Santo and H.R.Hobil, J.Bact. 100, 1020 (1969).

MITOCHONDRIAL SYNTHESIS OF GLYCOPROTEINS AND SURFACE PROPERTIES OF MITOCHONDRIAL MEMBRANES

H.Bruce Bosmann and M.W. Myers

Department of Pharmacology and Toxicology, University of Rochester School of Medicine and Dentistry, Rochester, New York, U.S.A.

Mitochondria have been firmly established as organelles capable of autonomous synthesis of certain macromolecules, including DNA, RNA, protein, and glycoproteins (see ref. 1 for review). The mitochondrial nature of this synthesis has been characterized by experiments with whole cells "*in vivo*" utilizing drugs to separate cytoplasmic from mitochondrial syntheses and also with isolated mitochondria "*in vitro*". Although recently *in vitro* mitochondrial protein synthesis has been challenged(2), the original work has been promptly defended (3), and both *in vivo* and *in vitro* experiments have contributed to our understanding of mitochondrial physiology.

Our laboratory has been concerned with glycoprotein synthesis and structure in mammalian mitochondria. In 1969 (4) it was demonstrated that isolated rat liver mitochondria could glycosyllate endogenous mitochondrial proteins; this observation was extended to cerebral cortex mitochondria (5). The fact that isolated mitochondria glycosyllate endogenous mitochondrial proteins has been independently confirmed by Ashwell and Work (1) and more recently by Morelis and Louisot (6). Subsequently it was demonstrated that the mitochondrial inner membrane contains some glycoprotein components and the neutral sugars of the mitochondrial structural protein fraction were identified (7). The enzymatic characteristics of the mitochondrial glycoprotein: glucosyl transferases and the mitochondrial glycoprotein: mannosyl transferases were described (8) and found to be somewhat dissimilar to other cytoplasmic transferases (9-11). It was demonstrated that some of the proteins glycosyllated by isolated mitochondria are also synthesized by isolated mitochondria by utilizing puromycin to inhibit mitochondrial protein synthesis prior to glycosyllation (12).

Several investigators have tried to determine the products of mitochondrial protein synthesis. Results have varied from indications of many mitochondrial proteins labeled (13), to individual peptides of complex mitochondrial proteins (14,15), to one or two specific types of protein to serve as components of other mitochondrial proteins (16,17). This laboratory has found that isolated rat-liver and cerebral-cortex mitochondria incorporate leucine into at least four major peaks, separable by sodium dodecyl sulfate-polyacrylamide gel electrophoresis (one peak at molecular weight of about 8,000); each of these protein bands also contained labeled monosaccharide (18). Thus the isolated mammalian cell mitochondrion seems to synthesize at least four distinct protein products, each of which is glycosyllated by mitochondrial glycosyl transferases.

Recently mitochondrial glycoproteins have been described in the intermembrane space (19) and intra-mitochondrial Ca^{2+}-binding glycoproteins have been partially purified (20,21). Glycoproteins on the mitochondrial surface have also been demonstrated both by techniques of lectin binding (22,23) and by particle microelectrophoresis (24). By this technique we have determined that both isolated mitochondria and derived mitoplasts contain terminal sialic acid residues on surface membrane glycoproteins which contribute to the negatively charged nature of the mitochondrial surface at physiological pH.

Because of recent interest in possible mitochondrial involvement in viral infection and oncogenesis (25-29) and suggestions of mitochondrial membrane alterations in neoplastic cells (30), we turned our attention to mitochondria of neoplastic cell models. Ini-

tially we reported that isolated SV-40 transformed
3T3 cell mitochondria exhibited a lower rate of protein and glycoprotein synthesis than did their normal
3T3 counterparts (31). In this paper we will describe
experiments which attempt to determine the nature of
autonomous mitochondrial synthesis in several neoplastic model cells.

Mitochondria were isolated from hepatoma, host rat
liver, and normal control liver and studied for patterns of macromolecular synthesis, protein and glycoprotein composition, and outer surface properties.
Reuber H-35 hepatoma (generations 83 to 87) was bila-
terally transplanted in ACI rat thigh muscle; non-ne-
crotic tumor was harvested 5-7 weeks later. Mitochondria from hepatoma and liver were isolated by homogenization of the tissue and subsequent differential centrifugation in a medium consisting of 0.3 M sucrose,
2 mM EDTA, 30 mM nicotinamide, 0.7 % bovine serum albumin, pH 7.4. This preparation of mitochondria was
found to have little plasma membrane and microsomal
enzyme activity and was greatly enriched in succinic
dehydrogenase and cytochrome c oxidase activity.

Mitochondrial acid insoluble macromolecules were
labeled by incubation in a protein synthesis supporting medium (13) supplied with the appropriate radioactive precursors--[^{3}H]thymidine (0.5 µCi/ml), [^{3}H]u-
ridine (0.5 µCi/ml), [^{14}C]leucine (0.1 µCi/ml), [^{3}H]-
amino acids from algal lysate (10 µCi/ml), UDP[^{14}C]-
glucose (0.03 µCi/ml), UDP[^{14}C]galactose (0.03 µCi/ml),
or GDP[^{14}C]mannose (0.03 µCi/ml). Media for DNA or RNA
synthesis were supplemented with the three appropriate
deoxy- or ribonucleotides (0.02 mM).

Incorporation of labeled precursors by hepatoma,
host liver, and normal liver mitochondria is shown in
Table I. Incorporation of both leucine and [^{3}H]amino
acids from an algal lysate was sensitive to 100 µg/ml
chloramphenicol but insensitive to 100 µg/ml cycloheximide and 500 µg/ml RNAse and thus indicates the
mitochondrial nature of the label incorporation. The
level of incorporation of DNA, RNA and protein precursors was similar for the three mitochondrial tissue

TABLE I.

MITOCHONDRIAL INCORPORATION OF LABELED PRECURSORS INTO ACID INSOLUBLE MACRO-MOLECULES.
Experiments were carried out as described in the text. Means from 9 independent observations.

Precursor	Normal Liver	Host Liver	Hepatoma
	(cpm/mg mitochondrial protein)		
[^{3}H]Thymidine	22,800	23,500	25,100
[^{3}H]Uridine	10,700	10,100	11,280
[^{14}C]Leucine	3,314	3,046	3,951
[^{3}H]Algal lysate (amino acids)	45,000	43,110	47,620
UDP-[^{14}C]Glucose	3,017	3,192	9,621
UDP-[^{14}C]Galactose	423	370	2,010
GDP-[^{14}C]Mannose	1,521	1,172	4,885

sources. However, there is a striking increase in the amount of labeled monosaccharide incorporated by hepatoma as compared to both normal and host liver mitochondria. This increase could reflect either increased transferase enzyme activity associated with hepatoma mitochondria or an increased availability of suitable intramitochondrial acceptor proteins. In order to investigate these possibilities, transfer of labeled monosaccharide onto exogenously added acceptor glycoproteins was studied using 0.1 % Triton X-100 extracts of mitochondria as transferase enzyme source. The exogenous acceptors utilized were collagen from which glucose had been removed and IgG from which sialic acid, galactose, N-acetylglucosamine and mannose had been serially removed (32). In a total volume of 100 μl the assay contained 0.5-1.0 mg of mitochondrial protein, 10 mM MgCl$_2$ or MnCl$_2$, 3 μCi nucleotide-monosaccharide-[^{14}C], 50 mM Tris, pH 6.8, and was incubated for 2 h at 30° C with or without the addition of 0.5 mg exogenous acceptor protein.

The data in Table II demonstrate the transfer of labeled monosaccharide onto both endogenous and exogenous acceptor proteins by enzyme sources prepared from hepatoma and host liver mitochondria. Hepatoma mitochondria transfer more glucose and mannose onto added exogenous acceptors than do liver mitochondria. These data indicate elevated levels of transferase ac-

TABLE II.

ENDOGENOUS AND EXOGENOUS GLYCOPROTEIN: GLYCOSYL TRANSFERASE ACTIVITY OF ISO-
LATED LIVER AND HEPATOMA MITOCHONDRIA.
Experiments were performed as given in the text. Means from 9 independent deter-
minations.

| Enzyme Source | Transferase Activity (cpm/h/mg protein) | | | |
| | With Acceptor | | Without Acceptor | |
	Glucose*	Mannose[†]	Glucose	Mannose
Liver mitochondria	317	406	97	115
Hepatoma mitochondria	822	1215	205	226

*Collagen minus *glucose*; [†]IgG minus sialic acid, galactose, N-acetylglucosamine, *mannose*.

tivity associated with hepatoma mitochondria. However, when exogenous acceptors were omitted, more monosaccharide was also incorporated by hepatoma mitochondria onto endogenous proteins in the extract. This fact indicates that hepatoma mitochondria contain more proteins suitable for glycosyllation than do liver mitochondria and suggests that hepatoma mitochondria may have an altered array of proteins and glycoproteins.

Polyacrylamide gel electrophoresis was performed on a preparation of insoluble mitochondrial proteins as described by Coote and Work (13). Mitochondria labeled with either $[^3H]$ amino acids or $[^{14}C]$ monosaccharides were extracted with pH 11.5 K-phosphate. The base-insoluble residue was dissolved in 1 % sodium dodecyl sulfate. This preparation contains most of the mitochondrial incorporated leucine label and most of the monosaccharide label. The material was electrophoresed on 5 % acrylamide, 0.1 % sodium dodecyl sulfate gels at 8 mA/gel in a 0.1 M Na-phosphate buffer, pH 7.1 for 6.5 h. Standard proteins in sodium dodecyl sulfate were used to determine molecular weight. Gels were stained with either Coomassie Brilliant Blue or periodic acid-Schiff (PAS; 33) and scanned for stain intensity in a Gilford Spectrofotometer. Identical gels were fractionated in a Savant Autogel divider and counted in liquid scintillation fluid.

Fig. 1 is a photograph of gels of hepatoma (A and C) and host liver (B and D) mitochondrial insoluble proteins stained with Coomassie Brilliant Blue (A and B) and PAS (C and D). As reported by Chang *et al.* (30),

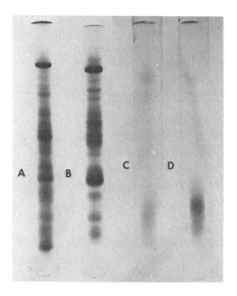

Fig. 1. Polyacrylamide gels of rat-liver and hepatoma mitochondria. Gels A and C are hepatoma mitochondria, B and D host liver mitochondria. Gels A and B are stained with Coomassie Brilliant Blue and gels C and D are stained with the PAS technique. Details are given in the text.

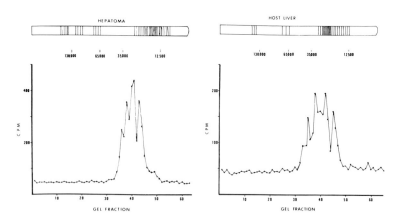

Fig. 2. Graphic representation of PAS-positive bands and radioactivity of gels of rat-liver and hepatoma mitochondria labeled with [^{14}C]monosaccharide. Details are given in the text.

Reuber H-35 hepatoma mitochondria contain an altered array of proteins. Several protein bands are missing or decreased relative to other bands in liver. However, a low molecular weight species is present which is absent from both host and normal liver. This band stains for glycoprotein. Three high molecular weight glycoproteins are present in hepatoma; two of these are missing or depleted from liver. For the same amount of protein loaded per gel, hepatoma material stains more deeply with PAS.

Fig. 2 shows the graphic representations of PAS-positive bands and [^{14}C]monosaccharide labeled glycoproteins. As in Fig. 1, it is evident that insoluble proteins of hepatoma contain more and different glycoproteins than do those of liver. However, those bands of radioactivity detected in hepatoma and liver gels are similar. No radioactivity was detected in the high molecular weight glycoprotein bands. Since the amount of monosaccharide present is very small, as indicated by the relative intensity of PAS staining, it is possible that these bands do contain monosaccharide transferred to protein by isolated mitochondria but undetectable by this radioassay. We are currently investigating this possibility.

Since some mitochondrial glycoproteins have been demonstrated to be present on the outer surface (22, 23,24), we have attempted to compare the surfaces of mitochondria prepared from hepatoma and liver by determining electrophoretic mobility in a dilute ionic solution containing 0.0145 M NaCl, 4.5 % sorbitol, 0.6 mM NaHCO$_3$ at pH 7.2 ± 0.1, as previously described (24). Table III illustrates that the electrophoretic mobility of hepatoma mitochondria toward the anode is significantly greater than that of liver mitochondria. Since negative electrophoretic mobility of mitochondria has been shown to be partially determined by sialic acid residues on the surface (24), we have determined the sialic acid content of hepatoma and liver mitochondria. Sugars were released from mitochondria by dilute acid hydrolysis (0.1 N H$_2$SO$_4$ for 1 hour at 80° C); neutral sugars were removed from the hydroly-

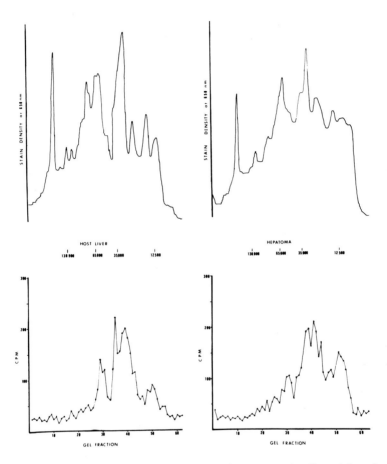

Fig. 3. Stain intensity of Coomassie Brilliant Blue and radioactivity of gels of rat-liver and hepatoma mitochondria labeled with [³H]amino acids. Details are given in the text.

TABLE III.

ELECTROPHORETIC MOBILITY AND SIALIC ACID CONTENT OF LIVER AND HEPATOMA MITO-
CHONDRIA.
Experiments were performed as given in the text. Values are means of 5 experi-
ments $\pm$ S.E.M.

Mitochondria	Electrophoretic Mobility* (µ/sec/v/cm)	Sialic Acid (µg/mg protein)
Liver	-1.82 ± 0.01	$6.4 \pm 1.0^{\dagger}$
Hepatoma	-2.03 ± 0.03	14.9 ± 1.8

*Measured at ionic strength of 0.015 g ion/liter; †Significant at 0.01 level.

sate by Dowex-formate ion exchange chromatography. Total sialic acids of the eluent were determined by the
Warren thiobarbituric-acid assay (34). The results
shown in Table III indicate that hepatoma mitochondria
do contain more total sialic acid than do liver mitochondria. What portion of this extra sialic acid contributes to electrophoretic mobility has not yet been
determined.

Related "*in vivo*" experiments have been conducted
using oncogenic virus transformed cells treated with
drugs that allow differentiation between mitochondrial
and cytoplasmic syntheses. Intact cells were incubated
in the presence of the antibiotics cycloheximide,
which inhibits nonmitochondrial protein synthesis (1),
and camptothecin, which inhibits nonmitochondrial nucleic acid synthesis (37). Experiments such as these
are complicated by the high dose of antibiotic needed
for inhibition of nonmitochondrial synthesis and the
fact that it is unknown whether the *in vivo* non-anti-
biotic treated mitochondria act in the same manner as
mitochondria in cells not treated with antibiotic. The
data of Table IV show that RVS-3T3 mitochondria synthesized slightly less protein than the 3T3 mitochondria while the MSV-3T3 and PY-3T3 mitochondrial protein synthesis was substantially elevated over that of
the 3T3 cells. RNA synthesis, as measured by the incorporation of [^{3}H]uridine in the presence of 10 mg/ml
of camptothecin, was greatly elevated in the virally
transformed cells; the highest elevation occurred in
the PY-3T3 cells. DNA synthesis, as measured by incorporation of [^{3}H]thymidine in the presence of 10 mg/ml

TABLE IV.

SYNTHESIS OF PROTEIN, DNA AND RNA BY MITOCHONDRIA IN INTACT NORMAL AND ONCO-
GENIC VIRUS TRANSFORMED CELLS.
Cells were cultured and harvested as described previously (36). Cells were incubated in complete Dulbecco's medium for 30 min at 37º with the indicated
concentration of antibiotic, then 10 µCi of [³H]thymidine, [³H]uridine, or
[³H]leucine were added to the cell suspension as indicated and further incubation of the suspension was continued for 1 h. Data are expressed as percentage of the 3T3 cell synthesis that the transformed mitochondrial synthesis
represents, with the 3T3 cells arbitrarily set at 100 %. Data are means + S.D.
for the transformed cells. For experiments with cycloheximide and [³H]leucine
the synthesis in all cells represented an average of 4.2 % of that without
any drug; for experiments with camptothecin and [³H]thymidine the synthesis
in all cells represented an average of 3.7 % of that without any drug; and
for experiments with camptothecin and [³H]uridine the synthesis in all cells
represented an average of 4.5 %. Macromolecular bound radioactivity per mg
cell protein was determined as given previously (32). Experiments were performed with 6 to 18 independent cell populations.

Drug:precursor	Cell Line			
	3T3	MSV-3T3	RSV-3T3	PY-3T3
100 µg/ml cycloheximide:[³H]-leucine	100	140 + 4	94 + 3	190 + 14
10 mg/ml camptothecin:[³H]-uridine	100	212 + 11	186 + 27	247 + 19
10 mg/ml camptothecin:[³H]-thymidine	100	189 + 32	216 + 41	289 + 44

of camptothecin, was essentially elevated two fold in
the oncogenic virus transformed cells compared to the
"normal" 3T3 cells; highest elevations were in the
PY-3T3 cells.

The data of this communication indicate differences in the mitochondria of "normal" and neoplastic
cells. The differences, however, are complex and not
clear-cut, as are many parameters related to neoplasia.
The results certainly call for more study in this potentially interesting area.

ACKNOWLEDGEMENTS

This study was supported in part by grants CA-13220,
GM-15190, GM-00032 from the United States National Institutes of Health. Dr. Bosmann is a Research Career
Development Awardee of the National Institute of General Medical Sciences. We thank Mr. K.R. Case, Mr. G.
Bieber, Mr. R. Ball, Ms. D. DeHond, and Ms. G.Z. Pike
for technical assistance.

REFERENCES

1. M. Ashwell and T.S. Work, Ann.Rev.Biochem. **39**, 251
(1970).

2. A.A. Hochberg, F.W. Stratman, R.N. Zahlten and H. A. Lardy, FEBS Letters 25, 1 (1972).

3. N.G. Ibrahim, J.P. Burke and D.S. Beattie, FEBS Letters 29, 73 (1973).

4. H.B. Bosmann and S.S. Martin, Science 164, 190 (1969).

5. H.B. Bosmann and B.A. Hemsworth, J.Biol.Chem. 245, 363 (1970).

6. R. Morelis and P. Louisot, C.R. Acad.Sci.Paris D 276, 2219 (1973).

7. S.S. Martin and H.B. Bosmann, Exp.Cell Res. 66, 59 (1971).

8. S.S. Martin and H.B. Bosmann, Biochim.Biophys.Acta 230, 411 (1971).

9. H.B. Bosmann, A. Hagopian, E.H. Eylar, Arch.Biochem.Biophys. 128, 470 (1968).

10. H.B. Bosmann, Eur.J.Biochem. 14, 33 (1970).

11. R.J. Bernacki and H.B. Bosmann, Eur.J.Biochem. 34, 425 (1973).

12. H.B. Bosmann, Curr.Mod.Biol. 3, 319 (1971).

13. J.L. Coote and T.S. Work, Eur.J.Biochem. 23, 564 (1971).

14. A. Tzagoloff, M.S. Rubin, M.F. Sierra, Biochim. Biophys.Acta 301, 71 (1973).

15. T.L. Mason and G. Schatz, J.Biol.Chem. 248, 1355 (1973).

16. B. Kadenbach, Biochem.Biophys.Res.Commun. 44, 724 (1971).

17. B. Kadenbach and P. Hadváry, Eur.J.Biochem. 32, 343 (1973).

18. H.B. Bosmann, Nature New Biol. 234, 54 (1971).

19. G.L. Sottocasa, G. Sandri, E. Panfili, B. deBernard, FEBS Letters 17, 100 (1971).

20. G.L. Sottocasa, G. Sandri, E.Panfili, B. deBernard, P. Gazzotti, F.O. Vasington and E. Carafoli, Biochem.Biophys.Res.Commun. 47, 808 (1972).

21. A. Gomez-Puyou, M. Tuena de Gomez-Puyou, G. Becker and A.L. Lehninger, Biochem.Biophys.Res.Commun. 47, 814 (1972).

22. G. Nicolson, M. Lacorbière and P. Delmonte, Exp. Cell Res. 71, 468 (1972).

23. R.H. Glew, S.C. Kayman and M.S. Kuhlenschmidt, J. Biol.Chem. <u>248</u>, 3137 (1973).

24. H.B. Bosmann, M.W. Myers, D. DeHond, R. Ball and K.R. Case, J.Cell Biol. <u>55</u>, 147 (1972).

25. J. Kára, O. Mach and H. Cerná, Biochem.Biophys. Res.Commun. <u>44</u>, 162 (1971).

26. F.E. Gaitskhoki, F.E. Erslov, O.I. Kiselev, L.K. Men'shikh, O.V. Zaitseva, L.V. Uryvaev, V.M. Zhadanov and S.A. Neifakh, Dokl.Akadem.Nauk.SSSR <u>201</u>, 220 (1971).

27. C.A. Smith and J. Vinograd, Cancer Res. <u>33</u>, 1065 (1973).

28. N.J. Richert and J.D. Hare, Biochem.Biophys.Res. Commun. <u>46</u>, 5 (1972).

29. A.J. Levine, Proc.Natl.Acad.Sci.U.S.A. <u>68</u>, 717 (1971).

30. L.O. Chang, C.A.Schnaitman, H.P. Morris, Cancer Res. <u>31</u>, 108 (1971).

31. M.W. Myers and H.B. Bosmann, FEBS Letters, <u>26</u>, 294 (1972).

32. H.B. Bosmann, Eur.J.Biochem. <u>14</u>, 33 (1970).

33. J.P. Segrest and R.L. Jackson, Methods in Enzymology, <u>XXVIII</u>, part B, 54 (1973).

34. L. Warren, J.Biol.Chem. <u>234</u>, 1971 (1959).

35. H.B. Bosmann, Biochem.Biophys.Res.Commun. <u>41</u>, 1421 (1970).

36. H.B. Bosmann, Biochim.Biophys.Acta <u>264</u>, 339 (1972).

MEMBRANE STRUCTURE AND MITOCHONDRIAL BIOGENESIS

Lester Packer and Lou Worthington

Department of Physiology-Anatomy, University of California, Berkeley, California 94720 and Physiology Research Laboratory, VA Hospital, Martinez, California 94553, U.S.A.

There is a consensus that membrane structure bears an important relationship to the biogenesis of mitochondria. Several papers in this volume have already made reference to the importance of this problem. In our laboratory we have been examining the structure of mitochondria from both yeast and mammalian cells by the technique of freeze fracture electron microscopy. Several points have emerged from these studies which may be relevant to the general problem of mitochondrial biogenesis.

Fusion between the outer and inner membranes. Examination of double membrane fracture faces of mitochondria from yeast (1) and mammalian cells (2) reveals the existence of double membrane fracture faces showing large regions where the outer and inner membranes are fused together. Indeed examination of mitochondria *in vivo* by phase microscopy and conventional staining electron microscopy reveals that the outer and inner membranes are generally fused. In isolated mitochondria this fusion is often lost as a result of preparation procedures, ionic composition of suspending media, etc., in both outer and inner membrane fusions and also in areas where the inner membrane system is fused together in the cristae. Cristae generally unfold after isolation. Such changes in the unique structural features of mitochondria must have extensive functional consequences.

Membrane lipids and proteins are mobile. Mobility of lipid and protein components occurs in the lateral and perpendicular plane of the membrane (3-4). The pattern of this mobility may be modified in fused membranes. The orientation of membrane particles and lipids (non particle containing areas) can be readily observed in mitochondria that undergo expansion and contraction in association with energized ion transport. Expansion or contraction of the inner membrane has been observed to change the characteristic cluster-like particle distributions seen in the outer membrane (3). Thus, the two membranes act cooperatively.

Relation of mtDNA to membrane structural components. In studies with mutants of both *S.cerevisiae* (1) and in *S.pombe* (carried out in collaboration with Dr. A. Goffeau), it has been found that the size, density and distribution of membrane particles is unchanged if alterations occur in DNA sequences or in ethidium bromide induced DNA-less mutants. Collaborative studies with Nolan, Katyare, Smith (5) and Worthington and Katyare (6) show similar results in human diploid lung fibroblasts (WI-38) treated with ethidium bromide or chloramphenicol under conditions that lead to growth inhibition and marked diminution of cytochrome oxidase activity. Some typical data for these parameters in outer and inner membranes of mitochondria are shown in Table I for *S.cerevisiae* mutants (after ref. 1).

It may be concluded from these studies that the main structural components of mitochondrial membranes are not coded for by mtDNA and further are probably not translated on mitochondrial ribosomes.

CONCLUDING REMARKS.

The structure of the membrane remains one of the main unsolved problems relevant to the understanding of mitochondrial biogenesis. The problem is not only to identify the components of the membrane and where they are synthesized, but also to determine the principle which directs their assembly and functional organization. Our studies with freeze-fracture electron microscopy suggest importance of membrane fusion and mobility of lipid and protein components to the pro-

TABLE I.

PARTICLE DENSITY/μ^2 IN MEMBRANE FRACTURE FACES OF WILD TYPE AND mDNA MUTANT YEAST MITOCHONDRIA*.

Mitochondrial Fracture Face	Wild Type	Ratio A/B	½ mDNA	Ratio A/B	No mDNA	Ratio A/B
Outer A	3270		3334		2909	
		2.06		2.50		2.48
Outer B	1589		1330		1175	
Inner B	3660		3433		4120	
		1.12		1.18		1.06
Inner A	4100		4050		4380	

*Experiment FE 94.

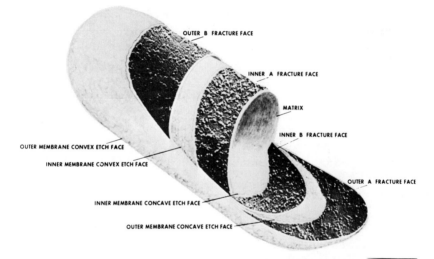

Fig. 1. Organization of membrane particles in mitochondrial membranes revealed bij freeze-fracture electron microscopy. The model shows the characteristic structural features seen in the half-membranes that arise as a result of fracturing of the outer and inner membranes. Non-particle areas in the fracture faces are presumed to be pure lipid domains. The B face of the outer membrane and the A face of the inner membrane are revealed as convex fractures, concave fractures show the B face of the inner membrane and the A face of the outer membrane.

blem of membrane assembly, functional coupling between the two membranes, and the possible attachment to the membrane surfaces of DNA, ribosomes and proteins.

ACKNOWLEDGEMENTS.

The authors would like to thank Susan Gooch for expert technical assistance with electron microscopy. This research was supported by USPHS HD 05875 and the Veterans Administration and the Musc. Dystrophy Ass.

REFERENCES.

1. L. Packer, M.A. Williams and R.S. Criddle, Biochim.Biophys.Acta 292, 92 (1973).
2. R.L. Melnick and L. Packer, Biochim.Biophys.Acta 253, 503 (1971).
3. L. Packer, Bioenergetics 3, 115 (1972).
4. H.M. Tinberg, L. Packer and A.D. Keith, Biochim. Biophys.Acta 283, 193 (1972).
5. L. Packer, J.S. McHale, S. Katyare and J. Smith, submitted for publication.
6. L. Packer, L. Worthington and S. Katyare, in preparation.

LECITHIN FORMATION IN PHOSPHOLIPASE C-TREATED RAT-LIVER MITOCHONDRIA

B.C. van Schijndel, A. Reitsema and G.L. Scherphof

Laboratory of Physiological Chemistry, State University, Bloemsingel 10, Groningen, The Netherlands.

The capacity of mitochondria to synthesize their major phospholipids phosphatidylcholine and phosphatidylethanolamine is generally thought to be very limited or is ascribed to microsomal contamination (1-6). To test whether the mitochondrial capacity to synthesize lecithin is indeed limited or even absent, we compared CDP-choline: 1,2 diglyceride cholinephosphotransferase activities in phospholipase C (PLC) microsomes and mitochondria. PLC action on membranes creates one of the required substrates, the diglyceride *in situ*. This was thought to be advantageous as compared to the addition of diglyceride dispersions which have been shown to stimulate microsomal but not mitochondrial lecithin formation (3,5). Table I*) shows a typical experiment of the incorporation of radioactivity from CDP-[^{14}C-Me]choline into lecithin in PLC-

*) incubation conditions and other experimental details will be given elsewhere (7).

TABLE I

LECITHIN SYNTHESIS FROM CDP-CHOLINE IN NORMAL AND PHOSPHOLIPASE C (PLC) TREATED MITOCHONDRIA AND MICROSOMES AFTER 30 AND 60 MIN OF INCUBATION.
Mitochondria and microsomes (40 mg protein in 2.0 ml) were treated for respectively 2.0 and 5.0 min with respectively 0.76 and 0.18 units of PLC. The reaction was stopped by addition of 2.0 ml 15 mM phenanthroliniumchloride, followed by two washes.

Cellfraction	% Phospholipid hydrolysis	Lecithin synthesis (nmoles/μmole phospholipid)	
		30 min	60 min
Mitochondria	–	1.9	4.2
PLC treated mitochondria	50.9	31.7	45.2
Microsomes	–	19.6	25.9
PLC treated microsomes	51.3	56.5	65.5

541

treated and untreated microsomes and mitochondria. The microsomal contamination of mitochondria, which is 5 % as calculated from the NADPH-cytochrome c reductase activities (8) can not account for the large increase in lecithin synthesis in the PLC treated mitochondria. Table II presents evidence that the mitochondrial and microsomal choline phosphotransferases are not identical enzym systems. The effect of increasing membrane phospholipid conversion to diglycerides by PLC on lecithin synthesis in both cell fractions is different. Mitochondrial activity is increasingly stimulated up to a maximum at some 25 % of PLC-catalyzed degradation whereas microsomal activity is continuously stimulated up to 85 % phospholipid hydrolysis. These results show that the mitochondrial cholinephosphotransferase can be very active under favourable conditions. Towhat extent the mitochondrial choline phosphotransferase activity actually contributes to the synthesis of mitochondrial lecithin during mitochondriogenesis and which factors would control such a contribution remains a matter of further investigation. An important question to be answered in this connection concerns the ability of mitochondria to synthesize and dephosphorylate phosphatidic acid and the possible involvement in these processes of other cell constituents (8).

TABLE II

EFFECT OF PHOSPHOLIPASE C (PLC) ON LECITHIN SYNTHESIS IN MITOCHONDRIA AND MICROSOMES.

Mitochondria		Microsomes	
% Phospholipid hydrolysis	Lecithin synthesis (nmoles/µmole phospholipid/30 min)	% Phospholipid hydrolysis	Lecithin synthesis (nmoles/µmole phospholipid/30 min)
no PLC	4.4	no PLC	21.1
zero time control*	0.9	zero time control*	7.0
4.8	13.8	-	-
12.8	29.2	12.7	22.2
24.2	33.6	21.1	44.1
47.3	26.6	46.3	58.5
75.9	3.3	85.0	73.4

*phenanthroliniumchloride was added before PLC and the mixture was not incubated at 37° C.

ACKNOWLEDGEMENT.

The authors wish to thank Dr. Zwaal and Dr. Roelofsen from the Laboratory of Biochemistry, University of Utrecht for ample supplies of pure phospholipase C from *Bacillus cereus*.

REFERENCES.

1. F.B. Jungalwala and R.M.C. Dawson, Eur.J.Biochem. 12, 399 (1970).
2. M.L. Williams and F.L. Bygrave, Eur.J.Biochem. 17, 32 (1970).
3. M.G. Sarzala, L.M.G. van Golde, B. de Kruyff and L.L.M. van Deenen, Biochim.Biophys.Acta 202, 106 (1970).
4. E.A. Dennis and E.P. Kennedy, J.Lipid.Res. 13, 263 (1972).
5. E.K. Miller and R.M.C. Dawson, Biochem.J. 126, 805 (1972).
6. J. Zborowski and L. Wojtczak, Biochim.Biophys.Acta 187, 73 (1969).
7. B.C. van Schijndel, A. Reitsema and G.L. Scherphof, article submitted to Biochem.Biophys.Res.Commun. (1973).
8. E.H. Mangiapane, K.A. Lloyd-Davies and D.N. Brindley, Biochem.J. 134, 103 (1973).

CONCLUDING REMARKS

C. Saccone[*] and A.M. Kroon[**]

Institute of Biological Chemistry[*], University of Bari, Bari, Italy, and Laboratory of Physiological Chemistry[**], State University, Groningen, The Netherlands.

The wide range of subjects dealt with in these proceedings and the variety of technical approaches to the different subjects make it to a difficult task to summarize the main points in a nutshell.Nonetheless we will try to give such a summary because at the one hand the various contributions clearly lead to a consensus of opinion on several points, which were quite controversial hitherto. At the other hand, from the papers and discussions, a number of clear problems have emerged which allow to trace out roughly the ways of future research in the field of mitochondrial biogenesis. In this respect these remarks attempt to be an outlook.

Mitochondrial DNA-dependent RNA polymerase. With regard to DNA-dependent RNA polymerase, agreement has been reached on the properties of the enzymes from *Neurospora crassa*, yeast,*Xenopus* and rat liver. In yeast a second enzyme with completely different properties is perhaps involved (Eccleshall and Criddle). The four enzymes mentioned are single polypeptides of about 60,000 daltons molecular weight, are able to transcribe mitochondrial DNA, but behave differently with respect to their sensitivity to inhibition by rifampicin and its analogues (Scragg, Gallerani and Saccone, Dawid and Wu).

The transcription products. The availability of the purified enzymes has allowed the *in vitro* transcription of mitochondrial DNA with its homologous polymerase. The RNA products transcribed by the yeast polymerase from yeast mtDNA (Scragg) and by the rat

polymerase from rat mtDNA (Gallerani and Saccone) show
S-values up to about 20S; in the rat system RNAs with
S-values similar to the mitochondrial rRNA's are found.
The presence of mt-rRNA sequences in the *in vitro*
transcripts of *Xenopus* mitochondrial polymerase was
demonstrated by hybridization-competition experiments.
The products of mitochondrial transcription in the
yeast system are, furthermore, capable to program an
E. coli cell free system to produce proteins with physicochemical characteristics of mitochondrial membrane proteins. The strand selection by mtRNA polymerase
has been investigated in *Xenopus laevis* (Dawid and
Wu). With closed-circular DNA both strands are transcribed *in vitro* but the heavy strand is preferred;
with open-circular DNA as the template heavy strand
preference is almost complete. On denaturation of the
template the strand preference is drastically altered
and this leads to the conclusion that the mtRNA polymerase recognizes the secondary structure of its template. The existence of mtDNA-coded messenger RNA species is shown by the isolation (Attardi *et al.*) from
HeLa cell mitochondria of two discrete poly(A)-con-
taining RNA molecules with sedimentation coefficients
of 7S and 9S, corresponding to 8.5×10^4 and 1.5×10^5
daltons molecular weight, respectively. It is interesting to note that the 9S component is coded for by the
heavy strand, while the 7S component is coded for by
the light strand, an observation reminding of the situation with the mitochondrial tRNA species. It is
tempting to speculate in this context about a possible
amplification of the informational content of animal
mtDNA via transcription of both strands. The presence
of genes for messenger RNA on mtDNA is most elegantly
shown in the studies with the "abn-1" stopper growth
mutant of *Neurospora crassa* (Küntzel *et al.*). From
these studies it follows that mtDNA of *Neurospora*
codes for a protein of 11,000 daltons molecular
weight. This protein is translated from a RNA molecule derived from a precursor transcript of 33S. This
interesting observation shows that one has to consider the possibility of extensive processing of messen-

ger-RNA precursors also in mitochondria. This processing of precursor RNA also occurs for the ribosomal RNAs (Kuriyama and Luck). It is evident, therefore, that the genetic information available for final transcription and translation products cannot be computed directly by relating the final molecular weights to the genetic complexity of the mitochondrial DNA.

The translation products. Looking at the gene products at the level of translation it has been convincingly demonstrated for lower eukaryotes that 3 components of cytochrome *c* oxidase, 4 components of the ATPase complex, one or two components of the cytochrome *b* complex and a protein necessary for the assembly of cytochrome c_1 are genetically coded for on mitochondrial DNA (Tzagoloff *et al.*, Sebald *et al.*, Ross *et al.*, Mahler *et al.*, Weiss and Ziganke, Schwab, and Werner). Furthermore *one* ribosomal protein may be translated on mitochondrial ribosomes (Groot). On the basis of the assumed molecular weights of the various polypeptides and assuming that these are all genetically localized on mitochondrial DNA, the minimum amount of organelle DNA involved in coding is 6×10^6 daltons. It is clear that one cannot accomodate much more information than that necessary for a similar set of proteins on animal mitochondrial DNA. Unfortunately for mammalian systems the details of information are still lacking but we can expect these to be available in the near future. An intriguing, although not very likely and still open question is whether the mitochondrial translation products are deficient in about 8 of the amino acids normally present in polypeptides. Attardi *et al.* have suggested this; if so it could explain the limited number of tRNA species (about 12) coded for on animal mitochondrial DNA. However, one should realize that 12 tRNAs are sufficient to incorporate 12 amino acids if there is no or only a slight degeneracy of the genetic code in mitochondrial messenger RNA in animal mitochondria, whereas extense degeneracy has just now been shown for *Tetrahymena* mitochondria (Chiu *et al.*).

An original approach to solve some of the questions

547

concerning the biosynthesis of mitochondrial membrane proteins is that of Mahler *et al.* They use radioactive formate to identify the mitochondria as the site of synthesis of polypeptide subunits of isolated enzymes derived from the inner membrane. Formate is incorporated into mitochondrial formyl-methionyl-tRNA, which as such is responsible for chain initiation in the mitochondria. Furthermore, using a temperature-sensitive yeast mutant blocked in polypeptide chain initiation on cytoplasmic ribosomes at restrictive temperature, they were able to demonstrate that mitochondrial protein synthesis is subject to direct translational control by a product of cytoplasmic protein synthesis. Also Ross *et al.* investigated the assembly of cytochrome aa_3 into the mitochondrial membranes with the aid of yeast mutants in which the biogenesis of cytochrome c oxidase was blocked by either nuclear or extranuclear mutations. The concerted biochemical and genetic experiments led these authors to conclude that mitochondrially-synthesized cytochrome c oxidase subunits regulate the assembly of the subunits of cytoplasmic origin. It is well appreciated at present that the products of mitochondrial translation are extremely hydrophobic (Michel and Neupert, Ross *et al.*, Sebald *et al.*). This hydrophobicity may raise the necessity of their synthesis close to their final site of integration and thus carry a teleological argument to justify why the mitochondrial protein-synthetic machinery has been retained. Furthermore the hydrophobic properties introduce special experimental problems *e.g.* with respect to release from the ribosomes *in vitro*, but also and especially for their identification by physicochemical and immunological means, etc.

Mitochondrial genetics. The favoured organisms for mitochondrial genetic studies still are the yeasts although also results with *Paramecium* are reported (Adoutte). Besides the few examples of yeast mutants used in the studies of mitochondrial biogenesis already mentioned, a large number of antibiotic resistant mutants appear valuable tools in the further unraveling of the mitochondrial and non-mitochondrial contributions to the final assembly of the fully active particles within the cell. The mutants resistant to antibiotics interfering with protein synthesis at the level of the ribosomes are the most popular. Apart from their applicability in studies of mitochondrial ribosomal functions (Grivell), they may be useful in searches for gene purification and characterization (Nagley *et al.*, Fukuhara *et al.*), just as other specific petite mutants (Rabinowitz *et al.*, Borst). Results of mono-, bi- and trifactorial crosses of strains characterized by antibiotic resistant markers have led Linnane *et al.* to propose that mitochondrial genetics has to be approached as a process comparable to phage genetics rather than to sex-linked bacterial genetics.

Resistance to other than the antiribosomal drugs has convincingly been shown to extend the insight obtained by the more "orthodox" biochemical methods (Linstead *et al.*, Griffiths *et al.*, Bandlow). Furthermore, the use of hybrid cell lines has now been introduced as a possible means to attack similar problems also in cells of higher organisms (Van Heyningen *et al.*, Dawid *et al.*). This approach opens nice perspectives for the near future.

The mitochondrial ribosomes. With respect to the miniribisomes complete agreement exists that these are less "mini" than their RNAs suggest and that, in fact, their size seems to be of the same order as that of other ribosomes known to date. This conclusion has been reached by several groups on the basis of different criteria and using different techniques (Kleinow *et al.*, O'Brien *et al.*, De Vries and Kroon, Greco *et al.*). It seems worthwhile to stress that in the electronmicroscope the dimensions of mitochondrial ribosomes from animal cells appear to be smaller in tissue sections (Kleinow *et al.)* than after isolation. The difference in the dimensions between isolated cytoplasmic and mitochondrial ribosomes is always less than between extra- and intramitochondrial ribosomal particles in tissue sections. The reason for this descrepancy may be the difficulty in the fixation of

these particles *in situ* owing to their special structure and chemical composition.

With regard to the ribosomes from lower organisms
it has been clearly established that the *Tetrahymena*
mitochondrial ribosomes display the same sedimentation coefficient (80S) as their cytoplasmic counterparts although the two types of ribosomes can be
clearly distinguished by analysis of their heavy and
light ribosomal RNA components, by their buoyant density in CsCl and by their fine structure (Stevens *et
al.*). An unusual property of the *Tetrahymena* mitochondrial ribosome is its dissociation into subunits
having identical sedimentation coefficients. These data, already known from the work of Suyama, have now
been firmly established (Stevens *et al.*). The sedimentation coefficient of the native mitochondrial ribosomes in *Neurospora* (73S and no matter of controversy until now) appears to need revision because the
particle actively involved in protein synthesis, the
monomer form in mitochondrial polysomes, sediments at
80S (Agsteribbe *et al.*).

Functional mitochondrial polysomes from mitochondria have been isolated from yeast (Cooper *et al.*)
and from *Neurospora* (Agsteribbe *et al.*). From both
types of polysomes messenger RNA was isolated. In the
case of *Neurospora*, the mRNA was able to direct protein synthesis when tested in a cell-free system from
E.coli. In yeast polysomes, Cooper *et al.* demonstrated the presence of a poly(A) containing mRNA confirming the observation (Attardi *et al.*) that poly(A)
sequences are present also in HeLa mitochondrial messenger RNA. The possibility to obtain mitochondrial
messenger RNA in combination with the availability of
sophisticated immunological techniques for the characterization of the translation products opens the
way to a more detailed identification of the mitochondrial gene products. Besides these interesting deve-
lopments it will be necessary to concentrate further
on the mechanism of synthesis, transport and assembly
of the mitochondrial polypeptides synthesized in the
cytoplasm (vectorial translation ?; *cf.* Kellems *et*

al.). Finally thorough attention should be paid in the near future to the (phospho)lipid part of the story as to make the picture more complete (Packer and Worthington, Van Schijndel *et al.*).